CW01276634

From Brooklands *to* Brize

A Centennial History of No 10 Squadron Royal Air Force

1 January 2015, exactly 100 years to the day from when 10 Squadron was first formed, a US F-18 Hornet receives fuel from a 10 Squadron Voyager over Iraq, supporting Coalition forces in the conflict against ISIL, the so-called Islamic State.

From Brooklands to Brize

A Centennial History of No 10 Squadron Royal Air Force

'Rem Acu Tangere – Hit the Mark'

by
Ian Macmillan
with
Richard King

Published 2015 by
The 10 Squadron Assocation

ISBN 978-0-9934646-0-7

First Edition
First Impression

A catalogue record for this book is available from the British Library.

Published in assocation with Colourpoint Creative Ltd, Newtownards

Designed by April Sky Design, Newtownards
Tel: 028 9182 7195
Web: www.aprilsky.co.uk

Printed by W&G Baird Limited, Antrim

CONTENTS

BUCKINGHAM PALACE

Number 10 Squadron was formed on 1 January 1915 and, as Honorary Air Commodore of RAF Brize Norton, it gave me great pleasure to mark its Centenary of service by presenting the Squadron with a new standard at a ceremony at RAF Brize Norton on 30 January 2015.

In my address following the dedication and presentation I recounted briefly the remarkable history of the Squadron and the diversity of its roles and operations. Later I enjoyed meeting not only those who serve today but others who had flown operational missions during the Second World War and in later years.

As the Squadron enters its second century it is clear that the commitment of its personnel today reflects that same dedication to the service of the nation as that shown by their predecessors across a century of war, peace, and humanitarian operations. This book records the deeds of former years, many of which are emblazoned as Battle Honours on the new Standard, and provides a comprehensive record to inspire future Squadron members.

Anne

INTRODUCTION

BY THE OFFICER COMMANDING No 10 SQUADRON

Wing Commander Jamie W. Osborne RAF
December 2013 – January 2016

It gives me great pride and it is my enormous honour to write the introduction for this fantastic book charting the history of 10 Squadron over the last one hundred years. I know it has been a labour of love for Dick King and Ian Macmillan but to have completed this in the Squadron's centenary year could not exemplify our motto better; they have truly Hit the Mark!

As the current Officer Commanding I am of course focussed on providing military capability with Voyager today much as all the former OCs were with their respective aircraft at the time. While technologies and the means of delivering airpower change there is one over-riding constant; our people. I have no doubt that the exploits of the current personnel could find parallels all the way back through to the first days of 10 Squadron with Major Shephard. We go where we are asked to go, we deliver what is required and we do it to the very best of our ability. This is a human endeavour and I, like Major Shephard, am amazed at what can be achieved by those who serve alongside me. In that sense, 10 Squadron has not changed at all and I am extremely proud to have commanded this historic unit in its centenary year.

The link to our collective history has also been critical to ensuring the current generation understand what they are a part of; my unreserved thanks therefore go to the 10 Squadron Association who have been thoroughly generous with their time and support be that for parades or after parties! To all past, present and future members of the 10 Squadron family; Rem Acu Tangere.

Jamie Osborne

DEDICATION

To the memory of some 900 members of 10 Squadron, fallen in action or in service since 1915

My tale was heard and yet it was not told,
My fruit is fallen, and yet my leaves are green,
My youth is spent and yet I am not old,
I saw the world and yet I was not seen;
My thread is cut and yet it is not spun,
And now I live, and now my life is done.

(From Tichborne's Elegy, 1586)

FROM BROOKLANDS TO BRIZE

PART 1

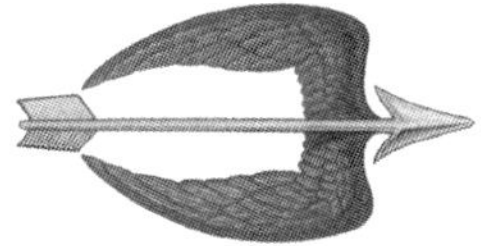

WORLD WAR 1 ~ 1914 - 1918

Royal Aircraft Factory BE2 Variants
Bristol Scout
Armstrong Whitworth FK8
Bristol F2B

&

THE INTER-WAR YEARS ~ 1919 - 1939

Handley Page Hyderabad & Hinaidi
Vickers Virginia
Handley Page Heyford
Armstrong Whitworth Whitley

CHAPTER 1

FINDING THEIR WINGS

"Ever since the war commenced, aircraft have loomed large in support of each of the nations—by the Allies as well as by the enemy—engaged in the present great struggle."

(Flight Magazine Editorial, 1 January 1915.)

No 10 Squadron was formed on 1 January 1915, one of a number of Royal Flying Corps (RFC) units to form in the first months of that year in the context of a wider expansion of the British Army. There was a pressing need. On the outbreak of war with Germany in August 1914 the British Expeditionary Force (BEF, but normally referred to at the time simply as the Expeditionary Force) had deployed to France, and the early expectation that "the boys would be home by Christmas" had quickly faded after the initial defeat at Mons and the retreat to the Marne. From there, a subsequent successful counterattack meant the failure of the German Schlieffen Plan to take Paris, but it was followed by the establishment of the trench lines on each side that would dominate the next four years. An uncertain horizon beckoned for the British Army.

In such circumstances, the much-needed expansion of a small force necessarily meant forming new squadrons in part from existing ones. Thus, there was an inevitable trade-off between current capability and future potential as a training organisation of Reserve Aeroplane Squadrons (RAS) was created. (Similar spreading of experience would occur in the late 1930s as flights from 10 Squadron were hived-off to form two further squadrons, and in 2013 Squadron personnel were cross-posted to 101 Squadron as it joined the Voyager force.) In a letter dated 28 November 1914 the War Office had authorised the Officer Commanding RFC at Farnborough to begin recruiting for a proposed No 10 Squadron, and it formed with personnel from No 1 RAS. Fifty-two other ranks were transferred to the new unit on 1 January 1915, a further twenty-two followed on 3 January, and the first six officers arrived the next day. These included the Commanding Officer, Major G S Shephard, who had already gained operational experience as a Flight Commander with 4 Squadron in France, for which the award of a MC was gazetted later in the month. With its initial personnel mustered, it was time for the unit to prepare for what lay ahead. So, a first BE2 aircraft was also transferred and, on 8 January, the fledgling squadron moved on to Brooklands, for all practical purposes its first base – and hence the title of this volume. Its strength increased and initial pilot training was carried out on a variety of types – Farman Longhorns and Shorthorns, Bleriots, and Martinsyde S1. (Looking ahead to deployment, men could have French lessons costing 1/6 for two weeks, and these became compulsory for Motor Cyclists. On which point, the last race meeting at Brooklands had been held on 3 August 1914, the day prior to war being declared, and on 30 September 'owing to the exigencies of military requirements… the Motor

No 10 Squadron's first commanding officer Major Gordon S. Shephard

Course and Flying Ground' had been closed to the public. Nonetheless, events were held later in the war, once the competition area had been secured.)

Some specialist training was undertaken elsewhere, as can be seen from an undated letter held by the Squadron from a Mr Gavin Dunn describing some of these earliest days:

"I had the good fortune to be selected for No 10 Sqn at its formation as a Rigger and my pal Louis Gray was the Engine Fitter, both of us being 1st AM at the time. We were sorted out at Brooklands racing track and hammered into shape for drill and guard mounting duties by a very strict but fair Warrant Officer whom we called the 'Old Man.' (1st AM was an abbreviation of Air Mechanic 1st Class.)

We were allocated to Flights and we were in B Flight and sent on detachment to Chelmsford in Essex ….. we had a large hangar for the plane which was a BE2b with a Renault air-cooled engine of 70 Horse Power. Louis Gray was a born mechanic and saw to it that it gave its full revolutions of 3000 per minute, while I had the airframe to see to. I worked from a manual that gave the angle of incidence as 3.5 degrees to the horizontal and had an instrument called the Abney Level to check this with. We had a Flight Sgt in charge of the other ranks who was a proper gentleman but we did no drill or parades during this 3 month period and when we got back to Brooklands, the 'Old Man,' our Sgt Major, was waiting for us and didn't he put us through the mill."

(Gavin Dunn was later promoted to Corporal, flew on many test flights, and was awarded the Meritorious Service Medal for disposing of a phosphorus bomb that separated from an aircraft on the airfield and ignited on 30 July 1916. He seized the bomb and had run around the aircraft into an open space to throw it away when it exploded in midair. In February 1918 he himself became the unit's 'Old Man' on promotion to Sergeant Major, before returning to England in September to train as a pilot. His friend Louis Gray later became a pilot flying RE8s on 6 Squadron, and was killed in action at Messines Ridge on 7 June 1917.)

During its time at Brooklands, the Squadron suffered its first fatality on 22 March. Captain J F A Kane was killed when, according to eyewitness reports, the Bleriot he was flying appears to have stalled and crashed. Capt Kane was an experienced officer from the Devonshire Regiment and his death at that point, when so much else was new, must have been a distinct loss to the new squadron. Then, too, in March, as the RFC continued to eke out its experienced resources, Major Shephard was sent back to France to command 6 Squadron in the field. (He later rose rapidly to the rank of Brigadier, commanding 1 Brigade RFC; was awarded the DSO in 1917; and was killed in a flying accident in France on 19 January 1918, the most senior RFC officer to die in the war. His father provided funds for the award of an annual Gordon Shephard Memorial Essay Prize that continued within the RAF for many years.) Command of 10 Squadron transferred to Major U J D Bourke, another officer with prior operational experience in France, who remained in command until June 1916. A further short stay at Hounslow began on 1 April, and was followed by a semi-permanent move to Netheravon from 7 April. There the unit began to be equipped with the BE2c aircraft that would later go to France. Basic wireless transmitters (but no receivers) were fitted to these and training with both pilots and observers began. At this point the Squadron was designated as a unit of No 4 Wing RFC, intended to move to France as a complete formation – a plan that would change.

The BE 2c: At the outset, the Army saw aircraft primarily as a more capable means of observation and artillery-spotting than any available hitherto. Their potential in such roles had been demonstrated in pre-war manoeuvres and their utility was proved in practice in the first days of the war, when vital information on German dispositions was obtained during the battle at Mons and in the ensuing retreat via Le Cateau to the River Marne. (The then Capt Shephard is credited with having flown one of the most vital sorties during those days.) So not surprisingly, many machines were designed as two-seaters providing a stable platform for the reconnaissance task, both visual and photographic. The Royal Aircraft Factory's BE2c (BE – Bleriot Experimental – signified use of a tractor or nose-mounted engine) and later slightly improved variants were built with this in mind. It had a 90hp engine giving a maximum speed of some 75 mph, with a service ceiling of 10,000 feet and an endurance of around three hours. The observer occupied the front cockpit where he had a

Lewis machine gun but, despite his being able to mount the gun in any of three positions, his field of fire was limited by his position between the wings. Added to its inherent stability and lack of manoeuverability, this made the BE2c particularly vulnerable to more capable German aircraft, to the extent that this vulnerability became a matter of parliamentary controversy during 1916. Nonetheless, with its need for aerial observation regarded as vital, the Army persisted and 10 Squadron flew BE2 variants until mid-1917.

A BE2c with the original black roundel Squadron marking, seen through the sights of a gun-camera used in training observers. When the RFC later adopted drab green aircraft colours, the roundel changed to white.

Typically, RFC squadron establishments in 1915 would build up to twelve aircraft, organised into a HQ and three flights, with a total of some 209 personnel (28 officers, 20 SNCOs, and 161 Air Mechanics). Deployment to France commenced on 22 July when unit transport sailed from Avonmouth, followed by personnel and some aircraft shipped from Southampton. Other aircraft were flown across the Channel on 25 July, when the initial destination was the RFC HQ at St Omer. The unit's arrival was reported in the first of the RFC communiqués issued for internal distribution for the remainder of the war:

> "Arrivals: On the 25th instant, 10 machines (BE2c) of 10 Sqn and two complete flights (Vickers Fighters) of 11 Sqn arrived at GHQ."

The Squadron was placed under command of No 1 Wing RFC, attached for reconnaissance purposes to 1 Brigade of First Army. (No 1 Wing was commanded at that time by a Colonel H Trenchard, later the RAF's first Chief of Air Staff.) It moved on to Aire for some days before arriving at its operational location on 7 August. This was to be an airfield by the Chateau de Werppe at Gonnehem, near Chocques, in the Nord Pas de Calais, where the remainder of the unit groundcrew caught up with the aircraft. The airfield had been created by No 3 Squadron in November 1914, when a beet field was trodden and rolled into a marginally useable surface, and later given repeated applications of cinders to permit regular use. 3 Squadron moved out on 1 June 1915, to be replaced by 16 Squadron till 18 July, and 10 Squadron would prove to have the longest stay there of any unit, moving out in November 1917. There were a number of HQ units in the general area, together with No 1 Casualty Clearing Station. (The officers lived in the Chateau with the de Sars family – "sans trop de heurts" according to one source – and a later squadron document suggests that Béthune, the nearest town of any size and relatively safe at the time behind the lines, had a reputation for "its greeny-mauvy oysters which seemed to be devoured with great gusto.")

Ira "Taffy" Jones, who ended the war as one of the RFC/RAF fighter 'Aces,' began it as an airman Radio Operator with 10 Squadron and, in his book "Tiger Squadron," described his arrival at Chocques after three days on the road from Le Havre:

> "Over a bank of trees behind the aerodrome shone the white turrets of an ancient chateau - the first I had seen outside picture books. This, I discovered later, was to be Squadron H.Q. and officers' mess. The lorries rolled on past the aerodrome, turned right down a narrow lane and came to rest in the yard of a tumbledown farm. The farmer, his wife and their pretty, sixteen year-old daughter waved and cheered from the door of the house as we pulled in. The daughter's greetings were returned with particular fervour.
>
> "I had visions of a nice quiet bed in a room under that lichened overhanging red roof. Instead, we were marched off in flights to the different outbuildings. My billet that night was a palliasse on the floor above the cow byres. It was clean enough, but somewhat strongly scented. Dumping

Above: *10 Squadron's base at Chocques, near Bethune, France from August 1915 to November 1917. The Chateau de Werppe, off frame to the right, was used as a Squadron HQ and the Officers' Mess. It exists still today.*

Left: *A pre-WW1 postcard of 'Le Chateau du Werppe', in the Gonnehem commune of the Pas de Calais.*

our kit, we got busy immediately on the job of going into occupation. The aircraft flew in from another aerodrome within a couple of hours of our arrival. By nightfall, 10 Squadron was operational."

In fact, the unit spent some time working-up and gaining familiarity with the area and it was not until mid-August that its first operational sortie was flown and, from the outset, a range of reconnaissance tasks was undertaken. Work with artillery batteries was particularly significant – "Registration" (direction of friendly fire or 'Shoots'), and counter-battery Artillery Patrols (later termed Flash Reconnaissance) during which crews set out to spot and report muzzle flashes from German guns. Given the essentially static situation on the ground that had developed, height had become critically important in gaining a picture of what was happening on the enemy side. Thus, photographic sorties were important to keep up to

date with changes in German trench lines, normally using a vertical camera mounted on the fuselage side, beside the pilot. Conversely, 'Long' or 'Distant' Reconnaissance sorties were also tasked for information relevant at Army or GHQ level. In time, Contact Patrols, flown to try and keep abreast of raids or advances by friendly forces and Counter Attack Patrols, seeking out German advances, would be significant developments - with much to learn about the practicalities of making contact with troops reluctant to draw attention to themselves, and many visual and aural devices were trialled. Patrols were also flown to seek out enemy aircraft and crews became increasingly exposed to the reality of aerial combat.

The reconnaissance camera used by the pilot in the BE2c.

Indeed, from the outset of the war it had been apparent that, with both sides using aircraft, the use of airspace over and beyond the lines would be contested. (The RFC was directed to be particularly aggressive and the majority of contacts would occur over the German lines.) The lesson was soon learned and it is noteworthy that No 11 Squadron deployed to France with 10 Squadron as the RFC's first dedicated fighter squadron, equipped with two-seat Vickers FB5 'Gunbus' Fighters. Although things had moved on from the earliest days when, for lack of aircraft armament, combat had often involved exchange of small-arms fire, the BE2c's single Lewis gun with limited arcs of fire offered only limited protection. (Ira Jones' book has a story from his time as a trainee observer with 10 Squadron. During his first air combat, flying with Captain O'Hara Wood, a German fighter got underneath their aircraft. He was so eager to continue the engagement that he lifted his Lewis gun from its mounting, leaned over the side of the aircraft and fired a burst at it. Unfortunately, the combined effects of recoil and slipstream pulled the gun from his hands, and it fell to earth! Happily, the aircraft got home safely where, presumably on account of the aggression shown, he was forgiven and no more was said.) Nonetheless, the Squadron's first casualty in France was not suffered in combat - an observer, 2Lt E R Nagel, was injured when his aircraft hit a tree on takeoff and crashed on 23 August. The first extant 10 Squadron Combat Report is dated 31 August 1915, and was submitted by 2Lt Fairbairn who described diving on a LVG aircraft whilst on Long Reconnaissance, south west of Douai, well beyond the lines. His observer, 2Lt James, fired as they flew to within some 50 – 100 yards, probably hitting the German observer who was seen to fall away from his gun. They gave up the chase at 5500 feet as their engine was missing badly.

In September, the Squadron was provided with a Bristol Scout fighter to provide a measure of organic fighting capability and Captain Gordon-Bell, a notable pre-war pilot, used it to achieve an early victory in driving an enemy aircraft to the ground. On 13 September a Squadron aircraft was surprised and badly damaged by a Fokker aircraft but managed to recover to base. In his report, 2Lt Quinell noted that his lower petrol tank had been pierced, he had sustained damage to all wing surfaces and the tailplane, with some armoured plating punctured in two places and an engine bearer shot through.

Then, on 21 September, the first crew was lost in action. 2Lt Caws and Lt Sugden-Wilson, the observer in the 13 September incident, were attacked by two Fokker aircraft during a general reconnaissance sortie and shot down by Leutnant Max Immelmann who would go on to become the first German 'Ace.' 2Lt Caws was killed, and was not only the first Squadron death in action but also the first Canadian airman to be killed. As their aircraft began to spin out of control, his observer managed to gain some control before it hit the ground but it caught fire on doing so. Lt Sugden-Wilson, himself badly injured, tried unsuccessfully to extricate his pilot's body from the burning wreckage before being taken prisoner. He subsequently wrote from Germany to his parents:

"We had a great fight lasting fifteen minutes. My back and legs are progressing as well as can be expected. Up to the present I have been in four different hospitals, and am as cheerful as my position permits."

(2Lt Stanley Winther Caws was born in England, served in the Boer War and became a prospector in Alberta. An early member of the Legion of Frontiersmen, he joined the 19th Alberta Dragoons in 1914 before transferring to the RFC in February 1915.)

Another squadron pilot, Lt H A P 'Pat' Disney, was one of those who had flown to France on 25 July and copies of a number of letters to his parents are held in Squadron archives. On 14 September he wrote:

"The weather has been very hot and fine lately and we have been as busy as can be. For instance, the day before yesterday I was up for three solid hours without a stop, during which time I blew up a railway and, I hope, a Bosch battery by directing four heavy 60 lb guns on them. The railway was a great touch. They gradually crept up to it with their shots and then about the sixth shell burst plumb between the rails. When the smoke cleared away I could see a great cavern broader than the track. I don't know how many hours flying I have done the last few days, but it must be somewhere round 15-20. There is lots more to come too ...... My engine gave up the ghost yesterday and I subsided gracefully into a potato crop but it is all right now."

RFC effort over those days in September was largely devoted to preparation for 'The Big Push' that began on Saturday, 25 September, later to be known as the Battle of Loos. A major French assault had been planned, centred on the Champagne region, the British had been asked for support to the north and six Divisions were committed. The squadrons of No 1 Wing were heavily involved and 10 Squadron had carried out extensive reconnaissance in advance, including two artillery ranging or 'Registration' successes mentioned in communiqués. In the battle, it was tasked to support the Indian Corps and its Meerut artillery in the north. Two Flights were used for this purpose and, with the heaviest fighting occurring to the south, the Squadron was able to use its third for strategic reconnaissance to assist First Army HQ, penetrating as far forward as Valenciennes, some 45 miles away, no doubt in association with bombing raids flown by aircraft of Nos 2 and 3 Wings. In a letter dated 25 September, Pat Disney reported what he could of the day's activity:

"As a matter of fact we were flying until the rain came down so hard ....I had two hours of it myself this morning, and drizzle most of the time. I wish I could tell you all that has happened since dawn this 25th[h] of September, which I shall never forget, but it would only be cut out by the censor if I did..... I have had awful bad luck with engines lately. It was a new one this time after Wednesday when my engine smashed inside, right over the trenches, but I was very high and made the aerodrome all right. On Monday I had a great strafe. I had to go up and direct a battery of the RGA onto an important house. They fired two shots and I signalled down the code correction, they realigned the gun and the next shot sent the house flying. It must have been a store of some sort as it was burning merrily 48 hours afterwards. It was a great sight seeing it blow up and the gunners were awfully cheered."

On what was probably the following day he added:

"The air was a wonderful sight this morning. We had about a dozen machines up over our bit of front and it was the same all along. We all had machine guns and not a single Hun machine has showed its nose all day."

On 29 September, he wrote briefly:

"There has been the hell of a battle out here as you know by now. We have had the devil of a lot to do but the weather has been against us. It is pouring today and the RFC is practically useless. I went out this morning but was driven in by clouds and rain. I got up to the trenches and I could not get higher than 1300 feet and could see nothing, so I beat a hasty retreat..... There is no news that I am allowed to tell you."

(Pat Disney fell ill some days later and was invalided back to the UK. He was eventually permanently grounded, served till 1919 in the Equipment Branch and went on to be a senior

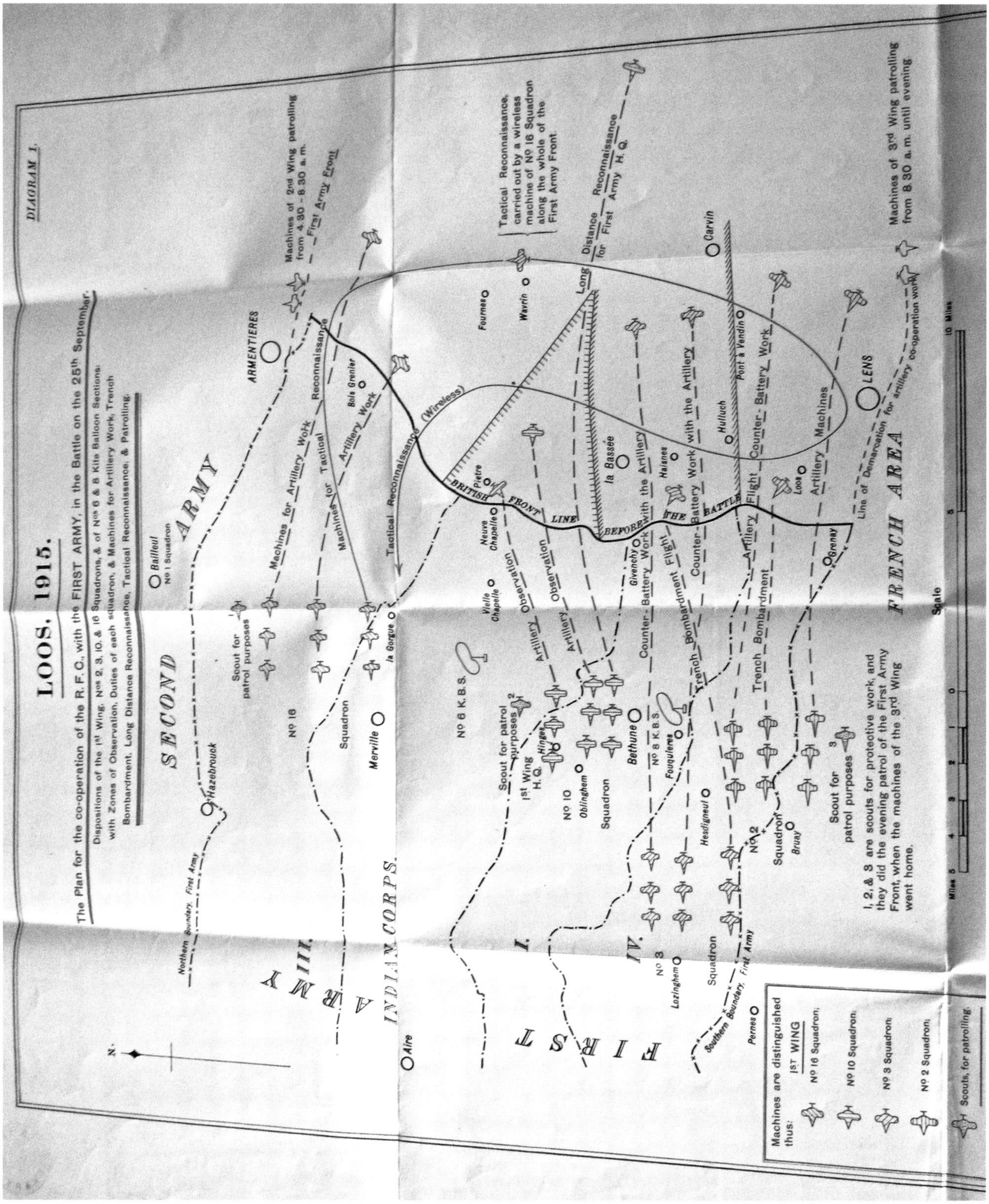

Battle of Loos September 1915, plan for cooperation between the RFC and the First Army, showing 10 Squadron's task.

Civil Servant in the Air Ministry. He rejoined the RAF on the outbreak of WW2, as did his four sons, and retired as a Group Captain.)

The Battle of Loos went well for the British on the first day but an inability to exploit that success meant that, in the end, only minor gains were made. Nonetheless, the RFC had performed well and, on 4 October, Field Marshal Sir John French, CinC British Army in the Field, issued a Special Order of the Day expressing his gratitude to all ranks for

> "the valuable work they have performed during the battle which commenced on the 25th September. He recognises the extremely adverse weather conditions which entailed flying under heavy fire at very low altitudes. He desires especially to thank pilots and observers for their plucky work in cooperation with the artillery, in photography and the bomb attacks on the enemy's railways."

More generally, in the edition dated 1 October 1915, the Editor of Flight magazine, having quoted enthusiastic reports by newspaper correspondents near the Front, concluded:

> "These measures of praise are but a faint reflection of the changes of moment which aviation has brought into the methods of warfare. As the years go by these changes will be still more pronounced. We look forward to our administrators seeing in the future that, as on the sea, there shall be no scintilla of doubt as to the command of the air being with us."

Unfortunately, even as he wrote, command of the air was being put in doubt as the Germans fielded more Fokker E1 Eindekker monoplanes. These combined agility with a gun that, thanks to the invention of a synchronisation gear, could fire through its propeller disc and this capability made it a distinct threat to Allied aircraft of the time. In comparison to others, 10 Squadron did not suffer too badly during the months later known as the 'Fokker Scourge' but, as the front to be covered grew to some 10 miles in length and, as more Eindekkers were deployed, squadron aircraft were involved more frequently in aerial combat as they went about their business. Crews were fortunate to survive some encounters with only aircraft damage and, on 10 October, having taken a hit whilst fighting one off during a Registration sortie, 2Lt Hallam, observer with 2Lt Fairbairn, reported having then fired four shots at a RFC Morane aircraft, before realising it was British.

Around the time of the Loos campaign, Squadron pilot, Lt John Lloyd Williams, had written home to a friend:

> "... it is now two months since we flew over from England and we settled down to work ....So far, we as a squadron have been very lucky, only having lost one man since we came out. Machines must be very hard to hit; sometimes machines are surrounded with shells bursting, but it is seldom any damage is done. My machine has only been hit on three occasions, and even then not damaged enough to cause me to come home until the job in hand was finished..... the Infantry are the people who are having the rough time in this war and by far the most trying ..."

A month later John's luck ran out as he became one of the Squadron's next casualties in a remarkable encounter on 26 October. During a photographic sortie near Lille, he was attacked by a Fokker and 2Lt Hallam, his observer, was hit in the hand, preventing further use of his Lewis gun. Whilst manoeuvering, Lloyd Williams was also hit and began to lose consciousness. As the aircraft began to spin, Hallam realised what was happening, climbed back between the struts and managed to turn off the fuel supply and gain sufficient control of the machine to land it behind French Army lines. The machine turned over and Lloyd Williams was thrown out, badly injured, but Lt Hallam managed to climb out and both men were passed on to the French Red Cross. A 2 Squadron observer on a liaison visit to a nearby artillery battery saw what had happened, reached the aircraft, and was able to salvage the Lewis gun and instrument panel. Two days later, after dark, 10 Squadron Scout pilot, Captain Gordon-Bell, took a party of four men and recovered the engine, still serviceable, under fire!

(Captain Charles Gordon-Bell joined the RFC as an experienced pilot. He had qualified for the Royal Aero Club's Aviator Certificate No 100 in July 1911 and later became a senior test pilot for the Short Brothers company. He deployed to France in August 1914 with the Air Park element of HQ RFC and, although injured, had survived an early forced landing near Mons. His time on 10 Squadron was cut short when ill health led to his being returned to England at the end of 1915. By that time he had amassed five victories, with opponents either forced to land or destroyed, and he has a legitimate claim to be recognised as one of the RFC's earliest 'Aces.' He remained with the RFC/RAF and died in an aircraft testing accident in France in July 1918.)

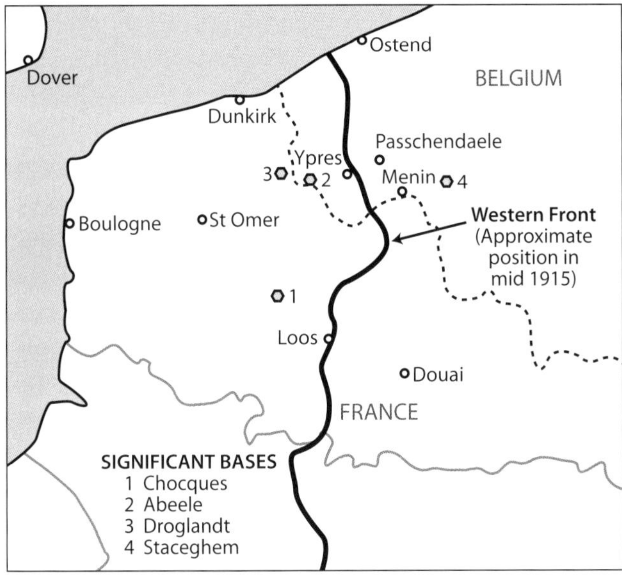

The next squadron losses occurred on 7 November when Capt Gordon-Bell, flying as escort in his Bristol Scout, was unable to protect a crew from two attacking aircraft. His report was given in full in a RFC communiqué:

> "I accompanied Lt Le Bas and Capt Adams who were in a BE2c 1715 doing long reconnaissance. At about 2.45 pm over Douai a Fokker and an Aviatik were seen. I was about 2000 feet higher than the BE2c and about the same height as the German machines. I engaged the Fokker and got off about 1 drum at him; he dived as if hit. I then turned to the Aviatik and got off ¾ of a drum at him when it dived and broke off the fight. I now went to look for the BE2c and saw it doing a steep left-handed spiral, with the Fokker circling down round it. The BE2c apparently hit the ground nose first near Quiery. The fuselage appeared to be broken. The two hostile machines then landed close beside it. My engine was only giving 900 revs, and I was at 4000 feet so made for home, crossing the trenches at about 1000 feet and landed near Verguigneuil, the machine turning over on its back in the plough."

Although the BE2c crew were both killed in this action, Captain Gordon-Bell survived his forced landing. His aircraft was recovered, with the records showing him in combat during two sorties on 16 November. His colleagues were also regularly engaged in air combat in the last weeks of 1915 – they all survived and the only casualty reported over the period was observer 2Lt Moss, to AA fire on 8 December. They were generally aggressive encounters with all ammunition being expended according to the RFC communiqués whilst, on other occasions, fights had to be broken off once the observers' Lewis guns had jammed. An action reported at some length took place on 14 December, when Captains Mitchell and Bruce were escorting colleagues Lieutenants Sison and Goodson who were on a First Army recce task. The recce aircraft was attacked but fought off the enemy machine, which was last seen diving steeply to ground. The escort then took on a second attacker, an Albatros, and drove it off eastwards. An unfamiliar aircraft now appeared. It attacked the recce aircraft, which returned fire, and the newcomer appeared to head away. Captain Mitchell gave chase and a close battle ensued, where it became apparent that the enemy biplane had three in-line cockpit positions for two gunners, with a pilot between them. Fortunately, the front gun jammed, but the rear gunner kept up continuous fire as the aircraft headed away. (This was thought to be the first sighting of the Goth G1, an aircraft with a limited and undistinguished service record.)

A new task was added to the Squadron repertoire in late November and early December when, after some practice sorties, crews participated in bombing raids on the railway junction and station at Don, southwest of Lille, mounted by large groups of No 1 Wing aircraft. (Bombing was often carried out by pilots alone, using a very basic sighting device.)

On 19 December, German aircraft were reported as having been particularly active in attempts to limit British observation, with some forty-eight encounters recorded. Captain Mitchell, now solo in the Bristol Scout, was engaged

in two of these, driving off enemy aircraft in each one. Nonetheless, despite these fortunate outcomes, the last weeks of 1915 had shown that effective reconnaissance had become almost impossible for lone aircraft. Although 10 Squadron was relatively fortunate in having only one pilot wounded by anti-aircraft fire in early December, losses elsewhere were mounting. Accordingly, in January 1916, RFC HQ issued an instruction that every reconnaissance aircraft was to be escorted by at least three fighting machines.

1916

In January 1916, as it continued to increase in size, the RFC was reorganised. (There would be 27 RFC squadrons in theatre by mid-year; just a year previously, 10 and 11 Squadrons had been the ninth and tenth to arrive.) Each Brigade now had a Corps Wing, commanding squadrons with a primary reconnaissance role, and an Army Wing commanding its fighter squadrons. Thus, 10 Squadron was now a Corps unit of 1 Brigade, still under 1 Wing RFC and, since November 1915, working with XI Army Corps. This meant, in general terms, that it could be tasked with a variety of duties, largely in the area of interest of the Corps to which it was attached. Whilst it would be tasked from time to time with bombing raids, its main role was Corps Reconnaissance, a term that grew to encompass a variety of activities – detailed recce coverage of the corps front, extensive photography, artillery registration and patrols of various kinds. However, where bombing missions were tasked, it is interesting to note that the RFC communiqués frequently describe these as having been 'in retaliation' for German bombing attacks on British military locations. For example, on 23 January, aircraft of both Nos 2 and 10 Squadrons were tasked on one such raid using the large 112 lb bomb, but had to return as the target was hidden under fog. And the rail facilities at Don were attacked once again – a mass raid of twenty-six 1st Brigade aircraft armed with the 112 lb bomb returned there on 26 February.

Aerial combats continued to occur. Indeed, the redoubtable Capt Gordon- Bell got involved in aerial scraps whilst patrolling on the first two mornings of the year. On 1 January his gun jammed, ending the fight. On the next day he attacked a 'large Albatros,' firing off a drum of ammunition. Having reloaded, in his eagerness to get tightly underneath the German aircraft to train his gun upwards, he got too nose high and stalled, turning over on his back and needing some 2000 feet to recover!

Detailed Squadron records only survive from February 1916 and looking at a sample week gives a good impression of the task at that time. Nine pilots and twelve aircraft are

On a foggy morning in 1916, 10 Squadron parades at Le Chateau de Werppe, Chocques.

shown as having been available at that time:

Sunday 6 February: One completed morning Artillery Patrol – no hostile flashes observed, but three train movements reported to First Army. Two hours later, a second crew returned after ten minutes on hitting cloud at 800 feet.

Monday 7 February: Morning – cloud defeated a Registration sortie, but a quick flight check of an engine was flown. The weather must have picked up after lunch as four Registration sorties were flown for four different Heavy batteries. Trenches and other targets were fired on and train movements were reported. Two Artillery Patrols were also flown, with gun flashes seen and reported – one crew saw two hostile aircraft over La Bassée, but could not get near enough to engage. Capt Gordon-Bell got airborne in the Bristol Scout after another hostile aircraft was reported but was unable to find it.

Tuesday 8 February: No flying until the afternoon, when 2Lt Evans flew a solo training flight. Capt Gordon-Bell in the Scout and Capt Kelly, with Lt Fairbairn, got airborne after reports of hostile aircraft were received. The others did not see a hostile, but did report a muzzle flash. Five Registration sorties and Artillery Patrols were flown, working with different batteries – hits were observed, including one on a trench that produced a yellow flame for about a minute, and many train movements were reported.

Wednesday 9 February: Two practice landing sorties and two training cross-country sorties were flown that morning, and two Registration sorties were flown. These had no success due to cloud, although other observations were reported. In the afternoon, a Recce training sortie was flown, together with two Photo sorties completed in cloudy conditions. Six Artillery Patrol and Registration sorties were flown with varying results – one appears to have failed on account of the answering letter put out by a battery on a green background that rendered it virtually invisible.

Thursday 10 February: Cloud defeated two morning Artillery Patrols and a recce sortie to Valenciennes with four escorts. Later on, a half-hour wireless training flight was flown, and four Artillery and Registration sorties were launched, but with cloud obstructing two of them.

Friday 11 February: No flying possible.

Saturday 12 February: No morning flying, and a weather recce flight was launched at 1225 hours. In just ten minutes, it cannot have gone far but it was enough to result in the recce to Valenciennes being attempted once again. Cloud cover forced all five aircraft to return within thirty minutes. Two Artillery Registration sorties were also attempted, but both were beaten by cloud. Two short training trips at base proved possible.

It is also worth noting that the records generally stated whether observations were made by the observer, or the pilot, or both. At that time, an observer's main task was to watch for and report enemy batteries opening fire and to look for hostile aircraft, whereas registration would be done by the pilot for, to quote a HQ RFC letter from later in 1916, "the pilot only is able to place his machine as he requires it at any moment to give him the view he desires." This had been written as a comment on proposals that Royal Artillery officers should take over the spotting role. The letter went on to state that more Artillery officers would be welcome, "but they must form part of, and live with, the squadron, and must be trained as pilots, or they will lose most of their value." Registration was a specialist process, carried out using an agreed clock-face grid, with distance rings from the target, and very much subject to the communication limitations of the time. Squadrons detached radio operators forward to artillery batteries for lengthy periods, equipped with a very basic wireless receiver attached to a 60-foot wire antenna rig, but no transmitter. (Ira Jones' book includes recollections of some of his detachments forward.) Conversely, the aircraft crews had a basic battery-powered Morse spark transmitter, with a hand-cranked streaming aerial but no receiver, and would receive acknowledgements via ground signal panels. The effective transmit-receive range was around 10 miles. (Towards the end of the war, a few 10 Squadron aircraft had Wireless Telephony equipment fitted, permitting voice transmissions.)

Later in the month, on patrol on 29 February, 2Lts Leggatt and Howe spotted an Albatros north of Bethune, dived into an attack and it flew off. A second Albatros was then seen and an attack on that commenced. Firing began at 150 yards and continued, the enemy's propellor was damaged and he

was seen to force land and turn over behind Allied lines, whereupon both crewmen were taken prisoner. Squadron personnel subsequently salvaged parts of the aircraft for souvenirs.

In the early months of 1916, the Scout aircraft with Corps Recce squadrons such as 10 were withdrawn and consolidated into fighter squadrons. The Fokker's supremacy in the air also ended around this time as both the RFC and the French fielded more capable fighter aircraft. So it is ironic that 10 Squadron suffered its next casualty during this period. On 26 April, whilst out on artillery observation, 2nd Lieutenant Milner, an observer, became the first squadron member to be killed by anti-aircraft fire (known colloquially to the RFC as 'Archie').

On 21 May 1916, the Germans attacked Vimy Ridge, on the IV Corps front. On the following day, 10 Squadron was tasked to assist 18 Squadron, the IV Corps squadron, when Souchez was heavily shelled and the bombardment of Givenchy was intensified. With fighter cover provided by 25 Squadron, both Corps units were commended on their artillery cooperation, with all observers locating active guns. Elsewhere, the communiqués continued to report actions involving 10 Squadron crews. On 1 June, whilst on artillery duty, Lieutenants Gould and Pearson attacked an enemy aircraft and, as fire was opened, they were themselves attacked by a Fokker. As it closed, Lieutenant Gould banked away; his observer fired at point blank range and the enemy aircraft tipped over to port and nosedived to earth. On 4 June, a crew successfully directed fire on to two anti-aircraft gun sites; on the next day, fire was directed on to a train that was set on fire; a convoy was scattered by ranged fire on 18 June; and other lorries were destroyed on 21 June.

Significant internal changes occurred over this period. On 2 June, having been promoted to Major, Captain Mitchell took command of the Squadron. In addition, the unit's aircraft establishment increased from 12 to 18 and, over the month, it had two new BE12 single-seat fighters on charge in succession for assessment. The second flew a few operational sorties before being returned to the Depot in mid-July (As the aircraft was basically a single-seat BE2, it had little manoeuverability and did not prove to be a success as a fighter.)

Saturday, 1 July 1916, is a day inscribed on Britain's national consciousness as the day on which the British and French Armies launched a joint offensive on the Somme with, by day's end, some 60,000 British casualties incurred. The offensive was preceded by a week of heavy artillery bombardment, some indications of which can be gleaned from the RFC communiqués – 198 targets ranged by aircraft or kite balloons on 22 June, 161 on 23 June, 154 on 25 June, 161 on 26 June, with unsuitable weather encountered on other days – and as is now well known, much of this effort was ineffectual. Whilst the main area of the battle was to the south of 10 Squadron's location, it was tasked, together with adjacent RFC Units, against recurring small scale attacks on its Army's front.

The first anniversary of the Squadron's arrival in France came around on 25 July 1916, and the records for that day show nineteen pilots on strength, with seventeen aircraft held in a mix of BE 2c and BE 2d variants. The day started with a weather checker airborne at 0625 hours for twenty minutes, followed by two crews who flew to 18 Squadron's airfield for a bombing mission on which they would be escorted by that unit's FE 2B fighters. Similar missions had been scheduled on previous days but nothing was to happen that day, and both aircraft were back at base by 0915 hours. An Artillery Patrol went up for two hours and reported "No flashes seen. Line very quiet. Clouds at 1200 feet." In the afternoon, a short Contact Patrol was flown during which messages were received from the unit tracked. A test flight of a new Mk II Compass was flown, and it was recorded as having a "very long period of oscillation." An Artillery Registration sortie had one success but was cut short by deteriorating visibility. (It was nonetheless able to record a train shunting at a Sugar Factory, the object of regular reporting by Squadron crews at the time.) Other Test flights were flown, with local visibility reducing markedly, and a later Registration sortie came home after its wireless set shorted out, "probably hit by a machine gun bullet." Another Artillery Patrol was attempted at 1930 hours, but the crew landed back after twenty minutes, reporting cloud at 1200 feet and ground mist precluding all observation. Conditions were better for an Artillery Patrol flow by 2Lts Woodley and Leckie on the evening of 29 July, and their report provides some good examples of the detailed reporting that could be carried out when conditions were favourable. They were airborne at 1845 hours and things were "very quiet up to 2000 hours." This permitted them to observe a new trench being cut out from Lorgies crossroads, quite straight and possibly for cable or drainage. Another trench showed fresh work and was continuous between two map positions, and Engineer stores were being dumped at a map position alongside a light rail or trolley line. A trench between two given positions now had what appeared to be circular pits at each end, and possibly unloading loops to the trolley lines running to these points. Then, at 2010, a British trench mortar fired about eight rounds, all of which fell in No Man's Land. The Germans retaliated and muzzle flashes were spotted and reported to a Section of 136 Heavy Battery, but no bursts were seen in the darkness and haze. However, at 2030, more muzzle flashes were seen and a different Section of '136 Heavies' engaged. Then, at 2045, both Sections were passed the coordinates of a point on a road, and a call to fire was made as four lorries reached it. Bursts were seen in the area but the crew lost sight of those vehicles. Nothing daunted, five others were spotted and their position was passed to both Batteries but, as it was by then too dark to see whether fire was effective, the crew called the shoot off and returned to base, landing at 2120 hours. All in all, a fair evening's work that had been accompanied by erratic anti-aircraft fire, "probably a moveable gun."

During the Somme campaign, the Squadron incurred a number of casualties and losses to anti-aircraft fire. 2Lt Bickerton and 2Lt Harris, both observers, were injured on 31 July and 13 August whilst, on 22 August, 2Lt Marnham and Lt Tetlow were shot-down and killed during an Artillery Registration sortie. The war came close to the Squadron's base at Chocques on 7 August when positions in Bethune were shelled. Gun flashes were roughly located and Capt Rodwell and 2Lt Bickerton quickly returned to action. The gun's position was later located precisely by another crew, and subsequently bombed. Bombing raids were also carried out by day and by night on enemy communications, particularly on parts of the rail network radiating from Lille. This pattern continued into September when it was intensified to disrupt the transfer south of German reinforcements. On 3 September, whilst they were flying as escort to colleagues on such a mission, Pte Philips, observer for 2Lt Gregor, was wounded by attacking aircraft and subsequently had a leg amputated. A significant attack took place on 17 September when twelve aircraft of 25 Squadron, plus seven each from 10 and 16 Squadrons, dropped twenty-two 112 lb bombs and seventy-five 20 lb bombs on Douai station. Sections of line plus sheds and trucks were destroyed, and the station buildings were set on fire. Similar results were obtained by another joint attack on 22 September on a rail junction at Somain, with further attacks on the St. Sauveur station in Lille and at Seclin on the two following days.

Writing in the 1950s, the practicalities of night landings were recalled by Lt P R Mallinson:

> "Landing at night on the primitive airfields of those days with their cinder tracks was a hazardous task. The 2c's skimmed over sunken roads and ragged poplars that so frequently obstructed the approach, the observer fumbling in a darkened cockpit for the Verey light pistol. A red rocket signalled the aircraft's approach and in response a searchlight threw a guiding beam across the airfield from the top of a Leyland power plant. The pilot thrust back his goggles and side-slipped in with the propeller just ticking over and flattened out. Immediately the undercarriage wheels rumbled on the cinder track, the searchlight went out like a snuffed candle lest it attracted the attention of a marauding enemy night bomber."

He went on to add:

> "Always the mechanics waited to attend to their charges, men whose eyes were heavy with lack of sleep, their faces sometimes grey with exhaustion. During the height of the Somme battles the fitters and riggers and armourers worked day and night, snatching food and rest where they could. No history of the RFC would be complete without a tribute to these loyal devoted men. When their particular aircraft failed to return and they had to turn away empty-handed from the last machine to land, their fatalistic expressions and muttered curse often masked sincere grief."

The Squadron's BE2c aircraft were progressively replaced during the second half of 1916 by a number of improved versions – BE2d and e and then BE2f and g. Perhaps the most significant change in the 2d version was the addition of dual controls; most were then used as trainers, with only a few flown operationally. The BE2e incorporated the dual controls, but with more substantial design changes. A shorter lower wing meant that only a single interplane strut was needed, and a strut connecting the upper and lower ailerons was introduced, with alterations to the shape of both tailplane and vertical stabiliser. The outcome was an increase of some 10 mph in top speed and an improved rate of climb. However, these marginal improvements were more than matched by more capable German aircraft being introduced at the same time and, as armament, crews had still only a single Lewis gun. In October 1916, a standardisation instruction was issued whereby all modified BE2c aircraft would be designated BE2f, with rebuilt BE2d machines designated BE2g.

As these newer BE2s came on charge, the Squadron's operating pattern of artillery observation, aerial photography, and occasional bombing continued for the remainder of 1916. That said, enemy air activity increased noticeably from around mid-October. For example, on 20 October, when the Squadron was on a bombing sortie attacking camps and billets at Petit Santay, the formation was attacked by nine hostile aircraft of which the escorting squadron, No 25, was able to shoot down two. Similarly, two days later, during a raid on Seclin station, six Roland Scouts attacked the 10 Squadron bombers, but were again engaged by 25 Squadron FE2 fighters that succeeded in downing three attackers. Between these actions, the first BE2d to fly with the unit was damaged by anti-aircraft fire on 21 October, during an artillery observation sortie. The pilot, 2Lt J A Simpson was hit and, although he was able to land safely, saving his observer, he died of his wounds on the following day.

An interesting footnote around this time was the arrival of four Portuguese officers to fly with 10 Squadron. Germany had declared war on Portugal in March 1916 and the creation of an Expeditionary Corps was put in hand for later service with the British Army. A number of officers were sent to both France and England to train as pilots and four of these joined 10 Squadron for some months into 1917 before moving to join their colleagues on French squadrons.

The Squadron CO, Maj G.B. Ward MC & bar, took but three years to be promoted from private to major. He was killed in action on 21 September 1917 aged 26 yrs, fighting several enemy aircraft single-handedly.

As the winter progressed, flying conditions deteriorated and fewer sorties could be flown. That said, on the morning of 11 December, whilst escorting a photographic mission, 2Lts Dampier and Barr (a US national) were shot down by an enemy aircraft and killed. Later that month, a change of command occurred when Major Mitchell was replaced by another highly experienced officer, Major G.B. Ward MC, on 17 December.

1917

With so much RFC activity at the time dependent on the ability of crews to see the ground and report accordingly, the weather over the front line areas was critical to obtaining results. For example, visibility on 22 January was particularly poor so very limited operational flying was possible. Nonetheless, a number of training sorties were launched and, unfortunately, whilst returning around noon, two aircraft collided over the airfield. Capt Foster, a pilot, survived with injuries. However, his observer, 2Lt

Above: *The Flares.*

Left: *There was a time for relaxation: the programme cover of the Squadron 'variety' show group, 'The Flares', who performed at Chocques.*

Right: *Christmas Day 1916 : a full programme of entertainment at Chocques.*

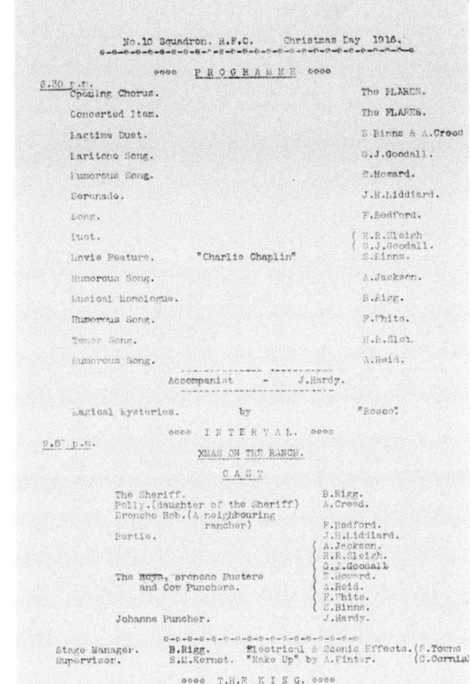

No.10 Squadron. R.F.C. Christmas Day 1916.

oooo PROGRAMME oooo

6.30 p.m.

Opening Chorus.		The FLARES.
Concerted Item.		The FLARES.
Ragtime Duet.		S Binns & A.Creed
Baritone Song.		G.J.Goodall.
Humorous Song.		E.Howard.
Serenade.		J.H.Liddiard.
Song.		F.Bedford.
Duet.		H.R.Sleigh G.J.Goodall.
Movie Feature.	"Charlie Chaplin"	S.Binns.
Humorous Song.		A.Jackson.
Musical Monologue.		B.Rigg.
Humorous Song.		F.White.
Tenor Song.		H.R.Slei.
Humorous Song.		A.Reid.

Accompanist - J.Hardy.

Magical Mysteries. by "Rosco"

oooo INTERVAL. oooo

9.0 p.m.

XMAS ON THE RANCH.

CAST

The Sheriff.	B.Rigg.
Polly.(daughter of the Sheriff)	A.Creed.
Broncho Bob.(A neighbouring rancher)	F.Bedford.
Bertie.	J.H.Liddiard.
The Boys, Broncho Busters and Cow Punchers.	A.Jackson. H.R.Sleigh. G.J.Goodall E.Howard. A.Reid. F.White. S.Binns.
Johanna Puncher.	J.Hardy.

Stage Manager. B.Rigg. Electrical & Scenic Effects. (S.Towns (C.Cornia
Supervisor. S.M.Kernot. "Make Up" by A.Finter.

oooo T.H.E KING. oooo

White, survived to die the following day, whilst both 2Lts Woodley and Kellett in the other aircraft were killed in the collision, the most serious incident of its type to affect the Squadron during the war. Better weather was experienced over most of the week in early February that can be compared with same week in 1916, mentioned earlier. The aircraft in use by this time were all BE2e, BE2f or BE2g.

Sunday 4 February: Five Artillery Patrols flown but heavy mist prevented observations all morning, and nothing was seen in the afternoon. A successful photographic sortie was flown over midday, with six plates exposed over three areas of interest. Two engine tests were carried out and a practice bombing sortie.

Monday 5 February: After a short weather check in early afternoon, six pilots set off on a bombing attack on the airfield at Douai. One returned with engine trouble, but the others were able to complete the attack. Each carried two 112 lb bombs and, dropping from 9000 feet, all saw bursts on the airfield and close to hangars. Later in the day, 2Lts Crow and Varcoe dropped two 20 lb bombs on a communications trench whilst on Artillery Patrol.

Tuesday 6 February: First up was an Artillery Patrol – no hostile activity was seen but a number of possible gun positions were reported. Rigging and engine tests were flown, with two training flights, and a new aircraft was flown in from the Depot. Other Artillery Patrols and Registration sorties were flown but with only partial success thanks to the weather. All dropped the two 20lb bombs carried, with one crew claiming bursts in a trench. (The records suggest that it was now standard practice to carry these bombs on patrol sorties for use against opportunity targets.) Operations continued after dark. Four pilots flew practice landings and three set off with pairs of 112 lb bombs to bomb Provin airfield. Lt Sisley's report is interesting in that he shut off his engine and glided down to 1500 feet to make his attack, whilst Very lights were being fired that suggested his was a German aircraft. Two other crews attacked rail targets. (Lt Max Sisley was the eldest of three brothers from Toronto who flew with the RFC. In WW2, he joined the RCAF in October 1939 and was a Group Captain at war's end.)

Wednesday 7 February: Sorties were flown from after dawn to after midnight, together with various test and practice flights. A morning Registration sortie reported its battery shooting wide, possibly due to frozen ammunition, and the crew saw several hostile aircraft and had a fight with one. Other Patrols had better results and dropped bombs on a variety of targets. After dark, some night landings were flown whilst six other pilots flew bombing attacks on airfields at Provin and La Pouillerie, and two Artillery Registrations were carried out.

Thursday 8 February: Seven Artillery-related sorties were flown, with partial success due to poor visibility obscuring the fall of shot, but two photographic sorties were more fortunate. 2Lt Chamier spent the afternoon flying short practice Contact Patrols, each with a different soldier from 112 Brigade in the observer's seat.

Friday 9 February: Operational flying began with a morning Artillery Patrol off at 0745 hours - 2Lt Ure and AM Bale reported on the weather (Visibility fair), no gun flashes were observed but an enemy aircraft was seen near Lille, heading north as three Sopwiths flew towards it. Train movements were noted and targets found for the two 20 lb bombs. Five practice flights and two air tests were flown. Eight Artillery sorties went up, some more productive than others – two aircraft had to give up with wireless problems – and Capt Robeson took off alone on a ninth at 2335 hours and at 0015 hours found a target for his twelve 20 lb bombs. After landing at 0055 hours, he recorded the sortie as Unsuccessful for, having ranged a battery on a target, he was only able to observe two bursts from eleven salvos fired.

Saturday 10 February: There were two mid-morning sorties – photography and Registration – both reported as Successful. In the afternoon, seven aircraft were sent to bomb an entrenched Infantry HQ whilst, related to this attack, two crews went up to range guns and take photographs of the area in question. Two other highly successful Observation patrols were flown, with the last up having to return as the light failed. Unfortunately, a slightly earlier Artillery Patrol, possibly aligned to the bombing attack, was hit by Anti-Aircraft fire at 1545 hours and came down west of the Lines. Both Lt W A Porkess and 2Lt E Roberts were killed.

On 17 February, 2Lts Ure and Manuel of 10 Squadron flew the only RFC sortie possible that day, a Contact Patrol monitoring progress of an XI Group raid into enemy trenches. No RFC flying at all was possible during the following week, with only limited flying carried out in the next. Nonetheless, RFC HQ staff had by now some experience of the effect of winter weather on flying and, in November 1916, had issued a comprehensive list of reminders to commanders on how breaks in operations might be used to good effect in training. This started with 'Bomb Dropping' and continued through to 'Pitching and Packing Aeroplane Tents.' It went on to acknowledge the changing reality of the Corps in that

> "Many of the Squadron Commanders have little or no experience of regimental soldiering, and there are numerous other points which they are likely to overlook."

So a further list followed, from 'Drill' to 'Improvement of Sanitation etc.'

On the ground, the Somme campaign had ended in November 1916 with Allied forces advancing less than 10 kilometres. The Germans then held their ground into early 1917 and, in February, withdrew some 50-60 kilometres to the newly constructed and fortified Hindenburg Line – a move that shortened the front to be defended. In early April, the British launched what would be known as the Battle of Arras, a 5-week campaign to breach the German line that called for extensive artillery liaison. 10 Squadron worked with XI Corps artillery and carried out photographic sorties and general reconnaissance. In addition, early in the year a flight of six aircraft was placed at the disposal of HQ RFC for bombing raids in the Douai and Lille areas that continued throughout the year by day and now also by night. Night bombing posed particular challenges as there had been no opportunity to train for the role. (This move was at the instigation of Brigadier Shephard, who had been the original Squadron CO in January 1915. Now the commander of 1 Brigade RFC, he had become a particularly strong proponent of bombing and a proportion of each of his corps squadrons was now devoted to that role.) The Squadron's next loss in action occurred on one such raid, in the early hours of 14 April when a number of aircraft had been dispatched on bombing raids (normally recorded as 'Special Missions') to a variety of targets. These were all flown by pilots alone and 2Lt C W D Holmes, forced to reduce height by cloud, was brought down by machine gun fire that punctured his fuel tank. He crashed and was eventually taken prisoner. Capt Snow, who had the same target, Henin Lietard, reported being unable to reach it due to low cloud and mist and bombed an alternative target at Vendin Le Vieil.

At the same time, RFC squadrons entered a period in which their German opponents became significantly more aggressive, thanks to their having newer and more capable aircraft. The RFC would be much better placed in the second half of the year as better fighter aircraft became available, but its most difficult days would come first. As the need for ever deeper reconnaissance beyond Arras grew, RFC losses mounted steadily with April 1917 becoming known as 'Bloody April.' So many escorts had to be provided for each recce aircraft that the RFC's numerical superiority was reduced. No doubt thanks to the efforts of the escorting units, 10 Squadron was comparatively fortunate during this period. (April 1918 proved much more costly.) 2Lt A W Watson was forced to land, but without injury, after his aircraft was damaged by AA fire during a 'Special Mission' on 23 April. On the evening of 25 April, Lt Kann was wounded in combat with an Albatros during an artillery sortie but survived, as did his observer. However, on the afternoon of 26 April, Lt Roux and 2Lt Price were brought down by a German aircraft during a photographic sortie and crashed behind German lines. Initially listed as Missing, it was later confirmed that both had become prisoners and that Lt Roux had died of his wounds.

Having survived on 23 April, 2Lt A W Watson flew a morning artillery patrol on 1 May and, a few hours later, took off with six other aircraft for some formation flying practice. Unfortunately, he was then involved in a midair collision with one of the others that managed to land successfully, whilst he and a Canadian, Lt J C Hartney, were both killed. Although the Arras offensive was coming to an end, the Squadron lost another crew to AA fire whilst out on artillery work on the morning of 5 May – Capt Lomer

was killed and Lt Bruce, his observer, died of his wounds later that day. Then, at midnight 6/7 May, 2Lt R P O Weekes was returning alone from a night bombing sortie when he hit some trees on landing, crashed, and was killed.

The British ground effort moved north for a planned offensive that would become the Battle of Messines in June. (Over this period, the Squadron was still a Corps unit of 1 Brigade RFC, supporting the First Army, together with Nos 2, 5 and 16 Squadrons.) Happily, injuries and losses were not unduly heavy during what were to be the unit's last weeks with BE2 aircraft. 2Lt B Ord was wounded in action during a photographic sortie on 27 May but, almost a month later, on 24 June, 2Lt Eyton-Lloyd and Sgt Matthews were killed when their aircraft nosedived and crashed after hitting a RFC kite balloon cable during an afternoon on which other crews were reporting fair to good visibility. The last loss of the BE2 era occurred as Lt F G Pearson was landing back from a bombing raid around midnight on 5/6 July, when a bomb that had hung-up exploded on landing and killed him. Unusually, a copy of a night recce report exists in the records over this period. It was submitted after a two-hour flight over the First Army area on the night of 7/8 July revealed that "there are by far too many lights shown," and inferred that several towns and rail stations "could form excellent targets for a hostile machine."

Over a few days in early July 1917 the squadron replaced its BE2s with the Armstrong Whitworth FK8, an aircraft it would fly until the war ended. Known fondly as 'The Big Ack,' other squadrons had flown it for some months by that time, and it had shown that it was well-suited to the Corps support tasks for which it had been designed. A sturdy and dependable machine, at 160hp, its Beardmore engine made it more powerful than any of the BE2 series but its performance was not significantly improved. The pilot now occupied the front cockpit. The larger rear cockpit was provided with dual controls, and gave the observer a much better arc of fire, both in defence, and for strafing ground targets, a task that now became more frequent. The observer still had just a single Lewis gun, but the pilot now had a forward-firing Vickers gun, with an interrupter mechanism, mounted to the left side of the fuselage. Despite the aircraft's sturdy nature, the statistics would eventually show that 10 Squadron suffered significantly

An early AW FK8 or "Big Ack," probably with its delivery crew 2Lt Gordon and Capt Saul on 14 July 1917. Later versions had a simpler undercarriage and different engine exhausts.

more casualties in operations and accidents in eighteen months of flying the FK8 than it had in its two years with the BE2s. These began with crewmen suffering wounds from AA fire on 15 and 22 July, but with the aircraft recovering safely to base. A new pilot, 2Lt Parsons, was also injured on 22 July when, during a local flying sortie, he flew into some cables. Then, on 29 July, shortly after taking off on a night bombing mission, Capt O M Conran and Lt H Mitton were killed when they had to make a forced landing for some unknown reason and crashed. And for the record, the Squadron's second anniversary of arrival in France had proved something of a washout. The weather on 25 July was a mix of low cloud and mist that was repeatedly reported as unfit for Artillery work or photography. A new pilot, Lt Mitchell, was able to do some local flying, a short practice Contact Patrol was flown, and that was all.

A British campaign, later known as the Third Battle of Ypres, was launched on 31 July, but was concentrated well to the north of the Squadron's location and the unit was not directly involved. Nonetheless, in mid-August, a diversionary attack to draw off German forces was launched in the Loos/Lens area and 10 Squadron took part in several bombing raids on airfields, troop areas and rail junctions beforehand. And by now there is frequent mention in the Communiqués of RFC aircraft carrying out attacks on ground targets, whether on enemy trenches or troop formations moving along roads. On the night of 1 September, 2Lt Rawson dropped six 40 lb phosphorus bombs on support and second line trenches, causing several fires, and then went on to fire some 300 rounds at a road convoy. Then in the morning of 13 September, another Squadron crew (Captain Atkinson and Airman Watson, back in the air having been wounded by anti-aircraft fire in mid-July) spotted an artillery barrage that appeared to be a prelude to a German raid. They circled over the battery concerned and then dived repeatedly to fire at the trench system from which they believed the raid would commence. When the front Vickers gun jammed, and once Watson had expended all of his ammunition drums, Watkinson removed the remaining belt from the Vickers and passed it back to his observer who removed its rounds into one of his drums and resumed the attacks.

There were, of course, still combat losses. Lieutenants Gordon and Cameron were shot down by anti-aircraft fire on 14 August and then, on 21 September, the Squadron's CO, Major Ward, was killed on a photographic sortie. The Times later reported the action:

> "He was killed in an aerial fight with four hostile scouts over the German lines, but his machine fell in the British lines and his body was recovered. His machine was completely smashed, and he died without regaining consciousness."

His observer, 2Lt W A Campbell had tried to land the aircraft but could not prevent a crash, and later died of injuries sustained as a result.

Major K D P (Keith Day Pearce) Murray, a New Zealander with some 2 years operational experience acquired with Nos 2 and 52 Squadrons, then assumed command and led the Squadron for the remainder of the war. Lt Mallinson's 1950s memoir includes a colourful note:

> "With the saving grace of humour, Major Murray's hat became a symbol in the Squadron, a portent of peace or storm. If the CO's headgear was observed to have strayed a few degrees from top dead centre and rested obliquely on his right ear, trouble was brewing for someone. Whereas if his hat was correctly rigged and in alignment, the CO's mood was likely to be a tranquil one and peace reigned on the tarmac."

Losses in action and injuries sustained in flying accidents continued to occur. On 27 September, 2Lt Ord was hurt in a crash after taking off on a solo night bombing sortie. Two days later, whilst on an artillery observation flight, Lt Burrell and his observer, 2Lt E A Barnard, were set upon by five German aircraft and forced to land behind British lines. Burrell was wounded, but Barnard was killed. The Ypres campaign continued into early November, with the British effort badly hindered by weather and, in October, the focus moved to Poelcappelle and Passchendaele. The Squadron was mainly engaged on artillery patrols over this period as weather conditions permitted. Another aircraft was lost, with the observer injured, when it stalled, crashed, and burned just after takeoff for a photographic training sortie on the morning of 16 October. Conditions that day were not good – two patrols packed it in after 25 minutes,

recording "gale blowing" and "high wind" respectively.

The records for 27 October include a test flight using a double gun mounting, and that day provided a reminder that it was not only the unit's aircrew who were daily at risk in action. 1AM W M Goldsmith, a radio operator working forward with 350 Siege Battery, one of several batteries the Squadron worked with that day, was killed, the first of five radio operators to be killed with forward batteries over the following months.

On the morning of 31 October, a day of mixed results for observation, 2Lt W Davidson and Lt W Crowther were shot down in combat during a photographic sortie and, although flames were reported as extinguished around 1500 feet, both men were killed. Then, on the night of 6/7 November, 2Lt K Le Gai Mills was taking off for a night bombing sortie when he hit a tree on the airfield perimeter and crashed. Three Squadron officers saw what had happened and ran to help and whilst they were trying to extract the pilot from his damaged cockpit, the main petrol tank under his seat caught fire. They succeeded in getting him free and carried him to a ditch some distance away as his bomb load began to explode, but he was badly burned and died from his injuries on 11 November. A happier incident began that afternoon when 2Lt Pattern took Lt Frackelton to the Aircraft Depot at Candas to pick up a new machine. Frackelton got back safely at 1700 hours and, whilst it was known that Pattern had also left the Depot, there was no sign of him as the day's record was closed. All became clear at lunchtime the next day, when Pattern arrived back to report that he had spotted a DH4 aircraft that had force landed and whose pilot was firing off red flares. He landed alongside to discover that the pilot had lost his way and, having given him his bearings, it was then too dark to risk a takeoff from the rough field, so he pegged down his aircraft and stayed the night!

Shortly before the British offensive at Cambrai, the RFC carried out a partial redeployment of its assets. Thus, after over two years at Chocques/Gonnehem, a relatively long period when compared with many other squadrons, on 18 November, 10 Squadron moved some 20 miles north to Abeele, just across the Belgian border, by Poperinge, 10 miles west of Ypres. The airfield lay across the road from what is now the Abeele Aerodrome Military Cemetery, an area that had been part of the airfield domestic site. (Poperinge was sufficiently far from Ypres that it remained unoccupied for most of the war, and it offered something of a haven to troops and airmen. It was also home to Talbot House – known as Toc H, from the phonetic alphabet used at the time - with its attic chapel offering a refuge for men needing a break from the battlefields, and today a photograph of some 10 Squadron personnel hangs in the hallway.) This airfield had been home to 6 Squadron for over two years and also to 4 Squadron, now taking 10 Squadron's place at Chocques, since May 1917. 10 Squadron would now belong to 2 Wing RFC, within II Brigade, Second Army, working with II Anzac Corps. In the 1993 biography of his father, Air Chief Marshal Sir Ronald Ivelaw-Chapman, GCB, KBE, DFC, AFC, his son includes a letter home from an 18-year old Lt Ivelaw-Chapman, newly arrived at Abeele in February 1918, fresh from training in England. It paints a detailed picture of conditions for the officers:

Le Chateau de Werppe in 1917

> "Well I think I have struck oil this time. I arrived at this squadron on Saturday night and was posted to C Flight. Each Flight has a different Mess consisting of 20 officers. C Flight is the Headquarters flight and the CO and Ground Officers mess with us, so you can imagine we do ourselves pretty well. We invariably have a four course and generally a five or six course dinner in the evenings; soup, fish, meat, sweets, savoury and dessert. At lunch we get a course of at least six dishes; lobster mayonnaise, cottage pie, cold brawn, ham, meat, etc. The ante room, though small, is very

comfortable and includes a piano, gramophone, and all the latest periodicals and magazines. The daily papers – all – we get just one day late. The sleeping quarters are little huts, two in each, very nicely got up with a stove, and very cosy….. The Squadron is just a little colony and quite peaceful; one might just as well be in England."

Lt Ivelaw-Chapman had a successful tour on 10 Squadron. He was made Temporary Captain as a Flight Commander some six months later, aged 19, and awarded the Squadron's first Distinguished Flying Cross (DFC). He stayed on with the RAF after the war, had a distinguished WW2, was appointed Commander of the Indian Air Force on secondment post-independence, and retired as the Vice Chief of Air Staff in 1957.

The new location did not mean that the Squadron was any less active and, within days, it lost a crew in action when 2Lt Muir and Gunner Dunsmuir were shot down and taken prisoner on 24 November. Things went better on 29 November, when 2Lts Pattern and Leycester, who had survived being hit by machine gun fire the previous day, were attacked during a photographic sortie by the leader of a flight of Albatros scouts and managed to shoot him down. It later emerged that the pilot had been Leutnant Erwin Böhme, who had some twenty-four combat victories to his name, including one earlier that day, and who had been due to be awarded the prestigious German decoration, L'Ordre Pour Le Mérite, or *Blue Max*, for his exploits later that day. Years later, Pattern related the story to author Robert Jackson:

"Leycester had started to work his camera ….. suddenly I heard the clatter of his machine gun above the roar of the engine. I looked round to see what he was shooting at and nearly had a heart attack. Slanting down from above, getting nicely into position thirty yards behind my tail, was an Albatros. I immediately heaved the old A-W round in a tight turn, tighter I think than I had ever turned before. I felt a flash of panic as I lost sight of the Hun, but Leycester must have been able to see him all right as he kept on firing. My sudden turn had done the trick. The Albatros overshot and suddenly appeared right in front of me. Because of the relative motion of our two aircraft he seemed to hang motionless, suspended in mid-air. I could see the pilot's face as he looked back at me. I sent a two-second burst of Vickers fire into him. His aircraft seemed to flutter, then slid out of sight below my starboard wing. I was pretty certain that I had hit his petrol tank. Behind me, Leycester was still blazing away. He was using tracer, and it may have been one of his bullets that ignited the petrol pouring from the Hun's ruptured tank. When I caught sight of the Albatros again, it was burning like a torch and side-slipping towards the ground, trailing a streamer of smoke. For an instant I saw the German pilot, looking down towards the ground, then the smoke and flames enveloped him. I pushed the A-W's nose down and headed flat out for home, aware that the other Hun scouts were coming down after me. They would probably have got me, too, if some friendly fighters had not come along just in time and driven them away. To say that I was relieved would be the understatement of the century."

Some years after the war, Keith Murray wrote to a magazine to correct a photo caption that had misidentified Leutenant Böhme. After detailing Pattern and Leycester's encounter, he continued:

"The big A-W was slow, but my pilots liked it for the particular job they had to do, and never regarded themselves as 'cold meat.' Owing to the nature of their work, they were rarely in a position to attack, but when attacked, as they were frequently enough, they gave a good account of themselves."

In that context, a note to the Squadron CO from the Anzac Corps HQ, dated 3 December, is worthy of mention. This conveyed the Commander's congratulations on the photographic work carried out from 28-30 November, and made a concluding point concerning the situation on the ground after three years of near-static warfare that, all these years later, is worth bearing in mind:

"He fully realises the difficulties the Squadron experienced in taking over, in bad weather, a new front on which recognised land marks had to a large extent disappeared."

That said, early December was not a good time for the Squadron. Two aircraft on Counter-Attack Patrols on 3 December were hit and both observers were wounded. One,

Lt S W Rowles, died two days later. Then, on 6 December, 2Lt A C Ross and Lt C W M Nosworthy were attacked whilst on artillery work. Their aircraft crashed and both died later that day. Their colleagues, Lt Burdick and 2Lt Fenelon, were more fortunate two days later when, on a Flash Recce sortie, they were attacked by an Albatros. Although the aircraft was damaged and forced to land, both survived unscathed.

The matter of finding effective methods whereby troops and airborne crews on Contact Patrols might contact and signal to one another, ideally without betraying positions to the enemy, was under continuous investigation. All manner of aural and visual devices were investigated and, during December, the Squadron was tasked to fly trials with troops to compare use of the 'Watson Fan' with flares. The Fan was of canvas construction and its rapid rotation was intended to catch the eye of aircrew overhead. The trial, flown on a dull afternoon, showed that at 600 feet, Fans and flares were equally distinguishable; at 1000 feet more difficulty was experienced in picking out fans than flares, but that both were equally visible once the Fan had been located; at 2500 feet, above the height normally used for Contact work, flares were more easily seen and, again, Fans were distinguishable once located. Other trials were flown elsewhere and the Fans were used operationally in places, resulting in very contradictory Army views on their effectiveness.

1918

Any partying from the night before must have been quickly put behind both groundcrew and airmen on the morning of 1 January as the agenda was set for the year. Fifteen flights were launched over a four hour period before thick mist stopped flying around 1430 hours – two air tests; a ground signalling trial; a pair on patrol reporting heavier than normal German air activity over British lines, with less on the enemy side; six artillery patrols assessed as unsuccessful thanks to weather; three photo sorties; and a final Flash Recce sortie. The first Squadron casualty of many in the year occurred two days later when 2Lt Baker, an observer out with 2Lt Cochrane on artillery observation, was wounded in combat after they were attacked by five enemy aircraft. They landed back safely, but two colleagues were much less fortunate two days later. 2Lts Garbett and Raggett lost a wing in flight whilst out on a Flash Recce and both died in the ensuing crash. It would then be some two months before the next casualties were inflicted, but operations continued as the weather permitted and Lt Mallinson later recalled the start of a standard day around this time:

> "A day's operations commenced characteristically with a shirt-sleeved batman shaking a somnolent pilot out of his blissful dreams of leave. "Five o'clock, sir." Wriggling into an oilstained Sidcot and gathering helmet, goggles and gauntlets from the litter on an upturned crate beside the camp-bed, the pilot and observer scheduled for dawn patrol would clatter along the duckboards with the taste of chlorinated tea in their mouths. Their machine would be ticking over on the tarmac, the Beardmore fluttering spasmodically in the chill morning air. Across the airfield an ominous figure would appear in the doorway of a Nissen. Major Murray was looking at his wristwatch."
>
> (Given the style of that extract, it is perhaps no surprise to learn that Percy Russell Mallinson, a pre-war journalist, went on to write many short flying-related stories in his own name, and under various pen-names. First commissioned into the Bedfordshire Regiment in 1915, he transferred to the RFC in 1916 and trained as a pilot. Head injuries sustained in a crash with a RE8 aircraft in November 1916 ended his flying career, and he was posted to 10 Squadron as Adjutant/Reporting Officer in 1918. He rejoined the RAF in 1941 and served until 1954.)

A boost to unit capability came on 2 February with the arrival of two Bristol F2B aircraft, brought in from 20 Squadron to supplement the established FK8s. (These were attached until late September 1918, when two RE8 fighters were attached for the last weeks of the conflict.) The F2Bs were two-seaters, armed like the FK8 with a forward-firing Vickers gun and a rear-mounted Lewis gun, but with a substantially more powerful 250 HP Rolls Royce engine that gave much better all-round performance. Not surprisingly, there was some competition amongst crews as to who could fly these new machines, taken from existing fighter squadrons and given to selected Corps squadrons

specifically to provide an aerial observation capability for long range artillery. For that purpose, the aircraft were equipped with a new Type 54A Transmit/Receive radio that permitted contact with new ground transmitters that had a degree of mobility, enabling them to be moved between batteries. The aircraft were to work under the Corps Wing on artillery tasks well behind enemy lines set at Army level. The squadron used these with considerable success, including directing a number of 'shoots' on a 'Big Bertha' gun emplacement.

A Bristol F2b Scout of the Shuttleworth Collection. Originally designed by the French Farman Aviation they were also made in a variety of UK factories. The 'F 'which originally stood for 'Farman' was later changed to 'Fighting'.

However, there was little that could be done with them or, indeed, the FK 8s, during the week or so that followed. Continuing the comparison of operations during the first weeks of February each year, the 1918 tale unfolded as follows:

> Sunday 3 February: A crew took off at 0750 hours for an early Flash Reconnaissance (formerly Artillery Patrol) but soon found that looking into mist and a rising sun made observation impossible. Flying resumed later in the morning as a crew carried out an aircraft exchange, and a general recce and a practice were flown. 2Lt Slattery, who had collected one of the Bristol Fighters, took Capt Ryder on a quick 15 minute check ride, after which he went solo for fifty minutes, with a practice recce sortie that afternoon. An hour after giving his Bristol Fighter lesson, 2Lt Slattery with 2Lt Hooker took off for one of six Artillery Observation sorties (now referred to as Knock-out Shoots) flown by the Squadron that afternoon. These had some success but visibility by around 1530 hours prevented anything more useful being accomplished. A photo sortie flown by 2Lts Bacon and Dennis was more successful, with twenty-three plates exposed over sixteen areas of interest. They also reported having been dived on by five hostile aircraft that were seen off by the arrival of five Sopwith Camels. 2Lt Hood with Capt Heath flew an afternoon Flash Reconnaissance sortie, with observation interrupted early on by four German aircraft. Capt Heath fired a drum from his Lewis gun at one and, with nothing further reported, that appears to have discouraged them. The crew later used both Vickers and Lewis guns from 2000 feet for an attack on enemy support trenches.
>
> Monday 4 February: Eight Flash and general artillery support missions were flown, with cloud partially obscuring in the morning. Later, there were many reports of successful friendly shoots resulting in counter-battery fire, when several new gun positions were located and recorded, as were and many train movements. Two Practice Recce sorties were flown, plus three Engine Tests.
>
> Tuesday 5 February: The day was largely occupied with Artillery Observation with fair to good results until, in late afternoon, bad light brought things to a halt. One crew had to pull out as a fuel supply problem began, whilst another worked with a Battery with a faulty gun that led to many re-ranging shots being fired. A photo sortie over the enemy's 'Back Areas' was flown, plus a variety of Practices and Air Tests.
>
> Wednesday 6 February: An early Contact patrol reported cloud at 400 feet, then two crews went up together for camera Gun practice for the observers, and each managed to each take fourteen exposures. A practice flight for a Gunnery Course was flown and flying for the day was stopped at 1445 hours – presumably on account of the deteriorating weather that meant no flying at all was possible for the next five days. Some limited local flying proved possible on 12 February before the weather clamped once again. Some limited flying was again possible on 15 February, but it was only on 16 February that normal operations were flown once again.

After the enforced break, 16 February was a busy day. 2Lt,

now Captain, Pattern had flown two daytime photographic sorties before taking off at 1855 hours as the second of three solo pilots flying night bombing missions over Menin. He dropped his fourteen 20lb bombs and saw them burst in the town but, crossing the lines on his return, he was bracketed by anti-aircraft fire and suffered concussion. On regaining consciousness, he found that he was inverted and perilously close to the ground! He later recalled righting the aircraft, but nothing else until he saw the airfield light beacon at Abeele, where he initiated an approach to land, hit some trees on the approach and crashed. (John Pattern was eventually sent home and was able to resume flying as an Instructor at Old Sarum at the end of May, but he did not return to France.)

In March 1918, the Germans launched the first phase of a major Spring Offensive to break through the Allied lines. The focus of this attack around Amiens lay well to the south of Abeele and, whilst 10 Squadron was not involved, it nonetheless suffered casualties carrying out its daily Corps task. On 9 March, out on artillery work, Capt Maclean and 2Lt McGovern were attacked by four enemy aircraft. The observer was wounded but both recovered safely to Abeele. On 14 March, another radio operator, 1AM L B Cope, was killed when attached forward to 252 Siege Battery. Then, on the night of 26 March, a day on which the RFC had been tested to the full in all areas, Lt F J Westfield crashed during a bombing sortie. He had encountered heavy mist in an aircraft with an erratic compass, ran out of fuel and finally crashed southeast of Dunkirk. Initially posted as Missing, word reached the Squadron on 19 April that he had been injured and was a POW. This phase of the German offensive ended in early April with their having regained ground lost in the Somme battles of the previous year at very substantial cost.

As the war went on, back in London there had been considerable debate over the merits of creating an independent military air arm. Almost one hundred years later, echoes of that debate can still be heard but suffice it to note here that, on 1 April 1918, the Royal Air Force was formed as an independent Service from what had been the Royal Flying Corps and the Royal Naval Air Service. The Squadron marked the day with 'Business as usual' – the collection of a new aircraft, training for new crewmen, five artillery observations, varied recce sorties and two Counter Attack Patrols.

On 2 April, General Foch, CinC of the Allied Armies, issued a directive on the employment of French and British air forces. He stated that "the first duty of fighting aeroplanes is to assist troops on the ground by incessant attacks.... and ... air fighting is not to be sought except so far as necessary for the fulfilment of this duty." The protection of Corps reconnaissance aircraft appeared to have less priority than it had been afforded since early 1916 and, no doubt coincidentally, the Squadron's first RAF losses were suffered that day. Lt E D Jones and 2Lt W Smith were killed when their aircraft was brought down by machine gun fire whilst on a Counter-Attack Patrol (CAP). Then, on 7 April, 2Lts Jukes and Stanley, again on CAP, were hit by AA fire, their petrol tank was punctured and they were forced to crash land behind Allied lines. Both survived, somewhat shaken but not badly injured.

In the interim, a second phase of the German offensive was in planning, further north in Flanders, with the aim of pushing the British back to the Channel Ports (Battle of the Lys). This was launched on 9 April in an area held by a weak Portuguese corps that was about to be replaced, and weather conditions severely limited the RAF's ability to assist troops on the ground. Initial German advances were made, several squadrons were forced to move and, grounded by fog, 208 Squadron was compelled to burn sixteen of its Sopwith Camels before evacuating its base. Accordingly, for days the possibility of the Squadron having to evacuate Abeele hung over all unit activity.

The National Archives have an anonymous diary covering the first three weeks of April. Signed "Lt L," (very possibly from the hand of a Lt W Law who often signed off the daily records for the OC around this timed), it provides a youthful, sometimes jocular, insight into a worrying and, given the weather, a frustrating time:

> April 9. Camp awakened by the growl of cannon. Breakfast a meal of flying rumours ..... B Flight get ready to move under the pretence of spring cleaning their ante-room. A

and C stand firm ….. Nobody flies. The clouds are low and thick. We dine to the noise of cannon and lie down to sleep in a roar of detonations, but we have no fear.

April 10. The battle rages and we are not in it ….. Damn the mist. The large AWs are restless in their sheds …..

April 11. Third day of the battle. We are early astir. The strife waxes hot and comes nearer home….. At last, later in the day, the clouds and mist are dispelled, and we take a hand in the game…..

Flying was possible from mid-afternoon, with fourteen aircraft used to bomb and strafe enemy positions, mainly in Ploegsteert Woods, south of Ypres, from which there was considerable machine gun opposition. This resulted in just one casualty, Lt I S Black, an observer, wounded in the right leg. Lts Webster and Cameron and Lts Ferguson and Hirst appear to have been attacked by the same seven Albatros D5 aircraft over the Wood – both observers fired on them and they dispersed. Three CAPs and general Recce sorties were also flown. In addition, and probably as part of a more general reshuffle of the supply chain in face of a deteriorating situation, four aircraft were returned to No 1 Aircraft Supply Depot at St Omer that evening, with a further three following early the next day.

That next day, April 12, was to prove critical and General Haig issued an Order of the Day stating that there was no alternative to fighting and holding ground. The need to know what was afoot would have been pressing and two Counter-Attack patrols were sent out at first light. One reported the British Corps front as quiet but, on crossing the lines, soon came under heavy machine gun and AA fire - two rounds were reported as bursting within ten yards of the aircraft. By mid-morning, a general Recce sortie revealed British troops having to give ground, and nineteen CAPs and other Recce sorties were flown that day. (The RAF flew its greatest number of hours during the war on 12 April.) Lt L's diary described the day thus:

April 12. The battle comes nearer. Townley packs his toothbrush and our morale receives a jolt, but we preserve the stiff lip. It is our way in times of stress. No signs of mist today, so the 'lads of the ruined village' hie them hence on the ungainly Armstrong. Young Harrison gets lost and we speculate. Churchman, forgetful of the Ledeghem Viaduct, fairs forth in search. We lose them both and Chapman goes out into the unknown in quest. This is exciting – we speak in whispers. It thrills us, this super seeking, we are awed. Chapman returns like the dove to Noah's Reporting Officer, with less news than the twig in the fable. Then came Anderson's wire - our baby has a hole in his leg, and has landed near a forest. Churchman and his man Friday are still at large. News at last. Churchman has sideslipped his chariot into the ground close to the Babe's handcart. Both well…. Short brings home McLeod, wounded in the leg and chipped in the head. Both swear terribly – the observer with more spirit – not unnaturally. Chapman keeps flying backwards and forwards to the scene of the disturbance – says he loves it. Medical Officer earns his pay. We drag him panting from his hut to the aid of a RE8 observer – very seriously wounded. We are about to have the odd cocktail before dinner when the blow falls. We are ordered to "git" at 7.55pm. At 8.05 pm the air was full of a whining crowd of large AWs. At 8.15pm the first landings began and by 8.45pm diligent and anxious searchers gathered in the remains….. Capt Maclean and Lt Wright are missing. We hope for the best. At midnight the tired airmen were having tea and the Recording Officer was having a small rum, neat….

In fact, Capt Maclean and 2Lt Wright were later confirmed as having been killed whilst on a CAP during the afternoon.

(Alexander Murchison Maclean had been working in Chile at the outbreak of war, made his way across the Andes by mule with two others to Buenos Aires, and thence to London. He was commissioned in the Scottish Horse and saw action at Gallipoli, for which he was awarded the MC. Later seconded to the Black Watch at Salonika, he transferred to the RFC in 1916.)

Lt Harrison, pilot of the first 'missing' crew, was found injured by machine-gun fire, having force-landed by a forest. His observer was uninjured, but Capt Churchman's later attempt to rescue the aircraft was less successful than Lt L made out – his sideslip resulted in a crash about 100

yards from the first machine, but with no further injuries. Again, and as noted, 2Lt McLeod, out with Lt Short on another CAP, was wounded in the leg. The anticipated move did finally start that evening and it extended into the next morning as the Squadron withdrew a few miles west to Droglandt in France. The urgency of the situation may have been behind the handful of records showing damage on landing, with one crew returning as they could not find the new field in the failing light, and yet another touching down for the night with 54 Squadron. Lt L continued:

> April 13. We bring in 'the 13th' in our boots, so to speak…
> ..A day of 'shaking down,' old ideas of comfort being destroyed and new ones formed.

That morning was spent completing the move to Droglandt and the unit went back to work in deteriorating weather conditions that afternoon. "Line very quiet in Ypres Salient" said one report, and the general impression was of some slackening of the German advance, with British artillery considerably more active than their

Droglandt Airfield, on the French side of the Franco-Belgian Border, 14 miles west of Ypres. Having left Chocques in November 1917, 10 Sqn moved to Abeele until April 1918 when it then moved here.

10 Squadron aircrew on a Big Ack FK8 at Droglandt, Belgium in 1918 Only 7 of the 8 names are known: Mallinson, Sutherland, Ryan, Blackman, Shields, Chapman and Showler.

German opposite numbers. However, another observer, 2Lt Mackenzie, was wounded by ground fire after attacking German troops, prompting Lt L to record wryly:

> "...but at least two machines have landed in this week, bearing observers who are not wounded anywhere at all."

Crews were having to fly fairly low as the mist encroached and by 1700 hours it had become impossible to distinguish friendly from enemy lines from 500 feet. On 14 April, no flying was possible at all.

> (A cauld morning, and a damn cauld day an'awl. We do no work ....)

Conditions were sufficiently improved on 15 April for six replacement aircraft to be brought forward from 1ASD, but only a limited number of Counter-Attack Patrols was able to get airborne – these reported things as fairly quiet on the Corps front. A weather check at 0600 hours on 16 April reported cloud and mist at 200 feet but, by the end of the morning, conditions had improved to allow patrolling to begin. Having survived on 12 April, Lt Short brought back a second badly wounded observer, 2Lt Harker. He had been hit by anti-aircraft fire and died two days later. Counter-Attack Patrols continued as hail, snow and mist permitted over the next few days, with little enemy movement reported after the earlier advances. It would appear that the Squadron's former airfield at Abeele had not been overrun as Lts Middleton and Jackson were forced to land there for the night by a snowstorm on the evening of 19 April. They were thus able to miss the ineffective German bombing attack on Droglandt later that night, despite the weather. Flying on the next day appears to have been unaffected, other than by continuing poor visibility, and sorties included practice Recces for three new, replacement observers. One of them, Probationary Observer 2Lt A C Clinton, went out for another practice with Lt Burdick on 21 April when they surprised an enemy LVG aircraft and, using the pilot's Vickers gun, forced it to crash amongst some trees where they reported that French troops had arrested one of the occupants. (The pair seemed blessed – they later survived damage from ground fire on 29 April and 3 May.) Flying conditions were much improved that day – photography from 6000 feet was possible for the first time in over a week, and several sorties were flown to update HQs with pictures of the position on the ground. The improvement also meant that aircraft crews were better able to see others nearby and three Squadron aircraft were attacked by large German formations that afternoon, incurring only minor damage. Lt L mentioned that the Squadron had been warned to be 'mobile' and, interestingly, Maj Murray flew to visit Abeele during the afternoon of 21 April, very possibly to assess its condition should a return there become feasible.

Weather conditions remained poor over the following days but patrols continued as far as possible and some artillery 'shoots' were carried out. The weather was not the sole hindrance, however – Lts Short and Hirst reported 'interruptions' by three lots of enemy aircraft during the afternoon of 23 April, all of which retired when fired upon. No flying was possible on 24 April until the end of the afternoon, when Maj Murray carried out a weather check and, with cloud at 500 feet, declared conditions as fit for patrols. Three CAPs were mounted and a recently-arrived pilot, Lt Riley, continued his local checkout – unfortunately damaging the wingtip of another aircraft in the process. A patrol went out at 0515 hours on the next morning, with two others an hour later, and all three reported no movement of German troops although enemy aircraft were out in some force. Lt Short and 2Lt Cousins were attacked – Cousins fired 100 rounds and one of the German aircraft was seen to go down trailing smoke until it disappeared into the mist below. However, a little later in the morning, patrolling around 1000 feet, Lt H W Holmes was wounded in the ankle by machine gun fire. The aircraft crashed but his observer, Lt Hooker, was uninjured. Two unusual entries for the day suggest that Maj Murray was concerned about the urgent implications of what his crews were seeing - he made special flights late that morning and in the afternoon to drop messages at the HQs of 2 Wing and 22 Corps, and Capt Churchman was sent on another to 22 Corps that evening. On 26 April, flying was just possible for a part of the afternoon and, again, no movement was detected in forward areas, though heavy German artillery barrages were reported. A similar picture emerged the next day.

As the Squadron's patrols had suggested, the German

offensive had begun to stall and, after Allied counter-attacks, was abandoned on 29 April, with losses of around 110,000 killed or wounded on each side, again for little strategic gain but with the Allied position in Flanders effectively secured. The German focus moved away to the south once again, back towards Amiens, and as the German offensive petered out in late July, the Allies were now in a position to counter-attack. What would become known as the 'Hundred Days Offensive' began on 8 August, and the advances that followed would lead to the Armistice and victory in November.

(A footnote to April 1918 is that 53 Squadron, operating RE8 2-seaters and leading a somewhat peripatetic existence through 1918, moved back into Abeele alongside 10 Squadron on 6 April, just in time to evacuate on 12 April. In the mid-1960s, 10 Squadron and 53 Squadron would both re-form at RAF Brize Norton and share opposite ends of a new squadron accommodation building.)

The normal range of support tasks was carried out over the summer, with the Squadron supporting variously 22, 2, and 19 Corps. However, with activity still intense, regular losses were sustained both in action and in flying accidents. During a Flash Recce sortie on 7 May, Lt Hooker, an observer, was wounded in combat – the aircraft controls were badly damaged and the aircraft was forced to land, and his pilot, Lt Hughes, was uninjured. Two days later, pilot 2Lt Jackson was wounded by machine gun fire during a bombing attack at 500 feet but survived. On 19 May, another observer, 2Lt Rushton, was wounded during an artillery sortie. Then, on 30 May, 2Lt Clinton's early luck ran out. On artillery work with Lt Short in the afternoon of 30 May, their second sortie that day, their aircraft was hit by a shell and made a forced landing, with both men wounded. On the positive side, what the Squadron was doing over the period was not lacking in appreciation by the troops being supported. Amongst a number of congratulatory letters, a surviving note to the CO from the Commander of 53 Brigade, RGA, reads as follows:

> "There is a general expression amongst the officers of the Brigade of admiration for the splendid work the Squadron is putting in for us, and attention is particularly focussed on the exploit of the pilot and observer of the AW working for 188 Siege battery, as recorded in the Corps Counter Battery summary of the 19/20th, after ranging the battery and seeing it shooting well, went for and and brought down a Hun triplane who might have brought fire on to one of our batteries or 188 Siege Battery itself."

(The encounter referred to is mentioned in the Communiqué for 19 May, when it appears that Lts Hughes and Peacock were in fact attacked by two Fokker triplanes near Zillebeke. Lt Peacock fired at close range to bring one of them down.)

On 15 June, 2Lt H L Storrs, a pilot, was wounded and he died five days later. On 16 June, Capt E H Comber-Taylor

10 Squadron Aircrew in April or May 1918. The RAF was formed from the RFC on 1 April 1918 and this is probably one of the first photographs of No 10 Squadron Royal Air Force. It is with regret that the names are unknown.

and 2Lt G A Cameron were taking off for an artillery task in one of the Bristol F2B aircraft when the engine failed. The machine stalled and crashed, killing Comber-Taylor and injuring the observer. The Squadron's next loss was on 10 July, when wireless operator 2AM R J Gibson was attached forward to a battery. Then on the night of 17 July, returning from a Recce sortie and attempting a downwind landing, Lt N J Forbes and 2Lt C F Berry were killed when their aircraft hit some trees and crashed. During the following week, and after a busy morning's work, time was found for a Sports Meeting on the afternoon of 25 July to mark the third anniversary of the Squadron's arrival in France. After the running and jumping, the programme ended with Blindfold Squad Drill just before the presentation of prizes!

Back to business, a record number of 22 artillery shoots was supported on 29 July and a detailed example of one of these 'knock out shoots' can be seen from the records for 31 July, when Capt Whittaker and 2Lt Kime flew an early evening sortie in support of 223 Siege Battery, targeting a German battery. They were airborne at 1655 hours, established contact at 1705 and nineteen initial ranging rounds were fired from 1725 to 1800. Ten were estimated as falling within fifty yards or closer. Thirteen re-ranging rounds were fired between 1835 and 1850, with similar results. A second re-ranging salvo of twelve rounds was fired between 1920 and 1935, with a closer grouping, and another salvo of nine rounds was fired between 1950 and 2000. Based on the fall of the observed ranging rounds, 'Fire for effect' instructions were given after each salvo and 150 rounds in all had been fired by 2005. The German battery appeared to have been knocked out and the aircraft was back on the ground at 2025 hours.

The Battle of Amiens, effectively the 'beginning of the end' for the British, was launched by an artillery barrage early on 8 August 1918. A significant advance was made that day and the RAF was heavily committed in support. The Squadron's next loss occurred on the afternoon of 11 August when, during a photo sortie, 2Lt H W Sheard was killed by gunfire from attacking Pfalz scouts. His observer, 2Lt Goodwin, took control from his cockpit and got back to a field near Droglandt where the aircraft crashed, with Godwin surviving the impact. As advances on the ground were made, Contact Patrols became an important method of keeping HQs abreast of the progress being made and, on 31 August, the entire Squadron flew patrols on a shuttle basis to locate a new German line after an evacuation from Kemmel. However, early September brought more casualties. Patrolling during the afternoon of 2 September, 2Lt Coulthurst was injured when he and his observer, 2Lt Macpherson, made a forced landing after being damaged by ground fire. The aircraft was shelled but, that night, Lt Mallinson took Sgt Lovell and a fitter out to the wreckage from which they salvaged the new W/T radio fitted to selected FK 8s and the Lewis gun. During an early Flash Recce on 3 September, Capt Ivelaw-Chapman and 2Lt Mills fired on a number of lorries and drew AA and machine gun fire in retaliation. Mills was wounded in the thigh and they flew back to base and, just over three hours later, Capt Ivelaw-Chapman was airborne with another observer for a 3-hour 'shoot' with three separate batteries. Then, just after 0700 hours on 4 September, Capt Blackman had begun a Contact Patrol and was repeatedly calling for flares with no result. At 300 feet, the aircraft was hit by machine gun fire and when the observer, Lt King, was wounded, they returned to base.

Later in the month, on 20 September, 2Lts Reader and Airey survived after being forced to land after sustaining damage in an attack by seven Fokker D VII fighters. Then, following the advances being made on the ground, on 22 September the Squadron returned to Abeele, in position to support the British offensive that began with some surprise at dawn on 28 September. That day's efforts fill twenty-four record sheets. Seven aircraft were out on Contact Patrols or Flash Recces by 0700 and the pace was kept up well into the evening, with many messages dropped to Divisional HQs and ammunition boxes dropped to advancing troops, mixed with strafing and bombing enemy ground positions. Given the flying achieved and what was happening elsewhere, the Squadron was fortunate to incur only one casualty that day. Lt Lisle was wounded in the arm by machine gun fire when on Contact Patrol and crashed in a forced landing from which his observer emerged unscathed. Overall, significant advances were made that day, with the enemy driven from ground east of Ypres that had been fiercely contested in the

previous year. It should also be noted that the Squadron flew a mix of Bristol F2B, RE 8 and FK 8 aircraft at that point, and that at least two of the FK8s were now fitted with Wireless Telephony, permitting voice contact for the first time during patrols. The records show attempts to call 'Carrots,' but it is not immediately clear which HQ or organisation that may have been. The following day was also busy and, emphasising the need for HQs to know what progress was being made, it included two Special Missions to drop baskets of pigeons to advancing troops. One crew carried three, the other two, all dropped at five different positions. Unfortunately, in one case the parachute failed to open, and in another the basket broke from its chute some 50 feet above the ground. In those cases, the birds may have been more suitable for the kitchen than message carrying. Another crew was brought down by machine gun fire that afternoon, with both men uninjured.

Despite the progress made on the ground, air operations became no easier and the Squadron sustained additional casualties and losses throughout October. The Squadron was out early on 1 October, but it seems that 'Carrots' was still not answering at 0800 hours. That day's effort was largely split between Contact Patrols and photo sorties, and it was during one of the latter that Lt Bennett and 2Lt Kime, in one of the Bristols, were attacked by sixteen Fokker triplanes over Courtrai (Kortrijk). Both men were wounded in the action, but recovered to the airfield. (2Lt Kime died in England on 12 May 1919, and is recorded by the Commonwealth War Graves Commission as a war casualty.) The British advance continued, requiring a good deal of ammunition resupply by airdrop on 2 October - Capt Short and Lt Peacock were both wounded by machine gun fire on patrol that morning, but both survived the ensuing forced landing.

Some idea of the pace of events in the following days can be gained from a citation prepared in January 1919, recommending the award of a DFC to 2Lt Douglas Scott Carrie. He had joined the Canadian Army Service Corps in 1915 whilst a medical student in Toronto. He then served in France until he transferred to the RFC in October 1917, when he trained as a pilot and was posted to 10 Squadron in July 1918. It is not necessary to understand all of the technical terms used (and we might recall that an observer would have been involved in all that follows), but his citation provides a lively snapshot of squadron involvement in the entire gamut of Corps roles overhead the advances on the ground in the last weeks of the war:

2Lt Douglas Scott Carrie

> "Exceptionally gallant and continuous good work during the period from the British attack on September 25 up to November 11th, working at low heights in the very bad weather prevailing at that period and always under heavy fire from the ground.
>
> October 1: On Contact Patrol, he located our troops at 10 points and the enemy at 7 points. He fired 160 rounds into machine gun posts which were holding up our advanced troops. GF call sent on 3 MT and 2 GS wagons. One WPNF call sent. He also reported on the condition of bridges at Bousbecque.
>
> October 6: He took 18 photos of the Messines area, fired 100 rounds into Oostaverne battery positions from 2500 feet and dropped two 20 lb bombs on Hollebeke.
>
> October 7: He carried out a successful knockout shoot of 120 rounds causing 3 explosions and a fire in the enemy battery. He dropped 2 bombs on this same target.
>
> October 14: He went up to do a shoot but was unsuccessful owing to mist and rain so went over to a main crossroad 4000 yards over the line, firing 100 rounds at it and dropping 2 bombs.
>
> October 14: On Contact Patrol, he located enemy troops

in several places round Gulleghem – fired 50 rounds Lewis at these, causing casualties. He was fired at from 7 places on the enemy line – own troops located on a 4000 yard line. Reconnoitred the enemy trenches and shell holes in front of our line. Dropped 2 bombs on Bisseghem and reported a V by wireless.

October 15: He carried out a successful shoot and re-ranged, causing continuous large fires and explosions on his target and also in a neighbouring enemy position close by. He dropped 2 bombs near this target.

October 16: He carried out 2 successful knockout shoots in the same flight with one re-ranging causing 3 fires and some explosions. 2 bombs on Garrard crossroads.

October 17: A very excellent Combat Patrol report – located our own troops on a 10,000 yard front south of the Lys. Located enemy transport SE of Courtrai, engaging them with 100 rounds MG from 800 feet. Enemy troops found east of Courtrai. Reconnoitred a large area south of Courtrai in front of our own troops trying to locate enemy. Reported on bridges over the Lys. Dropped 2 bombs on farms south of Lauwe.

October 23: He carried out a successful shoot of 130 rounds, causing large fires and one explosion round the OK point. He fired 175 rounds into Oostaverne battery positions.

October 25: He went up to give general ranging of a long range gun on Gheluwe, observing good results. He reported 4 active enemy batteries and later attacked and fired 400 rounds at an enemy balloon which was hauled down.

October 27: He carried out a successful shoot with re-ranging on a hostile battery.

October 28: On Contact Patrol, he located our troops at 6 points – reported by wireless telephone 4 GS wagons, fired 400 rounds at more transport, and also 200 rounds at various parties of enemy troops seen.

October 29: On Contact Patrol, he located our troops at 13 points in shell holes. He also located the complete enemy outpost line. Observing a number of enemy troops in a trench, he and his observer attacked them with machine guns, causing several casualties, and chased the remainder towards our own troops, who captured them, the officer in charge waving in acknowledgement to the machine. More enemy machine guns and post were located and fired at. One GF call was sent and answered with good effect. GF call sent on a column of wagons. Height 50 to 1000 feet.

Lt Carrie's award of the DFC was gazetted in June 1919, by which time he had returned to Canada to resume his medical studies. He went on to become a noted surgeon, and died in 1984. His WW1 medal group was sold for £1900 at auction in June 2013.

The Squadron's casualties in the first days of the month were: 2Lt Hinde, an observer, wounded by AA fire, on patrol on 5 October; 2Lt Glidewell, a pilot, wounded in the leg by machine gun fire whilst attacking and finally silencing the source during an early patrol on 7 October; 2Lt Cope was injured in a flying accident at base the same day; 2Lt Walsh, an observer, wounded when out with Lt Shields and attacked by nine Fokker D VIIs on 8 October. Towards the end of October: 2Lt Davey and Lt Mackenzie were forced to land after damage from gunfire on 25 October; Lt Aulph, a pilot, wounded by AA and machine gun fire when on patrol on the morning of 27 October; after surviving on 25 October, 2Lt D A Mackenzie, an observer, was killed in combat with an enemy fighter on the morning of 28 October, whilst on a photo sortie; and on 31 October, Lt Reader and 2Lt Fletcher were both wounded in combat and an enforced crash landing whilst on Flash Recce duty. 2Lts Neill and Ryan had survived being brought down by ground fire that same morning. There had been other losses: 1AM C B Haigh, a wireless operator attached to 100 Battery, Royal Garrison Artillery, on 13 October; and 3AM J R Brown had been killed in action on 14 October.

As the Allied advance continued in the last weeks, it became possible to occupy airfields that had been used hitherto by the Germans. The Squadron moved to Menin on 22 October and then on to Staceghem on 5 November, a move that appears to have delighted Capt Ivelaw-Chapman. He wrote to his mother on 9 November:

"We have had another move and a rattling good one too. We are now lodged in an English Brewer's ex-mansion. In other words, a gorgeous jerry-built chateau complete with Turkish carpets, grand piano, oak panels, billiard table, high ceiling, white table cloths, oak chairs .... Long may we

stay here .... I do hope we'll be in this chateau for Armistice Night or Xmas night; oh what a night that would be."

As that letter made clear, the prospect of an Armistice was widely anticipated by that date. Far beyond the moving front line, as the German military and internal political situations deteriorated, defeat and the need to seek terms had become inevitable. A sailors' revolt led to the declaration of a republic and the abdication of the Kaiser and, to bring matters to a head, a German delegation was brought to France, to Marshal Foch's private railway carriage in the Forest of Compiègne. However, before the war ended, there were further Squadron casualties and losses. On 7 November, a pilot, 2Lt H Parsons, was wounded by machine gun fire during a patrol – the Squadron's last casualty in action of the war. There were two additional deaths from illness in service - 2AM I Morgan on 7 November, and 2AM A W Wells on 10 November.

After some three days of negotiation at Compiègne, an agreement was signed early on 11 November with an Armistice to come into effect later that morning at 'the eleventh hour of the eleventh day of the eleventh month,' and the Squadron holds a photograph that suggests there had been an anticipatory celebration involving a volley of flares at Staceghem on the previous night. Squadron crews were out on patrol from first light that morning. Visibility was poor but signs of the imminent victory were seen – for example, German troops preparing to evacuate a town, large numbers of British troops advancing northeast in open order, and French cavalry and armour also moving forwards. Nonetheless, there was still residual opposition in those last hours. In two cases hits were sustained that resulted in forced landings, albeit with no crew casualties, and the Squadron's last shots in anger after some three and a half years were probably fired around 0830 hours by observer 2Lt Walsh as he returned fire with fifty rounds from 800 feet at a party of German soldiers who had fired on his aircraft. Nothing of the sort occurred during a flight over the Armistice hour for, written in 1931 by Lt Mallinson for a newspaper, and included in the Ivelaw-Chapman biography is a splendid story:

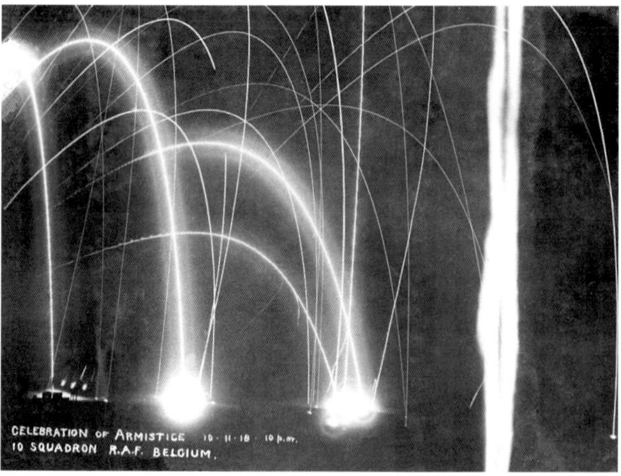

"We spotted our Flight Commander emerging from the recreation hut with a mysterious bulge beneath his leather flying coat.... At 11 o'clock we were over the advancing British and French line and it was then that the Flight Commander dropped his surprise from the skies. As a strangely moving silence fell over the battle area, the formation leader's streamered machine dived on our troops and from the cockpit spun ... a football. As it bounced gaily in front of them, the Tommies let out their first post-war cheer. In a moment a laughing, boisterous throng of khaki clad figures flung themselves on the ball and two rifles were thrust into the mud to act as goalposts."

The unit records for 11 November 1918 are somewhat indistinct and cannot positively confirm Mallinson's tale, although he was airborne with Ivelaw-Chapman on the next morning, ostensibly for an engine test, but it is heartening to think that something of the sort might have happened that day to mark the end of hostilities.

During the afternoon of 13 November, Maj Murray carried out a two-hour flight looking for potential airfields to the east of Stacgehem but, two days later, the Squadron moved westwards, back to Menen (Menin), now assigned to 81st Wing, 10 Brigade. On 1 December it moved again, to a nearby airfield at Reckem on the Belgian/French border, where it is said that surplus squadron funds were expended on a series of farewell dinners and the provision of a well-equipped recreation hut. Pilots kept in flying practice for some weeks – formation flying in groups of up to six became popular - and some had an opportunity to fly a Fokker D VII (F5284/18), a German fighter that had proved to be a very considerable adversary in the last year of the war. This was collected by Maj Murray on 1 January 1919 from the Enemy Aircraft Reception Park at Rely, not too far from the Squadron's first base at Chocques. Later in the month, the aircraft were flown either to Rely or, for the majority, into the temporary care of No 35 Squadron at St Marie Chapelle. The last FK8 (C 8572) was delivered there by Lt Carrie on the morning of 22 January, the Fokker went back to Rely, and the detailed Squadron record closed on the next day, showing fifteen pilots still on strength but no aircraft. After that, with the last after-action reports and medal commendations written, the Squadron was reduced to a small cadre of personnel that returned to Ford Junction in Sussex in February. Major Murray relinquished command on 16 February and went on to make his mark as an architect and ceramic artist whose work is still highly prized. The unit was formally disbanded on 31 December 1919.

And so, with "*The war to end all wars*" won and a widespread elimination of military capability taking place, the world's statesmen gathered at Versailles to dictate the terms of a peace that would ensure that no such conflict should ever again be possible in Europe ....

BATTLE HONOURS FROM WORLD WAR 1

Western Front	**1915-1918**
Loos	**Somme 1916**
Arras	**Somme 1918**

Peace at last.

A trophy from the Albatros aircraft forced down on 29 February 1916.

The abandoned and overgrown Chateau de Werppe development in August 2014.

CHAPTER 2

BETWEEN THE WARS

"The aim of the Air Force in concert with the Navy and Army is to break down the enemy's resistance. The Air Force will contribute to this aim by attacks on objectives calculated to achieve this end, in addition to direct cooperation with the Navy and Army and in furtherance of the policy of His Majesty's Government at the time."

(The war aim of the RAF as stated in Air Staff Memorandum S28279, October 1928)

In the decade that passed before 10 Squadron was re-formed, the Air Ministry was kept busy in maintaining the RAF as an independent service. Born and expanded in wartime, postwar demobilisation had left it a rump of its former self and a number of attempts were made to reabsorb the new entity and its assets into the two older services - echoes of such can still be heard today – but these were strongly resisted. And in doing so, the Chief of Air Staff (CAS) Sir Hugh Trenchard also did much that contributed to the future strength of his Service – the establishment of what would be the RAF College at Cranwell, starting the Apprentice scheme and founding a Staff College. However, after the rapid rundown after the Great War, and with the peace of Europe apparently assured at Versailles, service budgets were restricted under a Government policy of "No war in the next 10 years" and there was little that could be done to introduce marked improvements in equipment. Still, with the bulk of the RAF's squadrons serving in the Middle East and India, a concern over the Home Defence of the UK gradually gained ground until, in 1923, PM Stanley Baldwin was able to announce that, by 1930, the RAF at home would increase in strength by 34 squadrons to 52 in all. A unified command was created in 1925 called Air Defence of Great Britain, within which the Wessex Bombing Area would command a bomber force of some 35 squadrons (Trenchard had commanded the RAF's Independent Force created in May 1918 to carry out long-range bombing sorties into Germany), and it was in this context that 10 Squadron was reborn.

Whilst Chris Barnes' highly detailed 'Handley Page Aircraft since 1907' mentions 10 Squadron as reforming with the HP Hyderabad at Worthy Down in November 1927, that presumably takes account of a degree of administrative assembling at an existing bomber base before what the official record notes a few weeks later, for 3 January 1928:

> "No 10 Squadron formed at Upper Heyford under command of Wing Commander H R Busteed OBE, AFC. Establishment: HQ and one flight. Equipment to be Handley Page Hyderabads, and Squadron to be a twin-engined night bombing squadron."

Upper Heyford had first been opened in 1917 as a RFC, and later RAF, station and had closed in 1920 after the post-war rundown. It was selected in 1924 as one of the stations needed for the planned expansion of the RAF and initially provided a home for the Oxford University Air Squadron. It was reopened as an operational base on 12 October 1927, and 99 Squadron followed 10 Squadron there two days later, on 5 January 1928.

The HP Hyderabad was a large twin-engined biplane, with 4 crew positions (two pilots in tandem and two gunners), and a 1100 lb bomb load over a range of some 500 miles. It was thus a limited evolution of the HP 0/400 that the RFC's Independent Force had used in the later months of the war to attack Germany, and was first supplied to 99 Squadron from 1924. Eleven aircraft had been ordered for 10 Squadron in late 1927, and the first three were collected from RAF Hawkinge, near Folkestone, on 25 January

The Handley-Page Hyderabad prototype which first flew in 1923

Hyderabads first entered RAF service with 99 Sqn in 1925 but deliveries of the aircraft were slow, and the second squadron to be equipped was not until 1928 when 10 Sqn re-formed with them at Upper Heyford.

1928. A further two were collected in mid-May, and their essentially temporary nature as prime equipment was confirmed in early February when the Squadron was informed that it was to receive the essentially similar HP Hinaidi once Jupiter engines were available. However, some development of the Hinaidi wing design would intervene and it was an all-metal Hinaidi Mk II that was finally ordered to re-equip both 10 and 99 Squadrons.

Command passed to Wg Cdr F L Robinson in October 1928 as the Squadron increased in size to two Flights, but the narrative record of its history in the next years is sketchy in the extreme, with 1929 covered in only five handwritten entries, for example. In essence, it can be said to have provided part of a force in-being on which something more substantial might be built as opportunity and budget permitted. We know of Practice Camps for annual bombing and gunnery practice, air exercises, various competitions, and participation in Open Days – events that had begun to capture the public imagination. An indication of flying activity levels can be gauged from a 1930 entry giving July and August that year, the period of the annual Camp, as the highest since the re-formation – 413 and 372 hours, respectively. With the RAF List giving the CO plus twenty pilots on strength at the time (six were NCOs), that infers a maximum of 20 hours/month, with probably a good deal less in winter months, and there is no mention of what proportion – given the designated role – was achieved at night.

Wg Cdr A T Whitelock took command on 4 April 1929. He would remain in post for only some sixteen months, but he left a significant legacy – a Squadron crest, with motto, recorded on 11 June 1930 as having been officially approved, and with a separate record indicating that a copy or 'specimen' was lodged with the Air Historical Branch later that month. The story goes that, after watching an archery contest, Wg Cdr Whitelock (who was also prominent in the RAF Rifle Association) had the idea that a bomb had become the modern equivalent of an arrow. So, working with a Wg Cdr G L Robinson, he started with an arrow, gave it wings to indicate speed and delivery through flight, and chose a Latin motto "Rem Acu Tangere." A strict translation would be "To touch the thing/matter with a needle" or, more popularly, "To hit the nail on the head," a rendering that brings us close to today's free translation, "To Hit The Mark." In 1937, that crest and motto would become the basis of today's Squadron Badge when such badges began to receive Royal approval.

Wg Cdr P C Sherren MC took over as CO in September 1930 and, in December, the first Hinaidi (J 9300, one of six that had first been ordered as Hyderabads) was collected from the makers. A further fourteen aircraft (K 1910-23) would be collected by July 1931 and, whilst this was happening, a move to RAF Boscombe Down occurred at the

A rare airborne photograph of the Handley Page Hinaidi. 10 Sqn took delivery of them in 1931. It is believed that the 10 Sqn's WW2 sobriquet "Shiny Ten" may have been derived from the shiny, unpainted metal construction of the Hinaidi.

A Hinaidi 'heavy bomber' parked on the grass: concrete parking areas and runways were a rarity until WW2.

Designed at Brooklands in the 1920s the Vickers Virginia Mark X was made of duralumin and steel and 10 Squadron operated them from 1931 until 1935.

A Virginia shows off the unique twin stabiliser and rudders of the tailplane. Earlier models had three rudders.

start of May 1931, "by air, road and rail." The new all-metal aircraft appear to have been kept busy throughout July in a series of Air Defence Cooperation exercises, in others with the Observer Corps, and in an exercise over three nights as a Blueland unit. No doubt there was then a spell of Block Leave in August before the month-long annual bombing and gunnery Camp at Catfoss in Yorkshire.

(The origin of the Squadron's 'Shiny Ten' nickname that was well-established by WW2 remains elusive but it may well lie in the Hinaidi's all-metal finish – although other squadrons also claim to have been 'shiny' at various times.)

On 23 August 1932, news came that the Squadron would re-equip once again – with the Vickers Virginia MK X, the latest mark of yet another biplane bomber that had first entered service in 1924. It is hard to see it as any advance on the Hinaidi, particularly when the official RAF website now says:

> "The aircraft displayed no particularly remarkable qualities (3,000lb bombload, 108mph top speed (without bombs on later versions) and no defensive armament in early aircraft. It is a testament to the absence of strategic policy-making by the Air Ministry and lack of investment by the successive governments that the Virginia lasted as long as it did. (1938 in squadron service.) Nor was the aircraft particularly safe - the 124 airframes constructed suffered a total of 81 accidents."

The first (J 7430, an earlier serial number than any Hinaidi) arrived in September 1932. Elsewhere, on 30 January 1933, Adolf Hitler became Chancellor of Germany.

The next change of command came in early February 1933, when Wg Cdr H K Thorold DSC, DFC, AFC took over. And one of the 81 Virginia accidents occurred on 27 April 1933 when Fg Off Mathias and Plt Off Ross-Shore force-landed at Kempsford in Gloucestershire, and the aircraft had to be dismantled to allow its removal. Participation in a variety of tactical and cooperation exercises continued over the summer, but an interesting diversion is recorded for the

10 Sqn 'A' Flt circa early 1930s. L- R Standing: Harding, Oliver, Selby, Farrell (ranks not known) Seated: Flt Lt Dixon, Sqn Ldr Fulljames, Fg Off Baldwin Oliver was Australian and Selby came from New Zealand, pictured with their own uniforms and wings.

10 Sqn 'A' Flt photograph with ground crew in front of a Virginia circa 1935.

period from 8 May to 8 September 1933, when two aircraft were allocated from RAF Martlesham Heath, near Ipswich, (then the base of the Aeroplane & Armament Experimental Establishment that moved to Boscombe Down in 1939) for Service Trials. One was a Fairey Monoplane, the first of which had been built in 1928 to investigate methods of increasing aircraft range and, popular at the time, to establish some world records. The first had crashed on one such attempt in 1929 and a second was built in 1931. In February 1933, a crew from the RAF's Long Range Development Unit had flown this aircraft on a record flight from Cranwell to Walvis Bay in South West Africa. Regrettably, the Squadron record is silent on the nature of the trials carried out before the aircraft was returned to the manufacturers. It is similarly silent on what was achieved with the second aircraft, a Vickers Vanox. This was an aircraft designed to Air Staff specification B 19/27, seeking a Vickers Virginia replacement, and was another twin-engined biplane. However, as the HP Heyford was already on order as a Virginia replacement, any trials must have been somewhat cursory.

Wg Cdr C B Dalison AFC was appointed as CO in February 1934 as Wg Cdr Thorold was posted to command 70 Squadron in Iraq. His appointment was to be of short duration for, in August, he too was sent overseas and Wg Cdr M B Frew DSO, MC, AFC assumed command. Also, in June that year, the Squadron was informed that it would be re-

equipping with the HP Heyford Mk 1a, to an establishment of ten, plus two Reserves, and these were collected between August and November of 1934. The HP Heyford had also been designed in response to Air Staff specification B19/27 seeking a more advanced night bomber than the Hyderabad/ Hinaidi. Whilst still a biplane, it was notably different in design, with its two engines on the upper wings, close to the fuselage, and long undercarriage legs that gave it a marked nose-high attitude on the ground.

On 31 July 1934, a further expansion of the RAF had been agreed in light of what was happening in Germany, and further schemes would follow in subsequent years. It was these that provided the RAF with the aircraft, manpower, and new stations with which it would later face the outbreak of war. By 1939, front line strength had grown

10 Sqn's Heyfords were not yet a year old when King George V inspected them at his Jubilee Review at RAF Mildenhall on 6 July 1935.

'A' Flt , 10 Sqn: the 1935 Jubilee Flypast

from 42 squadrons and some 800 aircraft to 157 squadrons operating around 3700 aircraft, and a public display of this expanding RAF was given as part of King George V's Silver Jubilee celebrations in 1935. On the morning of July 6, he inspected a parade and static display of some 350 aircraft at RAF Mildenhall before being driven to RAF Duxford to watch them in a mass flypast that afternoon, led by Heyfords of both 10 and 99 Squadrons. It may also be a sign of awareness of a growing threat that the unit record became more detailed by 1935, with much more attention given to the scale and detail of exercises than hitherto – eg, in April, five out of six crews scored hits on a night exercise at what may have been a familiar target at Catfoss, but only two out of six did so on a different target at Porton on the next night. Days later crews were involved in a delayed Navigation and Signals Exercise, and in a concurrent RE Searchlight Company cooperation. July saw the annual Observer corps exercises, with up to five aircraft involved on some days, for a total of 86 hours flying. In the next week, Air Exercises 1935 took place, with 10 Squadron flying from home base for Southland. Weather hampered missions on the nights of 22 and 23 July when the Northland targets were, oddly, the Air Ministry and RAF Hendon. Nine aircraft were tasked for ops from Mildenhall on 24 July and, on 25 July, ten aircraft were able to carry out day attacks on Dagenham in two flights of five. In August, the Squadron acted as hostile for an Air Defence exercise over Portsmouth and Southampton, and in September went off for annual Camp at Catfoss.

E.V. Roe, Catt and McDermott in a Heyford, with the 10 Squadron badge adorning the aircraft's nose. Roe, killed in action in 1941, was the son of aircraft designer A.V. Roe.

Having drawn experience from elsewhere at its origin in January 1915, it was now the Squadron's turn to contribute to RAF expansion from its accumulated experience in the generation of two new squadrons. The first occasion was on 16 September 1935, whilst at Catfoss, when B flight was designated No 97 (B) Squadron, with its personnel and aircraft duly transferred. Six Heyford Mk IIIs were acquired in the following weeks to re-equip a new B Flight, and Wg Cdr Frew was initially in command of both squadrons.

Total flying for 1934 had been 1965 hours Day and 698 Night, and the totals for 1935 were 1853 Day, 452 Night.

Sergeants and commissioned officer pilots of 10 Squadron under the Heyford Bomber.

Handley Page Heyfords. It is not clear what has happened to the man on the ground – is he refueling or fallen from the cockpit?

The AOC later commented on results for 1935, expressing satisfaction that

"the long distance exercises show a marked improvement" and, as regards Air Gunnery,

> "the scores obtained by No 10 (B) Squadron ... reach a standard which he hopes other squadrons will make their aim."

There was a Winter Air Exercise in February 1936, with ten aircraft from the combined 10 and 97 Squadrons participating. Ten raids on a reservoir target were flown on the night of 18 February before high winds prevented more being flown. Unfortunately, in the early hours of 19 February, there were two aircraft incidents involving deaths, the first since re-forming in 1928. Three members of a Squadron crew were lost on their return from an attack when their aircraft crashed into a hill near Midhurst – Sgt McDermott, LACs Westlake and Adams. The pilot, Sgt Deakin, was injured. Then, some three hours later, a 97 Squadron aircraft ditched in the Channel. Three crewmen drowned, including Wireless Operator AC1 Watkin of 10 Squadron.

The standard mix of exercises continued through 1936, with now a first mention of

> "night cooperation with Fighter Command's ground and air defence organisation .... searchlights and sound locators."

Then, on 1 November 1936, came the second contribution to RAF expansion – 10 Squadron's fairly recently reconstituted B Flight was designated 78 (B) Squadron, but with new Heyford Mk III aircraft. Other highly significant changes lay ahead.

By 1936, with its expansion plan underway, the RAF had felt compelled to change its command structure and to arrange matters on a functional basis. As part of this, Bomber Command was created on 14 July 1936, initially at Uxbridge, with the move to High Wycombe occurring in March 1940 once new accommodation was ready. Next in the command chain, No 4 Group was formed on 1 April 1937 and initially had command of all night-bomber squadrons, including 10 Squadron. (Its first AOC was Air Cdre A T Harris, who became CinC Bomber Command from February 1942.) The Group was to concentrate its assets in Yorkshire and so it was that 10 Squadron moved north to RAF Dishforth, by the Great North Road, between 12 January and 8 February 1937.

10 Squadron was the first to operate the Whitley when it entered RAF service in 1937 . It was a significant improvement over its open-cockpit predecessor, the Heyford
Length: 69.26 ft
Width: 83.99 ft
Height: 14.99ft
Weight (Empty): 19,350 lb
Weight (MTOW): 33,501 lb
Crew: 5
2 x Rolls-Royce Merlin X V-12 piston engines
Max Speed:; 200 kts
Max Range: 1,501 miles
Service Ceiling: 26,001 ft
Rate-of-Climb: 938 fpm

A Whitley Cockpit

That done, a very significant rearming of the Squadron began and, between 9 March and 25 June, 10 Squadron accepted the first of the RAF's new Armstrong Whitworth Whitley Mk I aircraft, twelve on establishment, plus four

reserves. The Whitley was a twin-engined monoplane, with all crew spaces enclosed, derived from a 1934 Air Staff specification for a heavy night bomber. (Like the Whitley, the first Vickers Wellington and HP Hampden medium bombers also flew in 1936 but these were originally conceived as day bombers. The heavy four-engined types came later.) The Mk I was very much an early draft, and engines and armament were quickly improved to create Mk II and III versions. RR Merlin engines were fitted in 1938, resulting in a Mk IV that 10 Squadron received from May 1939. The Whitley had a crew of 5 – pilot, second pilot/navigator, two wireless operator/air gunners (WOP/AG), and a tail gunner. The mid-under retractable 'dustbin' turret was manned by a WOP/AG. The aircraft's performance was a distinct advance on the assorted biplanes of the previous decade, having a range of some 1400 miles at some 200 knots airspeed,

Whilst the re-equipment was in train, there was another change of command. Wg Cdr Frew took over the new 78 Squadron and was replaced on 17 April by Wg Cdr S Graham MC. Apart from noting that a formation of five Whitleys took part in the annual Hendon Air Day, the record is silent on the exercises carried out in 1937. The move to a new station and the change of aircraft no doubt combined to show a reduced flying total for the year – 1733 hours Day, and just 152 hours Night.

The situation in Europe began to deteriorate appreciably in 1938. Hitler was pressing the Austrian government to agree to a union of Germany and Austria and, in March, German troops crossed the border and the *Anschluss* was effected. Neville Chamberlain's government and the House of Commons deplored what had occurred but would do no more. In April, the UK recognised the Italian conquest of Ethiopia, and the next crisis would emerge as Hitler made his case for the ethnic Sudeten Germans areas of Czechoslovakia to be absorbed into Germany.

Meanwhile, back at RAF Dishforth, Wg Cdr Graham was posted to the Air Ministry Special Duty List in April. After two months with Sqn Ldr Steedman in charge of 10 Squadron, Sqn Ldr W E Staton MC, DFC arrived to assume command on 9 June, being duly promoted to Wg Cdr on 1 July. Observer corps and Home Defence exercises were scheduled over the summer, but were often

Wg Cdr W.E. 'Bull' Staton who was to command 10 Squadron from June 1938 into WW2 until Apr 1940 after it equipped with the Whitley.

limited by weather. Then, in the first days of September, a detachment of seven aircraft and crews went north to No 8 Armament Training School (ATS) at RAF Evanton, on the Cromarty Firth, for the annual Practice Camp. Meanwhile, in Czechoslovakia, the Sudeten crisis was increasing in urgency, with Hitler threatening annexation of the areas in question. Chamberlain made his first visit to Berchtesgaden on 15 September, to receive assurances that Germany had no designs beyond the Sudetenland. Then the Czech government fell and, in a second meeting with Chamberlain on 22 September, Hitler made greatly increased demands such that, on the afternoon of 24 September, the Squadron training detachment was recalled to base. Chamberlain's third visit to meet Hitler (plus Mussolini and Premier Daladier of France) in Munich followed on 29 and 30 September, resulting in the agreement that he described in London that evening as "peace for our time." The Sudetenland was duly absorbed and a virtually inevitable outcome was delayed. In the Commons debate on the matter on 3 October, Winston Churchill, who had very publicly opposed German expansionism over several years, told the House, "England has … chosen shame, and will get war."

Against this background, there was a tangible change in the tone of the unit record. Tactical exercises and operations under war conditions began and, at year's end, the 1938 flying total is given as 2000 hours Day - and 602 hours Night, a marked increase over that for 1937.

1939 began badly with an aircraft and crew of six lost during a navigational exercise on the evening of 23 January. The last message received placed them between RAF Manston, in Kent, and St Catherine's Point on the Isle of Wight and, despite a search by RN ships, no trace of aircraft or crew was found. (The crew was: Plt Off Miller, pilot, and Plt Off Miles, second nav; Sgt Cutts, nav; ACs Thompson and Lavery, WOP/AGs; AC Hanley, gunner.) Another aircraft was lost on the night of 2 May. It took off for a night flying detail with the elevator locks still in place and, unable to climb, it stalled, crashed and burned, killing both pilots, Sgts Donald and Fryer.

On 1 April 1939, the Air Ministry introduced a Municipal Liaison Scheme aimed at making the public better acquainted with RAF units, aircraft and personnel. Despite being Yorkshire-based, 10 Squadron was allocated Blackburn in Lancashire and Wg Cdr Staton, with his Flight Commanders and Adjutant, visited the town at the Mayor's invitation on 6 June. A reciprocal visit to the station was made later in the month and although occasional references to the Squadron as "Blackburn's Own" can be found, the outbreak of war soon after appears to have prevented the relationship from developing further.

Deliveries of Whitley Mk IV aircraft began in early May and participation in exercises continued. The annual Practice Camp was held in August once again at Evanton and included a successful test of new, classified W/T equipment with HMS Rodney. The new aircraft would soon be needed for, over in Europe, the situation was ever worse – Hitler had invaded other parts of Czechoslovakia in March and guarantees had been given to Poland should it be threatened. With that done, what followed was almost inevitable. Germany invaded Poland on 1 September 1939 and the Squadron record reads:

> "1 September: Squadron received orders to mobilise.
> 3 September: Operation Book closed."

A new and arduous chapter was about to be written.

FROM BROOKLANDS TO BRIZE

PART 2

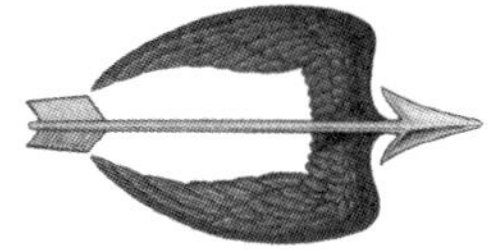

WORLD WAR 2 ~ 1939 - 1945

Armstrong Whitworth Whitley
Handley Page Halifax

&

1945 - 1950

Douglas Dakota

CHAPTER 3

NICKELS AND THE PHONEY WAR

"I have to tell you now that no such undertaking has been received, and that consequently this country is at war with Germany."

Neville Chamberlain, Prime Minister, broadcasting to the nation on Sunday 3 September 1939.

The war that had been long in coming was now declared. It had been averted over the 1938 German annexation of the Czech Sudetenland, that "quarrel in a faraway country between people of whom we know nothing," when the Munich Agreement had supposedly brought "peace for our time." However, the invasion of Poland on 1 September 1939 forced the issue – the United Kingdom and France had given pledges of support there that could not be honourably denied.

At 2200 hours that night, RAF Dishforth and No 10 Squadron received a Mobilisation Order by signal. In the hours that followed, armed guards were stationed, aircraft dispersed, gas masks issued, and blackout measures put in hand. Technical and equipment checks were carried out. In addition, suspecting that German Intelligence would be aware of the squadron identifiers then in use, Air Ministry instructions came to change the 2-letter identifiers used on all RAF aircraft. 10 Squadron dropped 'PB,' used since the introduction of such identifiers in late 1938, and adopted 'ZA,' used for the duration of the war and for some time afterwards. Unit badges were also removed from aircraft and by the morning of 3 September preparation was complete, with crews placed on Standby. (Interestingly, constituted crew lists now appeared for the first time in the unit records. This appears to have been an innovation and it was to remain standard practice in Bomber Command for the duration of the war and, indeed, afterwards.)

However, with our perceptions today largely coloured by the bomber offensive of the final three years or so of the war, what is less well remembered is that there were to be no immediate bombing raids over Germany. Both Britain and France had agreed to a call from President Roosevelt of the USA asking that combatant nations refrain from bombing civilian targets, and instructions were issued to Bomber Command accordingly. German naval vessels at sea or in harbours away from dockside areas could be attacked and bombers could fly over Germany to drop propaganda leaflets, or 'Nickels' as they would become known. Other restrictions would apply over potential routes – both Belgium and the Netherlands were still neutral, as was Denmark, and could not be overflown. As a result, direct routes were not possible and plans had to rely on diversionary routes to approach Germany from its North Sea coastline or from France, to the south.

September 1939

Bomber Command launched its first operation just hours after Chamberlain's broadcast on 3 September. A single Blenheim was sent on a reconnaissance flight seeking out German warships north of Wilhelmshaven and, later that night, Whitleys of 51 and 58 Squadrons flew the RAF's first Nickel sorties over Germany. 10 Squadron had crews on call each day, and on 7 September received an instruction requiring eight aircraft to drop leaflets and carry out general

reconnaissance over Lubeck, Kiel, Wilhelmshaven and Cuxhaven in the early hours of 8 September. Wg Cdr Staton, the CO, led the operation and all crews recovered safely to base, albeit one had to land first at RAF Manby to refuel. Only light opposition was reported, with searchlights active and very little AA fire. The first step along what would prove a hard and bloody road had been taken in comparative safety.

On 9 September, four aircraft were sent to Rheims, where they were to refuel and then proceed for Nickel drops that night over Nuremberg and Frankfurt before returning to Dishforth. (It would seem that only limited Host Nation Support was expected, as the instructions were that food should be taken for the period of operation.) Each aircraft would carry 20 packages, giving 240 bundles of leaflets to be dropped through the aircraft's flare chute, but things did not go entirely smoothly. Technical problems meant that two aircraft had to land at RAF Tangmere on the South Coast – one was able to continue and a replacement was sent for the other, to be taken on to Rheims once the Nickel load had been transferred. The operation was then postponed to the following night when, once again, opposition was slight with, perhaps significantly, no enemy aircraft encountered. And in the event, only one aircraft was able to fly home directly with the other three all needing to land first at different RAF stations. There was then a break in operations as the leaflet dropping policy was reviewed in London, and this allowed some low level bombing practice and fighter affiliation to be carried out as the weather permitted.

NACH HITLERS STURZ

Die Lebensfrage für das deutsche Volk ist heute:

SOLL DER KRIEG VERLÄNGERT ODER ABGEKÜRZT WERDEN?

Hitler will verhindern, dass der Deutsche dies selbst entscheidet. Er kann ihn aber nicht hindern, Fragen zu stellen. Hier sind sieben wichtige Fragen — und die Antworten.

Kann Hitlerdeutschland auf einen Kompromissfrieden rechnen?

Nein. In dem Vertrag von London haben England und Russland mit der Billigung der Vereinigten Staaten noch einmal erklärt, dass sie niemals mit dem Hitlerregime oder irgend einer anderen deutschen Regierung verhandeln werden, die sich nicht unzweideutig von allen Angriffsabsichten lossagt.

Bedeutet Hitlers Niederlage Deutschlands Zerstörung?

Nein. Immer wieder, zuerst im September 1939, zuletzt am 21. Mai 1942, hat die britische Regierung erklärt, dass sie zwei Ziele hat:
1. die Hitlertyrannei zu vernichten,
2. allen Völkern Europas, auch dem deutschen, nach dem Krieg den Aufbau eines Staates zu ermöglichen, der jedem Einzelnen unparteiische Gerechtigkeit, Rede- und Koalitionsfreiheit sichert und ihn vor Arbeitslosigkeit und wirtschaftlicher Ausbeutung bewahrt.

Hitlers Niederlage bedeutet also nicht die Zerstörung Deutschlands, sondern die Rettung Deutschlands vor der Zerstörung.

Wird England das deutsche Volk dem Bolschewismus ausliefern?

Die britische und die russische Regierung haben sich im Artikel V des Londoner Vertrags verpflichtet, sich nicht in die inneren Angelegenheiten anderer Staaten einzumengen, nachdem Hitlers Kriegsmaschine zerstört und eine Wiederholung von Angriffen unmöglich gemacht ist. Die Vereinigten Staaten nehmen dieselbe Haltung ein. Damit ist die „bolschewistische Gefahr" als das entlarvt, was sie schon immer war: ein Propaganda-Schreckgespenst.

Bedeutet Hitlers Niederlage Arbeitslosigkeit und Inflation für Deutschland?

In der Welt von heute hängt der Wohlstand eines jeden Volkes vom Wohlstand aller anderen ab. Es kann keinen dauernden Frieden und keinen Wohlstand geben, solange ein Volk versucht, sich zum Herrenvolk zu machen. In der Roosevelt-Churchill-Erklärung haben sich England und die Vereinigten Staaten in ihrem eigenen Interesse verpflichtet, nach dem Kriege keine wirtschaftliche Benachteiligung der Unterlegenen zuzu-

Leaflets dropped over enemy territory was known as 'Nickels'. Whitleys of 10 Sqn were the first RAF aircraft to visit Berlin in October 1939 on such a mission.

Close to 10 Sqn at Dishforth, RAF Topcliffe was often used as a temporary base. One of their 102 Sqn Whitleys is shown here prior to a leaflet-dropping ('Nickel') mission.

Operations resumed on the night of 24 September. Whilst leaflets would still be dropped, there would now be an emphasis on the reconnaissance possibilities that such sorties provided, and this emerges in subsequent mission instructions. Four crews were readied to fly but two sorties were postponed. The other crews were detailed for reconnaissance and Nickel raids over Hamburg and Bremen and to create a disturbance and test the air-raid precaution status, ie the effectiveness of blackout and the strength and

location of searchlights, over Berlin. Leaflets were dropped but Berlin was not reached on that night. The weather over the North Sea was bad, with severe thunderstorms and considerable icing. In addition, the High Frequency Direction Finding (HF/DF) bearings transmitted from both Leconfield and Linton-on-Ouse were found to be some 90° out. Similar circumstances would seriously affect Whitley operations quite often in the months ahead, sometimes resulting in navigational errors of considerable magnitude. Both crews flew for some hours over the North Sea before deciding to return to base – one made it directly, whilst the other landed first at Leconfield where there was insufficient fuel to onload before going on via Driffield. (The fact that all of these airfields are in Yorkshire suggests that the crews were operating to the limits of their fuel endurance, with little or nothing in reserve.) Already, within three weeks, a number of the significant factors that would affect operations in the months ahead were apparent. Nonetheless, some 18 million leaflets had been distributed in these first days and, as reported in *Flight* magazine's 5 October 1939 edition, the Ministry of Information was upbeat:

> "Since the war began night flights have been made covering thousands of miles over enemy territory; some of the raids penetrated into Germany to a considerable depth. The value of these paper raids has proved to be considerable. What the German High Command thinks of their moral effect is clear, for heavy penalties are imposed on persons seen picking up or reading the pamphlets. None the less, these pamphlets have given millions of people in various parts of Germany an opportunity to receive authoritative presentations of the Allies' case and of the reasons which have compelled them to take up arms."

On 28 September, two aircraft took all operational crews, less Gunners, to Linton-on-Ouse to watch Flt Lt Allsop practise 'ZZ landings,' a very early and rudimentary form of blind landing using DF bearings in which the ground controller would send ZZ in Morse to give clearance to land. Further general training was carried out before the next operation on the night of 30 September, when four crews were tasked for recce and Nickels over Hamburg and Bremen. Once again severe weather and false DF bearings were experienced. One crew returned to Dishforth safely and two found themselves misdirected by DF over the West of Scotland, but then recovered to base. However, the fourth aircraft was also misdirected over Scotland and, running out of fuel after some 9 hours 40 minutes airborne, its crew finally made a forced landing near Bolton, narrowly missing a farm house. The aircraft was badly damaged but there were no casualties. Unfortunately, this would not remain the case for long.

October – December 1939

The operation tasked for the night of 1/2 October would prove significant in two ways. Four aircraft were detailed for recce and Nickels over Berlin, once again led by Wg Cdr Staton. Despite adverse weather en-route, three crews reached the designated target, thus becoming the first RAF crews to fly over the enemy capital since the beginning of the war. There was only slight opposition reported but, unfortunately, only these three crews returned to base. The fourth crew is believed to have released its leaflets over Denmark, still neutral at that time, and failed to return. Flt Lt Allsop's crew were last heard on W/T at 0505 hours on the morning of 2 October, at an estimated position some 180 miles from St Abbs Head, and were posted as Missing at 1200 hours that day. Flt Lt Allsop, Plt Off Salmon, AC Bell, AC Hill and LAC Ellison were the first 10 Squadron casualties of the war, and it was assumed that they did not survive a ditching in the North Sea.

K 9018 Failed to return on the night of 1/2 October 1939 when 10 Sqn carried out the first raid of WW2 on Berlin, dropping leaflets. This Whitley was last heard on R/T when abeam St Abbs Head. A wooden 'Mouseman' plaque in the Squadron commemorates the event and the loss of the crew.

The official Air Ministry announcement on 2 October read:

> "Successful air reconnaissances by day and by night have again been carried out over Germany by the Royal Air Force. The night reconnaissance included Berlin and Potsdam."

However, beyond that bald statement, some UK newspaper reports also carried reports noting that Danish citizens had heard the sound of aircraft around 3am and had later found fields strewn with leaflets printed in German giving details of financial transactions by Nazi leaders. Similar leaflets were understood to have been found in Northern Holland. It was also reported that the Danish Government had ordered its Minister in London to protest against the dropping of leaflets, the text of which indicated a British origin.

Before continuing, it is worthwhile to consider the conditions in which the Whitleys of 4 Group went to war, graphically described by 'Larry' Donnelly in his book "The Whitley Boys." Later a pilot, he was at this point an Aircraftsman WOP/AG with Fg Off Bickford's crew on A Flight with 10 Squadron:

> "... our first raids were to reveal many deficiencies in equipment and training. In peacetime, flying in adverse weather conditions and flying at high altitude was rarely attempted. This inexperience was to become patently evident during the leaflet raids of the atrocious winter of 1939/40 when aircraft began to fly at great altitudes. The lack of suitable heating, oxygen facilities and adequate flying clothing was to cause unimaginable hardships for the crews and especially the gunners in their unheated turrets. Airframe de-icing consisted of a compound smeared on the leading edges of the wings, an antidote which was less than satisfactory."

It might also be noted here that two crewmembers were needed to handle and dispatch leaflet bundles down the flare chute, and that they would be doing so at altitudes of up to 20,000 feet, engaged in fairly vigorous activity, and alternating on a single oxygen point to stave off hypoxia. Larry continued:

> "The only aid to navigation was the M/F and H/F Direction Finding W/T stations. Their range and accuracy was limited to between 250 and 300 miles for the bearings and fixes and 100 miles for QDMs (Homings). Aircraft W/T equipment, the T1083/R1082 combination, was difficult to operate, where the main difficulty was the necessity of changing the coils every time a change in frequency was required. Outside the M/F and H/F direction finding range we had to rely on basic navigation, Dead Reckoning, which could prove to be fatal at night and in bad weather conditions. Despite the difficulties, the crews made the best of what they had .....The lessons learned by them during the leaflet raids were to prove of great value .... But this was learning the hard way."

Bad weather now caused an enforced break in operations until, on 15 October, nine crews were tasked once again on Nickel/recce over Berlin, Magdeburg and Hamburg. Things did not go well and only one of the Berlin aircraft was able to complete the mission. One did not take off and the others became unserviceable en-route or were unable to fix position by W/T or visually. Operations for the remainder of the month were often similarly blighted. Of six crews detailed on 18 October, only two succeeded in dropping, and one of these had to force land near Amiens, finally reaching Dishforth via Tangmere on 24 October. Three out of four were successful that night; the fourth returned to base with W/T failure, its leaflets were quickly transferred to another aircraft and crew that took off but was forced to return on losing the port engine. On 31 October, two aircraft left for recce over the Elbe and Weser estuaries. One was able to see Hamburg but nothing else, thanks to cloud cover that prevented the other aircraft from seeing anything at all. Both then had difficulty in finding Dishforth under low cloud, and the second had finally to land at Driffield having been airborne for 11 hours 55 minutes.

Nonetheless, despite the real challenges of operational flying that were daily becoming apparent, other aspects of Service life could not be denied and Squadron crews were paraded on the afternoon of 1 November when RAF Dishforth was visited by HM King George VI and senior RAF officers. HM discussed recent operations with crewmembers and, on having a look inside the Whitley's cockpit, was reported as having expressed amazement at the "enormous number of gadgets."

Whilst weather was preventing operations over Germany, a sortie was flown on 9 November to investigate the accuracy of D/F bearings from Leconfield, and a number of air gunnery

sorties were planned. (The bearings recorded were assessed to be "on the whole satisfactory," suggesting that they were taken in considerably more favourable circumstances than crews had been experiencing in the previous weeks.) It was also decided to improve operational range by detaching aircraft to the airfield at Villeneuve (codenamed 'Sister'), south of Paris, then in use as a base for Fairey Battles of the RAF's Advanced Striking Force that had gone to France to support the British Expeditionary Force. 77 Squadron went first and 10 Squadron's turn came on 20 November when three aircraft deployed. The aim was still to drop Nickels and carry out recce as possible, and it is noteworthy that the OpOrders now carried instructions that aircraft were not to fly within 10 miles of neutral territory and should be clear of German airspace by 0440 hours, before daybreak. The instruction was significant as vulnerability of other bomber types to German fighters during the day was by now painfully clear and daytime loss rates were high. (Two crews were partially successful that day, in that leaflets were dropped, but no recce over Hamburg or Bremen was possible. Things went better over Dusseldorf and Frankfurt on 21 November, after which the aircraft recovered to Dishforth.)

A very different operation was put in hand on 23 November. The German cruiser *Deutschland*, which had already sunk Allied ships, had been located southeast of Iceland. Bomber aircraft of 4 Group were to deploy north to RAF Kinloss, on the Moray Firth, under the operational control of Coastal Command's 18 Group, to attack it if spotted again. Wg Cdr Staton's crew and six others flew there on 24 November, bombs loaded, by a largely over-sea route. Aircraft and crews from other squadrons were also deployed. In the event, the warship was not located; no flying was tasked; and the Bombs taken north were downloaded for potential use by 99 Squadron which replaced 10 Squadron once it recovered to Dishforth on 28 November. (The special instructions in the OpOrder are of some interest, and clearly anticipated that any attack would have been made in daylight, in circumstances similar to those in which the day bomber force had incurred heavy losses. Attacks were to be made from the highest possible level and in no circumstances below 1000 feet, and aircraft were to resume a defensive formation on completion.)

HM George VI visited the Squadron twice within a month in November and December 1939. His first visit was to RAF Dishforth and the second was to the forward base at Villeneuve, France –codenamed 'Sister'.

December began with the weather once again preventing some air gunnery sorties at Acklington from being flown. Then on 3 December, an order came for another Nickel dropping detachment to 'Sister' (Villeneuve), and four crews plus a reserve duly departed that afternoon. The intention was that two aircraft should drop leaflets over Prague and the other two over Frankfurt, but yet again the weather intervened and no operations were possible. Nonetheless, crews were detailed to remain as HM The King was to visit the airfield on 8 December whilst in France. This visit appears to have gone well, with His Majesty recognising some of the personnel whom he had met at Dishforth the previous month. The crews, with leaflets, departed for home on 9 December, to be plagued again by weather – two got in with difficulty, two had to divert to Linton-on-Ouse and the fifth to Tangmere before reaching Dishforth the next day. In the meantime, two crews had managed to complete Nickel sorties from Dishforth over Hamburg and Bremen on the night of 6 December. Although both were by now relatively experienced in the role, they reported particular difficulty with the cold at altitude and one of the tail gunners had to be treated for frostbite on return. The aircraft also flew within 12 to 15 miles from the Danish border but were reported by German News as having violated neutral territory.

Crews were brought to readiness for daylight anti-shipping sweeps on 10 and 11 December but then stood down. On 13 December, two were sent to enforce blackout conditions at Baltic floatplane bases and to attack any

aircraft taking off or landing. None were seen, no bombs were dropped and both had difficulty in finding base under cloud. And to a large extent, this pattern of operations into the Baltic with indifferent weather pertained until the end of the month when, finally, on 31 December, Flt Lt Phillips' crew was able to drop a flare followed by a single bomb on a suspected flarepath near Cuxhaven – the Squadron's first bomb dropped in anger. (Significantly, an aircraft had been flown to RAF St Athan on 19 December, with five others following on 29 December, for the fitting of 'special W/T apparatus' that was, in all probability, the earliest form of IFF (Identification, Friend or Foe) that would respond to interrogation by the Chain Home radar system. Aircraft fitted with IFF were certainly identified in daily tasking orders from around that time.)

January – Early May 1940

Having ended 1939 with a bang, 1940 started quietly for 10 Squadron with four crews detailed for a sector sweep over the North Sea on 1 January that was cancelled owing to fog. However, on 4 January, two crews took off as a pair and managed to complete Nickel ops, albeit not without incident. Sgt Versage dropped over Hamburg and his crew was able to make a useful recce report on return, but Fg Off Paterson's crew had to return shortly after coasting out as an oxygen bottle was leaking. This was replaced, fuel was topped-up, and they departed once again to drop successfully over Bremen despite heavy AA fire. Unfortunately, on return the aircraft undershot on final approach, touched down heavily, broke the starboard undercarriage leg and finally slewed to a halt. There was, inevitably, some moderate airframe damage but the crew emerged unhurt. On the next day, four crews were readied but only one was required. Flt Sgt Cattell's crew patrolled for some hours in the Borkum/Sylt area and this was the last operation flown for some time. The weather at home and in likely target areas was unsuitable for days on end and only occasional training flights were completed. Matters got worse thanks to snow later in the month, when a large scale operation involving nine crews from 10 Squadron and six from 51 Squadron (also based at Dishforth since December) was planned to attack German warships. The force could not launch on either 27 or 28 January and the operation was cancelled. After an Air Raid Warning and All Clear, three crews managed to take off on 29 January to test airfield conditions – with the last ending in a snowdrift after its landing run. The squadron record notes laconically: *"After this the aerodrome was placed unserviceable."* And thus it remained. Possible operations and training programmes were published but cancelled, and the squadron record for 7 February notes drily that *"One P/O Jeremiah reported on posting during afternoon."*

During all of this, another saga was developing. On 20 January, Fg Off Paterson and crew had managed to get airborne for France to ferry stores to 'Sister.' They managed to land there with some difficulty, but were unable to get back, thanks to Dishforth being closed. Finally, on 13 February, it was decided that attempts should be made to get them home, but Dishforth remained closed! On 15 February, HQ 4 Group signalled to the Squadron that the aircraft needed an oil cooler and that one should be flown south to RAF Hendon for onward carriage. Sgt Johnson managed to get airborne with the part early next morning and delivered it - but fog had now closed Dishforth and, on return, he diverted to Waddington before finally landing back later in the afternoon. On 22 February, and after a month in France, Fg Off Paterson managed to start back but had to land at Abingdon as fog still precluded landing at Dishforth. Almost defying belief, it appears that on 23 February a stock of Nickels had to be taken to France and Fg Off Paterson headed back to 'Sister' with them, whence came news that he had damaged an airscrew after landing! On 25 February, word came that he now needed a hydraulic pump; and that one could be sent via an aircraft from Driffield from where operations appear to have been possible, so this was duly dispatched. On 28 February, an officer was detailed personally to take a spare oil pipe for Paterson's aircraft to Hendon as previous spares had been mislaid in transit, a precaution that seems to have done the trick. The crew finally made it to Dishforth on 1 March, with the record showing *"Trip quite uneventful"* – a remark that we can assume referred more to that day's flight than to all that had gone before.

Winter conditions persisted through February. Plans were written and cancelled and only a few light weight training and air-testing sorties were possible. On 18 February, another 15-aircraft operation with 51 Squadron was planned, once again with the object of attacking enemy warships. Crews were repeatedly brought to readiness until the operation was finally cancelled on 21 February. As the snow passed, a thaw with some rain induced flooding such that the airfield was still unfit for take-offs of loaded aircraft. Finally, on 26 February, the airfield was declared usable for essential operations and Sqn Ldr Whitworth's crew was able to carry out a Nickel and Recce sortie to Berlin, the first operational flight possible in well over a month. He landed at 'Sister' on return, to avoid bad weather over the North Sea, and finally recovered to base on 29 February.

On 1 March, two crews were detailed for Nickel flights and recce over Berlin. Flt Lt Bickford's crew returned after a relatively uneventful flight, with little opposition. Sgt Johnson's crew was less fortunate. They ran out of fuel on approach, both engines stopped, and the aircraft crashed in a field about a quarter of a mile out. The aircraft was badly damaged but there were no casualties amongst the crew.

Sunday 3 March was an eventful day. Another aircraft was badly damaged on landing after a training sortie; Sqn Ldr Beaman had to force land the unit's Miles Magister communications aircraft whilst en-route to Catterick; and the squadron's first five Whitley Mk V aircraft were ferried in. Unfortunately, on arrival these were found to be 'incomplete' - there are no details as to what was missing in the record – and on the next day word came that they were not to be kept as it would take too long to rectify the deficiencies found! They were duly ferried away again on 15 March. (Two aircraft were delivered later in the month without difficulty, on 28 March.)

The pace of operations was about to pick up and on that same afternoon, instructions for a significant Nickel/Night Recce operation were received. On the following day, three crews were to fly to 'Sister' in order to have the range to reach Prague later that night. Two did so, with the third used as a reserve, and Prague Radio was reported as having shut down for a time. Back at Dishforth on the same night (16/17 March), six crews were detailed for Nickel/Recce ops over the Ruhr. All completed their tasks, with UK weather requiring landings at 'Sister' on return. Weather precluded a return to base that day; eight aircraft recovered on 18 March, with the ninth landing at Amiens with engine trouble. (Despite the instructions in all OpOrders that aircraft should avoid neutral territory, one crew is reported to have overflown Luxembourg with another over Holland – further reminders of the difficulty that crews faced in maintaining an accurate navigation plot above cloud at this early stage in the war.)

The pace now quickened and, on the night of 19 March, eight aircraft (with seven from 51 Squadron) flew the unit's first bombing raid against a ground target, the seaplane base of Hornum, on the Frisian island of Sylt. This raid was authorised as a reprisal for German bombs dropped on Orkney Island during a raid on the RN base at Scapa Flow, killing a civilian and injuring seven others. (A further 15 Whitleys of 77 and 102 Squadrons and 20 Hampdens from 5 Group participated in the raid.) Following the Driffield-based Whitleys, the 10 and 51 Squadron aircraft were to attack at timed intervals over a 2-hour period, on 5 separate but converging axes, twelve using standard 250lb and 500lb bombs, and three with a mix of bombs and incendiaries. With the operation already well underway, the 10 Squadron crews reported the target as visible from many miles off, with both fires and searchlights. All crews also reported having dropped successfully, albeit two reported partial hang-ups, and there were no casualties from the AA fire. Overall, some 20 tons of high explosives and 1200 incendiary devices were dropped from 50 aircraft, for a single Whitley lost so, on the face of it, this first significant night operation was considered in London as having gone rather well. The press and newsreel companies were duly alerted but, once later target analysis was complete, it was found that an exaggerated success had been claimed. (Nonetheless, 10 Squadron tail gunner, Larry Donnelly, was one of those interviewed by a Movietone newsreel team and he records having *"played my new-found fame for all it was worth"* when back on leave shortly afterwards. Also, Driffield-based squadrons felt that 10 and 51 at Dishforth had won an unfair amount of the national publicity and organised the distribution of leaflets poking fun at the claims made, a jape that resulted in a retaliatory leaflet drop from Dishforth.) There would be no more bombing operations for the time being but, that apart, what

The YMCA tea-lady is given a 'helping hand' passing a cuppa to a Whitley's crew.

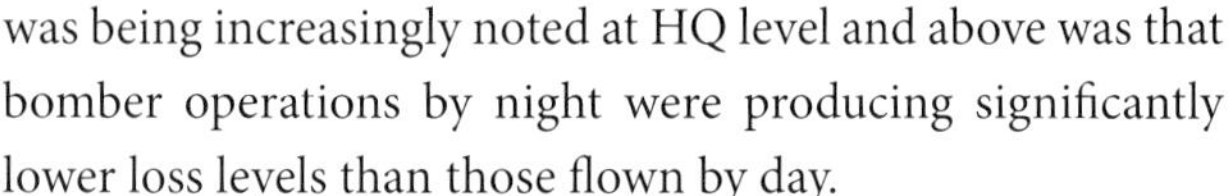

Elaborate 'Nose Art' was to come later. 10 Squadron's first bombing raid of WW2 was on the German seaplane base at Hörnum, Sylt on 20 March 1940, accompanying No 51 Sqn.

was being increasingly noted at HQ level and above was that bomber operations by night were producing significantly lower loss levels than those flown by day.

Weather and serviceability affected flying again. Some training sorties were completed and the last operation of the month was again with Nickels over the Ruhr and Recce along the Rhine on the night of 22 March. After some delays, six crews departed but two returned early on account of the severe icing encountered. Sgt Parsons was unable to climb above 7000 feet and his windows became completely obscured, and Sqn Ldr Whitworth gave up once his DF aerial broke away under the weight of ice. Recalling the difficulties that crews had already had with winter weather, it must be significant that all six reported that the conditions on that night were the worst they had encountered. On the following days, forecast weather over Germany led to the cancellation of operations although some local training was possible. Another joint operation with 51 Squadron was listed for 5 April, cancelled and re-programmed for 6 April. This was planned primarily as a Recce of canals and potential targets in the Ruhr, with Nickels as a supplementary task, and involved four 51 Squadron crews and two from 10 Squadron. The missions were completed against heavy AA fire, with one crew reporting that it had been too dark to spot activity on the ground, and both had eventful landings. Sgt MacCoubrey lost an engine in the circuit as a connecting rod broke through its crankcase; and Fg Off Prior saw his petrol gauges reading near-empty and force-landed near Grimsby, hitting a tree in the process. And yet again, fortune smiled, and no injuries were sustained.

Bombing Range sorties were planned for 8 April … and were cancelled that afternoon as crews were detailed for a pending operation. They were alerted and put on an hour's notice at 0430 on 9 April, and at 0700 news came that Germany had invaded both Norway and Denmark. Nickels were put aside and a first warning order was issued for attacks on German shipping near Trondheim, too far north to be within Whitley range. A new order was issued for shipping sweeps of the North Sea around Norwegian waters and crews found themselves in a cycle of stand-bys – 'standing by to stand by' – as plans changed. Eventually, on the evening of 11 April, six crews plus six from 51 Squadron were detailed for bombing attacks on shipping between the Kattegat and Oslo Fiord, with 10 Squadron operating in the north of the area. In the event, Sgt Johnson's aircraft suffered engine trouble and could not get airborne and the other crews had only partial success. Flt Lt Bickford bombed some lights, thought to be a ship, but no results were seen; Plt Off Cattell saw no targets, but had to jettison his bombs over the sea on return as an engine began to fail; and none of the others saw anything that justified an attack. Two aircraft had also been sent to the nearby Misson bombing range that night and demonstrated that incendiary bombs alone would not light-up a target sufficiently to guarantee a successful attack. Quite independently, 10 Squadron had begun to investigate the matter of target marking, a subject that would only become more pressing in the months and years that followed.

Stand-bys continued on the next two days, with no calls made on the crews involved. On 14 April, six crews were readied for an attack on the airfield at Stavanger but they were stood down later due to the adverse weather forecast for the target area. Three aircraft were also required for another operation with six from from 51 Squadron and twelve from other bases, but only two were available. This OpOrder is of interest in that, for the first time, it envisaged bombing attacks on rail and road junctions in Germany, west of the Rhine, near the Belgian and Dutch borders. Its objectives were stated as 'to delay enemy advance through the Low Countries," presumably en-route to Denmark, and 'reconnaissance of road and rail movement.' Very strict instructions prohibiting indiscriminate bombing were included: it must be possible to distinguish and identify the targets to be attacked; and there must be a reasonable expectation that damage would be confined to the objectives attacked and that the civilian population should not be 'bombarded through negligence.' The operation did not take place but, had it done so, it would have been a marked departure from the policy in force to that point and it was presumably a matter of considerable political concern in London.

The planned attack on Stavanger was rescheduled for the night of 15 April. Six aircraft got airborne but things did not go well. Two crews had to return with unserviceable intercoms – a subsequent investigation found that "two airmen had not carried out their work to the best of their ability." A third crew returned due to weather and a misunderstood message from base, and the fourth had engine trouble and landed at Waddington. A fifth crew had been unable to find the target after a series of unreliable DF bearings, encountered severe icing, and was compelled to jettison its bomb load over the North Sea. Only Sgt Johnson could report a successful mission. Previous aircraft from other units had started fires on the airfield, and he was able to drop his bombs on a runway.

An armed reconnaissance sortie against the Oslo airfields at Fornebu and Kjeller was planned for the night of 16 April, "to determine feasibility of air attack by night on aerodromes near Oslo, with a view to more extensive operations later." Wg Cdr Staton's crew flew the sortie and very severe icing was encountered over Norway, where snow and low cloud prevented any bombs being dropped. He reported that the number of small islands made identification very difficult and that accurate location of these targets would only be possible on a clear, moonlit night. Three crews were sent back to Oslo on 18 April, to attack ships in the fiord "provided these are lying well clear of the land or wharves," and with the two airfields as secondary targets. None of them could find targets under mist and low cloud. After a rest day, five crews and a reserve were detailed to return to Oslo on 20 April, together with four 51 Squadron crews. The primary 10 Squadron targets were the airfields, with shipping as secondary and, failing these, shipping and the airfield at Stavanger. Once again, cloud cover protected the Oslo area, but two crews were able to attack the Stavanger airfield and the four that had reached the target area all returned via a diversion to Leuchars. Ops for the night of 21 April were cancelled, as was that first planned for the next day. However, within hours, a further three crews were alerted, of which one was finally dispatched to Oslo fiord. Ironically, conditions were good but no permissible target could be identified, although the crew reported extinguishing "an annoying intense blue searchlight" with the aircraft's guns.

On 23 April, the primary targets for two waves were shipping in Oslo fiord and the airfield at Aalborg. (The relatively new OC A flight, Sqn Ldr D P Hanafin, who had first joined 10 Squadron as a new pilot in 1934, would fly the unit's first operational sortie in a Whitley V (N 1483) in the Aalborg raid.) From the first wave, two crews attacked an alternate target, the Kristiansund airfield, hitting both a hangar and the runway. Another was able to attack ships in the fiord, appearing to cause one to run aground; and the fourth bombed Fornebu airfield, causing some fires. In the Aalborg wave, one crew had to return immediately with an unserviceable airspeed indicator, and a 51 Squadron reserve crew took off to replace it, but the other two were able to carry

out attacks. Messages were sent directing all crews to divert to Kinloss or Leuchars on return as weather at Dishforth would be bad. Sgt MacCoubrey's crew did not receive it and had great difficulty in landing at base; the others duly diverted and could not finally reach Dishforth until the afternoon of 25 April. Weather precluded operations and interrupted training plans on the following days, and it was not until 30 April that another attack could be launched. This was to be on Fornebu airfield, using six aircraft each from both 51 and 10 Squadrons. For the first time, a mix of long-delay fuses in the bomb loads was detailed – 3, 6, 8, 12 and 36 hours – to prolong disruption of airfield operations. Unfortunately for this plan, three crews had to be recalled about an hour after take-off. Back at Dishforth, it had become apparent that the safety pins in their long-delay bombs had been left in place, so the weapons were not 'live.' (The armourer concerned was immediately placed under 'open arrest.') However, the weather over Norway was clear and the other three crews pressed home an attack against heavy AA fire but without casualties, causing damage to hangars and airfield facilities. On the night of 1 May, six crews attacked Stavanger airfield in clear conditions with a mix of HE and incendiary bombs in what was to be 10 Squadron's final operation of the Norwegian campaign. On 4 May, four crews flew patrols over the Frisian Islands to enforce blackout in German floatplane bases and to attack minelaying aircraft spotted on take-off or landing. No bombs were dropped.

Armstrong Whitworth Whitley Mk V with the Nash and Thompson Type FN4 rear turret.

The unit's conversion to the improved Whitley V got underway in May – having received the RAF's first Whitleys in 1937, and the Mark IV in 1939, 10 Squadron was the last 4 Group unit to receive the latest version. (A major change affected the aircraft's armament – the retractable mid-under turret was removed and the single tail gun was replaced by a new 4-gun turret; rubber wing de-icing boots were installed; and the DF loop aerial was encased in an aerodynamic housing of a type that would be familiar on aircraft for the next 20 years or so.) Two aircraft had been received in late March (N 1482/83), one of which Sqn Ldr Hanafin's crew had flown on operations. Five new aircraft were delivered on 4 May and, on the next day, HQ 4 Group stood-down the squadron from operations for seven days to allow conversion to the new aircraft. Some training was carried out and the tail gunners, in particular, were given range firing at Filey Bay to gain experience with their new, four-gun turret. Additional aircraft arrived on 8 May, another two on 10 May and, with thirty two aircraft now on charge, it was decided to disperse ten of them to Topcliffe.

More significantly, on 10 May Germany invaded Belgium and the Netherlands, having occupied Luxembourg on the previous evening. France deployed an Army Group north to assist in defence and what had become known as the "Phoney War" was over. The strategic offensive as we now know it was about to begin and RAF Bomber Command would soon be involved in the type of operations over Germany envisaged in pre-war plans.

The remains of Whitley V P5094 that crashed at Dishforth early on 9 September 1940, as a result of hydraulic failure. It came to rest across the A1 Great North Road adjacent to the Yorkshire base. Because the subsequent fire and explosion only occurred some 10 mins after the aircraft came to rest, the only crew injury was to the captain, Plt Off Cairns, who suffered a broken leg. He was rescued by his WOp, Sgt Nicholson.

The Whitley production line at Armstrong Whitworth's Baginton factory, near Coventry. After Whitley production ceased in June 1943 the factory switched to making Lancasters.

The Whitley won few prizes for its looks but it was the first heavy bomber of the 1930s to enclose its crew out of the elements.

CHAPTER 4

THE BATTLE BEGINS

"Let us therefore brace ourselves to our duties, and so bear ourselves, that if the British Empire and its Commonwealth last for a thousand years, men will still say: This was their finest hour."

Winston Churchill, 18 June 1940.

The monthly 10 Squadron Operations Record Books (Forms 540, or F540) for the Whitley era of WW2 give a fairly detailed insight into each day's activity – much more so than in the later Halifax years - and the entry for 10 May 1940 conveys something of the excitement and confusion that must have prevailed that day as it became apparent that the war had moved into a higher gear:

0730 Information received that Belgium and Holland had been invaded by German Forces; maximum dispersal of our aircraft ordered.

1000 Instructions received that stripping of Whitley IV aircraft to cease. Later this was cancelled and the preparation of the Whitley Vs 'P' Series to continue at utmost dispatch.

1400 Four aircraft were collected by Ferry Pool pilots and flown to Abingdon. Two more Whitley Vs 'P' Series have arrived. Plt Off Parsons was sent to Harwell for some camouflage paint. Flt Lt Bickford was flown down by Plt Off Parsons to Hendon to attend a conference at Harrow but this was cancelled and he returned with Plt Off Parsons. Information received that Fg Off Phillips and crew would be attached pending posting to No 97 Squadron. Crew consisted of Plt Off Wood, Sgt Donald, AC Nicholson and AC Mathews. Fg Off Phillips would be second-in command of a Flight.

1615 Crews released for recreational leave. Eight aircraft were dispersed at Topcliffe.

Weather was warm, 4/10 cloud at 2000 feet, no rain and visibility 6 miles.

The German attacks in the Low Countries were successful in drawing French and British forces north and in outflanking the defensive Maginot Line whilst German forces moved across the River Maas (Meuse) and into the French Ardennes. Advances were made on all fronts, eventually leading to the evacuation of the British Expeditionary Force from the beaches at Dunkirk that began on 27 May. At the outset of the Battle of France, the resources of Bomber Command were committed in support, albeit that targets were initially restricted to locations on the west of the Rhine.

On 12 May, the first attack on German towns was launched by a force of 18 Whitleys and 19 Hampdens. 10 Squadron contributed six crews from nine first alerted, to attack road and rail targets close to the Dutch border at Kleve and München-Gladbach. (The unit was now using only Whitley Vs on operations.) Five found targets but the sixth could not make a positive identification and flew over the area for some two hours encountering moderate AA fire and considerable searchlight activity before having to land at Manston on return with fuel running low. And, for the first time, a returning crew reported being fired upon by British AA guns over the Thames - no doubt a sign of increased nervousness on all sides. No operations were called for 14 May. However, it was on that afternoon that the Luftwaffe bombed the city of Rotterdam, killing some 900 civilians and making some 30,000 people homeless – an attack that

was to be critical for Bomber Command and its squadrons.

On 15 May, as a consequence of the Rotterdam bombing, the War Cabinet abandoned the policy of bombing restraint and directed that attacks on the German heartland should begin. Some of the plans that had for so long kept Air Staffs busy could now be implemented. But that night, as would inevitably happen again in the years ahead, there were competing priorities for the limited bomber effort available and that was immediately apparent in the 10 Squadron targets allocated. A maximum effort had been called for and the unit was able to generate thirteen aircraft. Six crews flew against industrial targets in the Ruhr, with a seventh aircraft used by 51 Squadron (HQ 4 Group had forbidden Wg Cdr Staton to fly on that night) and a mix of primary, alternate, and military targets of opportunity was attacked without loss. At the same time, efforts had to be made to stem the German advance on the ground, and a further six crews were detailed for attacks on road bridges and river crossings at Dinant and Turnhout in Belgium.

Those first raids on Germany that night marked the start of the strategic bomber offensive as we now know it, and set the scene for the next five years. The scale of operations would, of course, increase and a comparison of gross numbers gives an indication of how Bomber Command operations would change over the next years. Responding to the max effort call, a total of 111 bomber aircraft was involved on 15 May 1940. The 1000 Bomber Raids were far ahead and were exceptional in scale but on 25 April 1945, the date of the last raid in which 10 Squadron participated, 482 aircraft would be launched.

Keeping up the pace, on 17 May all 14 currently operational crews (this time including Wg Cdr Staton) briefed for an attack on an oil refinery at Bremen. That did not mean that the Whitleys were any more biddable than before and the starboard undercarriage leg on Sqn Ldr Hanafin's aircraft collapsed on take-off. Twelve of the remaining thirteen completed attacks with a mix of HE and incendiary bombs, and using a variety of attack profiles. One crew had considerable navigational trouble and, failing to locate the refinery, attacked a military airfield, an alternate target permitted by briefing instructions. A number of aircraft incurred flak damage but all returned to base safely, albeit one had to make a forced landing as a tyre had been punctured by shrapnel.

Operations were now flown as often and as intensively as weather and serviceability allowed, against targets normally shared with 51 Squadron. As might be expected, these operations were interspersed with aircraft incidents and accidents. In that context, and serving to make a point of how fortunate the squadron had been thus far, the first wounded crewmember since hostilities had commenced was AC J P Atkinson, a tail gunner, hit by flak shrapnel on the night of 21/22 May over Julich rail station – and he was not badly wounded. In noting this, it is also worth underlining that

10 Squadron possibly taken in 1939 at Dishforth

only a handful of minor encounters with fighters had taken place in these first months during which searchlights and flak had been the major concern for attacking crews. In one of these, near Utrecht on return during the night 27/28 May, Sqn Ldr Hanafin's tail gunner, AC Stan Oldridge, is believed to have shot down the first Luftwaffe night fighter of the war.

However, there is another incident from that night that needs to be recorded in more detail. On 27 May, eleven crews were briefed for attacks on marshalling yards or self-illuminated targets in the Ruhr, with the German-occupied airfield at Flushing nominated as the last resort target for the night. The first aircraft departed at 2018 hours, with the others following at approximately one minute intervals, heading for the coast at Great Yarmouth, and with thunderstorms on the way. The third aircraft suffered a lightning strike that fused the WT aerial, causing the pilot a severe electrical shock conducted through his intercom leads. His second pilot took over and landed at Bircham Newton in Norfolk. However, nine of the other ten aircraft reported making attacks in the designated target area. The last aircraft took off at 2029 and also encountered severe electrical activity on the early part of its route – a crew member later recalled lightning flashes streaking off the guns in the nose turret and his pilot making several course alterations trying to steer out of it. On reaching the Ruhr at 2351, this crew was unable to identify a target due to industrial haze and searchlight glare and, as instructed at briefing, set course for the airfield at Flushing at 2357. Some two hours later, an airfield was seen and attacked and bombs were observed to burst on an illuminated flare path. (The operational record has the aircraft over the target at 8000 feet, from 0200 to 0210.) The crew continued home to base but were eventually given a fix putting them well to the west, over Wales. They finally landed at Dishforth at 0420, rather later than many of the other aircraft. About an hour later it became known that RAF Bassingbourn, near Cambridge, had been bombed at around 0200 – and it was to become clear that it was this crew that had done so. The subsequent investigation revealed that the aircraft's magnetic compass was some 40° out of true, a result of the early lightning activity. There are some time and distance questions from all of this that cannot be answered without the log and chart for the sortie, but these were available to a Board of Inquiry. As a result, the navigator, Sgt Donaldson, was exonerated. However, the experienced captain, Plt Off Warren, was demoted to Second Pilot on the grounds that he should have made more use of his gyro compass – but both were back in the air within days. (And as a footnote to this unfortunate incident, the Air Staff became alarmed by the very limited damage done at Bassingbourn by the 250 lb bombs in use at the time.)

By late May, the Battle of France was close to being lost and Operation Dynamo had begun to evacuate what remained of the British Expeditionary Force, together with French forces in the area, some 340,000 men in all, from the beaches and harbour at Dunkirk. On 4 June, Winston Churchill made his historic "We shall fight them on the beaches" speech. On 18 June, in the equally inspirational "Finest Hour" speech, he added that, "I expect that the Battle of Britain is about to begin." Britain, with the Commonwealth, now stood alone against Germany, with the aircraft and men of Bomber Command the only weapon to hand capable of carrying the fight to the enemy heartland. Reflecting the uncertain circumstances, on the afternoon of 31 May a precautionary dispersal of aircraft was ordered. All serviceable aircraft were flown out to RAF Topcliffe and operations would now be flown from both there and Dishforth, with crews bussed to-and-fro as required. In addition, Home Defence Exercises were begun and given sufficient priority that a proportion of bomber effort was devoted to them. 10 Squadron was tasked with flying up to three aircraft over the Midlands on these on a number of nights in June.

On 1 Jun, in what may have been the first trial of such a procedure under fire, an attempt was made to mark a target, the refinery at Homburg, using flares. This was defeated by weather over the area, but Sqn Ldr Bickford's crew with another from 51 Squadron were able to do so on the following night. In addition to a bomb load, each aircraft carried 30 flares to be dropped at intervals so that the target would be continuously marked for the ten following aircraft. Consequently, these crews were exposed to severe AA fire for an extended period, and Bickford's tail gunner was severely wounded in the process. (Accurate target marking remained an issue in bomber operations and eventually led to the formation of the Pathfinder Force in August

1942, commanded by Gp Capt DCT Bennett, a former 10 Squadron CO.) The Homburg refinery was the target once again on 3 June, when crews found it still in flames from the previous attacks. Following that attack, in the early hours of 4 June, the unit's first fatal accident occurred when P 4963, captained by Flt Lt Phillips, crashed at Battisford, close to RAF Wattisham, having lost an engine over Germany. Plt Off Fields, the tail gunner, was killed, and only the Wireless Operator escaped without injury.

In marked contrast to the miserable winter weather, high pressure prevailed at base in early June – indeed, by 7 June, the F540 described it as 'confoundedly hot.' And in the midst of this there was a VIP visit by the RAF's founding chief, Lord Trenchard, on 7 June. There was also some good news for the wireless operators and air gunners – they were to be promoted to Sergeant, a rank that would be awarded in future on completion of training, and the new policy can be seen taking effect in crew lists over the following weeks. This improved the pay and conditions of those involved, albeit there remained a gap between the pay rates for these trades and those for NCO pilots and navigator/observers. However, the weather did not prevent accidents – there was another crash on take-off at Topcliffe on 8 June, when Sqn Ldr Bickford's aircraft developed an uncontrollable swing, leading to an undercarriage collapse, severe damage, and minor injury.

On 10 June, after some delay but now calculating on an imminent collapse in France, the Italian Dictator, Benito Mussolini, declared war on France and Great Britain. As the day went on, information was received that another maximum effort would be needed on the following day and that aircraft were to be fitted with auxiliary fuel tanks by the morning. Something special was afoot. On the morning of 11 June, eleven crews and thirteen aircraft were declared to Station Ops. Nine crews including a reserve were called and briefed, and eight departed for Guernsey that afternoon. Another twenty-eight Whitleys from 4 Group also deployed there, where they briefed for an attack on the Fiat engine and vehicle factory in Turin that night. In the afternoon, Wg Cdr Staton demonstrated that a takeoff from the short grass strip was possible and later that evening the force departed. Although the aircraft were carrying just two 500 lb bombs, they had full mains and auxiliary fuel loads. This weight, and the severe icing that many crews experienced in thunderstorms en-route, limited their ability to climb. Some twenty crews had to return to Guernsey, including five of those from 10 Squadron. (Fg Off Smith's crew was then unable to find Guernsey and continued home to Dishforth.) Wg Cdr Staton reached Turin and spent an hour over the target area, largely covered in thick cloud. Flares were used to try and identify the factory and an attack was finally made, hitting the target and an adjacent rail siding, but Sgts MacCoubrey and Johnson elected to attack facilities at the port of Genoa, an alternate target. All the crews returned to Dishforth later that day, to discover that another five squadron crews had launched the previous night for attacks on targets in France to dislocate German army movement. Two had to divert en-route with mechanical problems, two located and attacked targets at Amiens, and the fifth, Sgt Keast and crew, unfortunately failed to return. Sgt Keast, Plt Off Braham, Sgt Myers, Sgt Black and LAC Nuttall were, therefore, the first 10 Squadron crew to be lost in the main bombing offensive.

As other operations continued, on 12 June instructions came to fit a number of aircraft with special carriers for 'W' Bombs – a type of mine devised for use against river and canal traffic, to be released within a limited height and speed band to prevent the devices braking apart on hitting the water. The navigational challenge posed in attempting to do so by night, possibly in indifferent weather, was substantial, the more so as instructions were given that no weapon was to be dropped on land where they might be recovered and examined! An operation was mounted on the night of 14/15 June, using twelve aircraft of 10 and 51 Squadrons, in an attempt to disrupt oil barge movement along the Rhine, between Mannheim and Bingen. It was not a great success – one 10 Squadron aircraft was unable to get airborne; another had to return with radio failure; and two crews found the Rhine but were unable to complete an attack within the set parameters. Another two did release their mines, but not without incident. In one case, a container fell with its bombs inside, whilst the other had a partial hang-up and had to return with several bombs. Seven crews set off on 17 June for the same stretch of the Rhine, again with mixed results – three attacks were completed, whilst combinations of weather, AA and searchlight activity prevented the other crews from achieving a suitable release. On 19 June, word came that the 'W' Bomb carriers were to be removed from those aircraft to which they had been fitted – the short-lived experiment was at an end. Whilst the squadron record gives only the bare instruction, Larry Donnelly's book makes clear that the decision had the full approval of all the aircrew personnel who had tried to deliver the mines. Not all of the experimental weapons tried in WW2 proved as successful as Barnes Wallis's 'bouncing bombs.'

Whilst some crews had struggled with the 'W' Bombs, their colleagues had continued with operations over Germany and France. In the early hours of 20 June, Fg Off Smith's crew had to abort a mission over Amsterdam as they lost power from the starboard engine. Headed for RAF Honington, they found the lights out on account of an air-raid warning and, after attempting two circuits to land, crashed in trees in the undershoot. The aircraft broke in two and was destroyed in the fire that followed. Fg Off Smith was killed, and two colleagues suffered from burns. (An air raid warning had also been in force at Dishforth that night, with AA fire in the vicinity and unidentified aircraft heard overhead.)

Operations for 10 and 51 Squadrons were now settling into a pattern, with sorties still being flown from Dishforth and Topcliffe and, as far as possible, crews had a night off between flights. Targets were generally oil refineries, industrial plants, marshalling yards, and airfields – but besides attempting to destroy as many of these as possible, a regular supplementary object set for crews was 'to create maximum disturbance.' The one hundredth Station OpOrder was reached with instructions for an attack on the docks of the inland port at Duisburg on the Rhine, on the night of 27/28 June. However, given the considerable navigational difficulties that crews faced at that time, at night and in weather, how effective all of this effort was is open to question. At this stage, they were attempting precision bombing of point targets – area bombing was some way ahead – often spending extended periods over target areas whilst trying to identify a particular factory or whatever in extensive built-up areas through industrial haze and searchlight glare, and usually amidst AA fire. It is hardly surprising that many had to give up, in accordance with their instructions to avoid what we now call 'collateral damage.' Either that, or they dropped as best they could in the circumstances, and Wg Cdr Staton's early attempts at target marking suggest that there were already misgivings at some senior levels over what was being achieved. It is not surprising, therefore, to learn that Blenheims were now to be sent by day to targets attacked at night to extend the disruption as much as possible, and to take photographs of the night bombing damage. (After the war, it became known that the apparently random spread of bomb damage over this period meant that the German authorities had often been unaware of the intended targets.) This would all come to a head in another year's time, when a comprehensive study of the available photography would prove how few targets had been hit accurately. In the meantime, recalling once again that the crews and aircraft of Bomber Command remained the only means of striking at the enemy, those crews had to continue to give of their best – and they did so. Fortunately, 10 Squadron losses thus far had been relatively light.

Summer 1940 - A move to RAF Leeming and the Battle of Britain period

As the warm summer of 1940 wore on, the Battle of Britain that Churchill had anticipated was fought over Southern England, a period that earned RAF Fighter Command a special place in our national memory. Bomber Command also played a part, if largely behind enemy lines. And whilst a real risk of German invasion persisted, attacks were made on the shipping mustered in French ports and, each day, crews were placed on standby for operations against any landing force. At the same time, as Bomber Command began to grow in size, Operational Training Units (OTUs) were being formed and experienced squadron personnel began to be posted to some of these – No 19 OTU at RAF Kinloss and No 10 OTU at RAF Abingdon – and, whilst all of this was in hand, significant organisational changes were afoot for 10 Squadron.

A new airfield at RAF Leeming, some 15 miles up the Great North Road from Dishforth, had just been completed as part of the late-1930s RAF expansion. As such it had five of the large C-type hangars and substantial brick-built accommodation blocks and technical buildings, with

concrete runways to be added from late 1940. This was to be 10 Squadron's new home. An advance party moved there on 6 July, as did some aircraft, with the main move taking place on 8 July. However, operations were still to be controlled and flown from Dishforth, with crews and aircraft going there beforehand. Wing Commander Staton was appointed as Station Commander on 10 July, on promotion to Group Captain, and he appears to have made an impression in his new role. Sgt Jim Bassett arrived at Leeming as a new Observer later in the year and recalled his early impressions for a Squadron Association newsletter:

> "What a shock to see RAF Leeming! 'Kong' Staton was the Station Commander and he ran it as a peacetime unit. The SWO walked around – sorry, marched – with a pace stick under his arm, plus a waxed moustache……The Gp Capt did not regard Observers and WOP/AGs as Sgts and therefore we lived in barracks, with bunk and room inspection daily. I once had the nerve to protest to the Squadron Adjutant, and was told that the Pilots needed rooms for study purposes. The only thing I saw Pilots study at Leeming was the bottom of a tankard at The Green Dragon."

Sqn Ldr Whitworth assumed temporary command of 10 Squadron pending the arrival of the new CO, Wg Cdr S O Bufton, who arrived on 21 July. (Bufton flew initially as Second Pilot to Sgt Robinson to gain experience on the Whitley, and went on to be a significant figure in the development of bombing operations in the Air Ministry – *inter alia*, arguing the case for the creation of the Pathfinder Force in the face of opposition from Air Chief Marshal Harris at Bomber Command.)

Amidst these administrative changes, Invasion Standby and operations continued as weather at base or over Germany permitted. Attacks were planned on two different targets at Kiel. On the night of 8/9 July, the designated target was the Howaldts Shipyard. Five crews were detailed, but only one reported a successful attack and two were unable to attack any target so brought their bombs back to base. Sqn Ldr Whitworth elected to attack an alternate target at Eckenforde and, on approach, his aircraft was hit by flak – a fragment penetrated the cockpit and struck his parachute harness before striking him on the face. His rudder controls were cut and there was other airframe damage but he was able to recover to base. Flt Lt ffrench-Mullen's crew did not return. They too had abandoned the mission circa 0145 hours and were later shot down over Heligoland, to survive as POWs. (This was confirmed via The Times on 12 August.) Ops were briefed then cancelled on account of weather on subsequent nights until 12/13 July, when six crews were detailed for a raid on the Krupps Works in Kiel, a target destined to become familiar in the years ahead. One returned immediately after takeoff as the front turret

Wg Cdr W.E. 'Bull' Staton DSO & Bar.. … and dog.

Gp Capt Staton, Wg Cdr Bufton and Squadron officers, at Leeming, 27 September 1940.

proved unserviceable, a problem that could not be quickly rectified, so the sortie was cancelled. Two crews claimed to have bombed the target area in poor weather conditions that defeated the three others, both over Kiel and over potential alternates.

On the next night, three crews were detailed for an attack on aluminium works at Monheim. All reported bomb releases in the target area but with weather preventing any assessment of results. Diversion messages were issued to all three but a W/T problem prevented Sgt Hillary's crew from receiving theirs, and he managed to land at Dishforth nonetheless. On the next night, 14/15 July, the target was an air park at Diepholz, with four crews detailed. All claimed successful results, with some positive hits identified from a variety of level and glide attacks. No two crews adopted the same bombing procedure and all were diverted once again on the way home.

There was a weather break before the next Op could be flown and during that period the Squadron Ops setup moved to Leeming on 17 July, leaving a detachment at Dishforth, from where aircraft were still being shuttled to Leeming for Ops. That next Op required nine crews on 20/21 July, when the target was an aircraft factory at Wenzendorf. No reason is given, but the records show that Sqn Ldr Hanafin took off, landed back, and took off again in a space of seven minutes. Sgt Green's crew had to change to a reserve aircraft and were late on target as a result, incurring some flak damage to the rear turret, but with no bombs dropped. Seven crews reported completing attacks on the primary target, with no results visible, and once again using a number of techniques. However, on landing, it became apparent that a blown fuse, not detectable in-flight, had prevented one bomb load from being released. The ninth crew had been unable to identify the primary target or any authorised alternative. Another aircraft factory, this time in Bremen, was chosen for the Op on 22/23 July. Eight crews were detailed, and five claimed attacks with whole or part loads, once again with weather preventing any assessment of some results. Two crews elected to attack alternate targets and one was unable to identify the target, despite spending some time in the attempt. For what proved to be the last raid in the month, eight crews departed for the harbour at Bremen on the night of 24/25 July, with targets designated as the battleship *Bismarck* and two merchantmen, *Europa* and *Bremen*. One crew did not get away and another had to return quite soon with engine trouble, whilst the others were defeated by a combination of dense cloud and icing conditions that persisted to the target area. Plt Off Parsons crew moved to the alternate, the *Tirpitz* in Wilhelmshaven, and believed that they released a stick over the port area, with no results visible. The battleship survived to be a target again on more than one occasion.

The twelfth month of the war began with an attack on an example of a target type that would recur right to the end. Oil refineries and synthetic oil plants were seen as essential to the German ability to fight and were repeatedly targeted till 1945. The target for the night of 2/3 August was the refinery at Salzbergen, near Osnabruck, one of the oldest in the world, having started the production of lamp oil in 1860. Seven crews were briefed, including that of Wg Cdr Bufton who was recalled to see Gp Capt Staton, the Station Commander, whilst preparing for takeoff. The other six took off as planned but unserviceabilities forced two to turn back to base. The remaining four all claimed to have bombed the target area, with Plt Off Prior spending some forty minutes over it to drop in three sticks. Nickels were also dropped later that night, as they would be on many bombing missions in the years that followed.

What transpired in the conversation between Gp Capt Staton and Wg Cdr Bufton on the evening of 3 August is not recorded, but it may be no coincidence that, on the next afternoon, Sqn Ldr Hanafin took off on a local flight "observing likely aerodromes," presumably (and prudently) spotting possible dispersal sites for contingency use. Luftwaffe attacks on RAF airfields and radar sites in the south began shortly after this, and attacks in the north would follow. In the meantime, on the night of 5/6 August, seven crews were briefed for an attack on the Dornier aircraft factory at Wismar. One crew returned with its bomb load having been unable to see the target, but the other six reported bomb releases through heavy cloud that made positive assessment of results impossible. Wg Cdr Bufton completed the mission that night but his crew could only get two bombs to release, and several crews also reported having to take evasive action on being held by searchlights.

> Razzle Pellets were an experimental incendiary weapon introduced in August 1940, to be dropped into forest areas such as the Black Forest where it was believed military stores were being hidden, normally after a bombing raid had been completed. They consisted of phosphorus pellets covered with gauze, carried in sealed cans containing water to prevent them drying out. The cans would be opened and the pellets dumped into the flare chute and as they dropped and dried out, they would ignite in the forest canopy. However, it was found that not all dropped as planned – some were caught in the airstream and adhered to the elevators and tailplane, causing fires as they dried! They were first used by 10 and 51 Squadrons on the night of 11/12 August over the Black Forest after a raid on another target, with both units reporting that fabric on control surfaces had caught fire. Their subsequent use appears to have been rather spasmodic.

The primary target for the night of 11/12 August was a synthetic oil plant at Gelsenkirchen, but the Object of that night's Op was amplified to include the destruction of "Forests by fire" in a designated area. Eight crews were designated but, as one crew had to withdraw with engine trouble at start-up, a reserve crew and aircraft quickly got airborne in its place. The target was largely covered by cloud and the records indicate that Wg Cdr Bufton attempted to mark it for the others using flares and incendiaries. This proved unsuccessful, and he brought the remainder of his bombs home, but five of the others reported having made attacks using two sticks. The two others were unable to see the target and both attacked alternates elsewhere. All reported dropping their Razzle pellets over cloud within the designated area, some experiencing partial hangups, and with Fg Off Henry noting that his had started to ignite on release, thus drawing flak onto his aircraft!

On the morning of 12 August, the Squadron was informed that ten aircraft would be required for operations over Italy on the following night. (51 and 58 Squadrons would also be involved on a similar scale.) Crews were briefed the following morning that their targets would be either of two Fiat aircraft factories in Turin, and flew to Abingdon that afternoon to refuel. (The unit F540s of this period were compiled in considerable detail – hence it is that we know that a duplicate Ration Return for HQ 4 Group had to be compiled at 2030 hours to replace that submitted at 1800 hours, but since lost.)

In the afternoon of 13 August, ten Squadron crews took off for Turin but two had to turn back en-route – one with engine trouble, the other on account of the rear turret being damaged as its door blew open. The other eight all claimed successful attacks against light flak. (The records show that some Nickels were to be loaded for release over Italy. There is no mention of this having happened, but a lack of reporting on the leaflet task appears to have become quite normal.) However, Plt Off Parsons' aircraft was damaged after the raid in an encounter with an Italian fighter and was on one engine from Geneva. Gradually losing height, an attempt was made to make a forced landing on a beach at Hythe, on the Channel Coast, when the starboard elevator dropped away, precipitating a crash into the sea. Plt Off Parsons and his Second Pilot, Sgt Campion, were killed and their bodies were finally washed ashore near Boulogne, where they were buried. Fortunately the crash was seen from the land and the other three crewmen, floating in lifejackets, were picked up. One, Sgt Marshall, the WOP/AG, was rescued by a Miss Peggy Prince in a canoe, an action for which she was later awarded a BEM. The crews landed back at Abingdon for interrogation and refuelling before returning to Leeming, but another Italian mission was in the offing.

On the next day, 15 August, news came that four crews would be required for a second raid on Northern Italy that night, again using Abingdon as a forward base. In addition, nine crews were to be detailed for an attack on Munich, using Honington as their forward base. These duly briefed and deployed but all returned that evening when it was decided that weather in the target area was unsuitable. The Turin raid went ahead but, thanks to low cloud, crews had to attempt attacks from as low as 2500 feet with Sqn Ldr Whitworth finally electing to attack a blast furnace in Genoa instead. Sgt Green's crew signalled that they had attacked Turin but nothing further was heard from them. A report forwarded through the US Embassy in Rome said that Second Pilot, Fg Off Higson, was at the controls as the crew baled out but he died in the subsequent crash. He was buried with full honours by the Italians, with his colleagues in attendance as POWs. (In early September, The Squadron heard from the mother of Plt Off Oliver, the Tail Gunner, that he was in an Italian camp.)

The next Op was called for the night of 16/17 August when, after a change of target, the destination was the Zeiss optics factory in Jena, beyond the Ruhr towards Leipzig. Nine crews were detailed and six reported attacks on the target, with Flt Lt Raphael's tail gunner (LAC Cowie) shooting down a Me 110. Three failed to locate it – one dropped Nickels and, accidentally, a container of incendiaries, another attacked an airfield target, and the third dropped incendiaries on a factory at Mellingen. Fg Off Nixon's crew had reported engine trouble en-route and that the primary target had been attacked. At 0324 hours it was reported as over the Dutch coast, at the mouth of the Scheldt, and nothing more was heard. Later that night the BBC reported that the German authorities were offering a reward for news of a British crew that had landed near Zevenbergen and, on 2 September, word came via the International Red Cross that all five airmen had been captured and were unhurt.

For 18/19 August the target was an aluminium factory at Rheinfelden and ten crews were detailed. Three attacked the primary target; five attacked a secondary, the airfield at Freiburg; one attacked an airfield target of opportunity; and one returned with its bomb load, having failed to identify any target. For some days following that, up to ten crews at a time were brought to varying degrees of readiness and later stood down as plans changed, but with the Invasion Standby still maintained. Finally, on 24 August, ten crews were detailed to deploy south to Abingdon, under strict radio silence, to launch from there for another raid on Italy, when the target was an electrical factory in Milan. Things did not go terribly well. Three crews had to return to Abingdon with technical problems, and a fourth did so as a result of a misunderstood radio message. Two crews attacked the primary target, and another the secondary, a factory at Sesto Calende. Cloud and haze made positive identification very difficult, and the remaining three crews attacked targets of opportunity in the area. As crews landed back at Leeming, via Abingdon, word came that some would be required on the next day, 26 August, for what turned out to be a second run at the previous target in Milan, but this time using Harwell as the forward base. Having been warned that ten crews would be required, in the event just six were called and these deployed south that afternoon. Only five aircraft took off and four reported having attacked the primary target using a number of runs. Sgt Howard's crew was shot down near Varese, north of Milan, and all five crew were killed.

It is worth recalling developments elsewhere around this time as they inevitably impinged on the pattern of bomber operations in the longer term. The Battle of Britain was still in full swing, with the Luftwaffe maintaining its offensive against RAF Fighter Command assets and associated industrial targets. (RAF Driffield suffered severe damage and evacuated its squadrons to nearby airfields, including Leeming.) However, on the night of 24/25 August 1940, in what is often held to have been an accidental attack, bombs were dropped in areas of east and north London. A first attack on Berlin on the night of 25 August was authorised as a result, but weather prevented it having any real effect in the city. Additional attacks were ordered in the following days and, in turn, these caused Hitler to order that the Luftwaffe's bombers should now concentrate on London and major British cities. What we now know as the Blitzkrieg (or The Blitz) began on 7 September. Other cities took their share, but London was targeted on a nightly basis for some two months. Whilst this change in German policy went a long way towards permitting Fighter Command to win the Battle of Britain, it also paved the way to the sustained attack on the German homeland that would occupy Bomber Command and, thus, 10 Squadron, until April 1945.

More immediately, further attacks were made on oil-related targets. On 28/29 August, four crews were tasked for raids on a synthetic oil plant at Dortmund and a refinery at Dusseldorf, with two aircraft tasked against each, and all four crews claimed to have attacked their designated primary targets. Whilst over the target, Flt Lt Raphael's crew also carried out a successful night photographic mission, the first such for No 4 Group. (Investigations continued with Fg Off Warren's crew carrying out an air test of a camera mounting for HQ 4 Group on 5 September and, in due course, all aircraft would have cameras, with pilots under instruction to maintain a steady heading after 'Bombs Gone'

until a photoflash picture was taken.) Then on 31 August/1 September, the target was a synthetic oil plant at Wesseling. Six crews were called and three reported attacking the primary target, whilst the others attacked the designated secondary in Leverkusen or targets of opportunity. (In relation to this latter mission and others like it, it should be noted that crews had authority to attack a variety of targets. The daily operation orders of the time would normally list Primary, Secondary, and Alternative Targets, together with generalised targets of Last Resort listed as SEMO and MOPA – Self Evident Military Objectives and Military Objectives Previously Attacked.)

On 3/4 September, after being warned to prepare for a mission over Italy, the target was later changed to a Power Station in Berlin, using seven crews. Some indication of high level interest in the changed target may be gauged from the fact that the CO, Wg Cdr Bufton, was then added to the night's crew list and that he took a 4 Group staff officer as 2nd Pilot to observe the operation. (They took a good lump of shrapnel through the windscreen over Berlin, but without injury.) Only one crew claimed to have bombed the primary target and, in extreme darkness and haze, the others attacked a variety of others in the area. The weather back at base was also poor - Flt Lt Tomlinson's crew could not locate the airfield and, running out of fuel, crash-landed near Northallerton. The aircraft was wrecked but there were no casualties. This Op was followed by another to an oil depot in Berlin on 6/7 September. Five crews took off; four claimed attacks on the primary target and the fifth attacked guns and lights in the area. Plt Off Thomas' crew signalled "Task complete" but did not return to base and probably crashed into the North Sea.

In the late afternoon of 7 September, the Invasion Alert status was raised. All crews were recalled from leave, and all aircraft were made serviceable and bombed up. Late that night, the alert status was raised to Alert No 1 and, whereas the normal Invasion Standby commitment had been running at 3 or 4 crews per day, it was increased over the following days from 9 to 12 crews and all serviceable aircraft, as other operations permitted. Meanwhile, as already noted, London would suffer the first night attack of The Blitz and amidst all this, during the morning of 8 September, the squadron crews were visited by MRAF Lord Trenchard. Later that day, as a consequence of the raised invasion alert, a planned afternoon's bombing training on the Misson range for eight crews was cancelled and an operational target for the night was nominated - barges and invasion shipping at Ostend. Six crews took off but only two claimed to have found the target. Weather over the target area beat the others, and their bombs were brought back to base. The first aircraft back, with Fg Off Cairns' crew, overshot the runway on landing and crashed through the airfield fence onto the Great North Road (A1). A hydraulics problem had prevented full flap extension. The crew escaped uninjured, other than Fg Off Cairns who suffered a left leg fracture and had to be pulled through an upper escape hatch by Sgt Nicholson, his WOP.

By now, some of the effects of having been at war for a full year can begin to be seen. 'Screening' of individuals after a number of operations had begun, with the personnel concerned usually leaving the Squadron to move to the training units that were now providing a growing number of new, non-regular replacement aircrew. For example, a new 10 Squadron crew list posted on 11 September showed sixteen crews in two Flights, up from the thirteen listed on 1 September, but ten of them were annotated as Not Yet Operationally Fit, largely on account of postings-in and promotions to Captain. After a burst of night circuits and cross-country flying, the list was posted once again on 13 September without the annotation. Also, the Record Book for 14 September 1940 has the first mention of a 'Nursery Crew' in the unit's daily returns to Station Ops, and specially tasked missions for such crews are mentioned in the following weeks. These appear to have been potentially less hazardous sorties, allowing the crews to gain some experience before joining colleagues in the main bomber offensive. At the same time, the Invasion Alert remained at its highest state and, whilst crews were released once not required, all were reminded of the need to remain within one hour's travel time of the station.

For the night of 11/12 September, the designated target was the naval armaments and shipyards in Bremen, with seven crews detailed. (Unusually, the record notes that all crews "were contacted by R/T prior to both takeoff and setting course." This stands in marked contrast to

procedures later in the war when, with knowledge of long-range German listening devices, all starts and departures were carried out under radio silence.) All seven crews claimed to have attacked the target area, with the first firing Red Very Flares to help mark for those following. Significant fires were created, large enough for the CO's Observer, Sgt Bessell, to use back-bearings on them for navigation on the way home. During a second attack run, Flt Lt Tomlinson had to take violent evasive action to break from searchlights and subsequently discovered that his Tail Gunner, Sgt McIntosh, had abandoned the aircraft, believing it to have been shot down. Elevator damage was discovered on landing, possibly caused as Sgt McIntosh bailed out. He was captured and became a POW. (It is also worth noting that crews had standing instructions to report on poor blackout locations spotted over the UK, and two crews did so that night regarding a blast furnace near Darlington.)

On 14/15 September, as the air battle over Southern England reached what is generally seen as its climax, the invasion threat remained high. Three crews were dispatched against shipping in Hamburg but, shortly after takeoff, were diverted by radio message to join another seven crews briefed to attack barges and trucks in Antwerp's dock area. Nine of the ten crews reported having attacked the dock area, hitting shipping and causing fires. On the return, Sgt Willis' crew spotted distress signals from the sea some 20 miles off Spurn Head. He circled the position and passed back a good fix that permitted a RN Minesweeper to pick-up Sqn Ldr Ferguson's crew from a dinghy. On their outbound leg to Hamburg, they had lost the starboard engine, jettisoned their bomb load, and ditched without casualties. They were landed back at Grimsby during the afternoon of 15 September, and returned to Leeming the next day to report that the aircraft had floated for fully five minutes after ditching and that they had salvaged confidential manuals and signals logs.

After a weather cancellation of operations planned for 16 September, on 17/18 September ten crews briefed for a return to significant targets on which unsuccessful attacks had been mounted in late July – the *Bismarck* at Hamburg (primary), and the *Tirpitz* at Wilhelmshaven (secondary). Engine problems prevented one takeoff, and blocked guns forced another crew to return to base. The weather over Hamburg was very poor and although the *Bismarck* could not be seen in the position given at briefing, six of the remaining crews claimed bombs dropped into the dock area, with the last crew spotting too late that the ship had moved to a new location within the harbour. Thanks to the weather, two crews were unable to locate either the primary or secondary targets and they went on to attack Last Resort targets.

For 20/21 September, after an early notification that the night's target would be Berlin, a change of plan included marshalling yards at Hamm (four crews), Soest (two crews), and Ehrang (four crews). In addition to HE bombs, five of the crews were briefed to carry and drop ten tins of Razzle pellets each. Crews reported successful attacks on Hamm and Soest, but only one could see enough to bomb the yards at Ehrang. with the others attacking opportunity targets. For the next night, 22/23 September, the target was an aluminium factory at Lauta, deep in Germany towards the Polish border. Ten crews were briefed, plus one for a nursery sortie over Brussels. However, weather conditions caused the cancellation of the latter sortie, and a reduction to four crews for Lauta. In the event, three claimed successful attacks, whilst the fourth was defeated by weather in the target area and attacked a railway yard. Two aircraft suffered lightning strikes, making their radios unserviceable and causing navigational problems on the way back to base as no DF bearings were obtainable.

For the night of 23/24 September, just three Nursery crews were called for an attack on invasion barges and harbour installations at Boulogne. Conditions were good over the port and all of the crews were able to confirm successful attacks in and around the docks and workshops. An altogether larger Op was planned for the next night when the Squadron provided half of the crews detailed to attack a power station at Finkenheerd, near the Polish border, with another at Charlottenberg, Berlin, as an alternate. Twelve crews were briefed for the task, which was to include a return via West Raynham, in Norfolk, for refuelling. One returned early with engine trouble whilst, in difficult weather, only one reported an attack on the primary target. The others attacked a variety of nominated and last resort targets over Germany and Razzle pellets were also dropped.

On the return, fuel shortage compelled Sgt Snell to land in a Norfolk field. The crew refuelled but, unfortunately, whilst taxying for a takeoff an elevator clipped a plough and was damaged. After repair, the aircraft was eventually recovered to Leeming on the afternoon of 27 September.

During a short break from Ops, a number of administrative issues were taken care of. New crews were detailed to go to HQ 4 Group to meet the AOC and, though Invasion Standbys continued, leave was permitted once again, for two crews at a time. Nonetheless, attacks on potential invasion ports were mounted once again on the night of 27/28 September, when ten crews were detailed for a raid on the U-boat base at Lorient, with a single Nursery crew destined for Le Havre. Conditions over Lorient were reasonably good and the first crew marked the target with flares for those following, each making two or three runs. Luftwaffe aircraft were also active over England that night and Sgt Towell's takeoff for Le Havre was delayed by an Air Raid warning. Later, on return from a successful attack, his aircraft was fired upon by AA crews around Coventry, Nottingham and Birmingham and, as the main group returned later still, their landings were delayed by fears of an intruder that turned out to be a Squadron aircraft. (The records also show that runway construction work at Leeming was affecting operations around this time.)

On Sunday 29 September, Gp Capt Staton, the CO and Squadron officers returned to the Station Church at Dishforth for the unveiling of a plaque commemorating Flt Lt Allsop's crew, lost almost exactly a year before during the first mission over Berlin. And on the next day, the commemoration continued insofar as the target for that night was the Air Ministry building in Berlin, with alternatives in the area. Ten crews departed, but one returned early with an unserviceable rear turret. The others reported attacks around the city despite severe weather conditions and seven landed back variously at Watton, Marham, and Pembray. However, one crew failed to return after reporting a successful attack – both pilots, Sgts Snell and Ismay, were killed, but the other three crewmembers survived to become POWs. Fg Off Wood's aircraft overflew the UK on its return, came down in the Irish Sea and the crew was picked up by a trawler off Waterford. All aboard survived and were landed at Holyhead on 1 October for transfer to hospital. Two Nursery crews also set off to attack invasion shipping at Le Havre that night, but one had to return early and force land at Abingdon after an engine failure.

The turnover of aircrew personnel in the latter part of September had continued and the consequential dilution of experience was clearly of concern to the Station Commander, and probably to all levels in the chain of command – messages from AOC 4 Group concerning Navigation and Crew Training were read to crews on 3 October. For example, with the Squadron Commander in hospital, Gp Capt Staton issued instructions on 2 October to the effect that three crews were not fit for full operational duties, and a crew list the next day showed seven out of sixteen crews with Nursery or Training status. This set a pattern for the months that followed, with some 50% of available crews categorised in this way. Over the same period, the number of crews increased to eighteen and nineteen, but the weather over the winter months did little to help the newcomers accelerate the gaining of experience. The records show many cancellations of both planned operations and of bombing range training sorties, together with a number of meetings of squadron executives to discuss the issue. However, as the war progressed, the comparative stability of both crews and experience inherited from peacetime in that first year would give way to a continual churn of personnel to replace casualties lost in action or posted out ('screened') after surviving a thirty-mission Op Tour. (From a strength of some 70,000 in 1938, the RAF had already grown to around 250,000 in 1940 and it would end the war at over 1,140,000.)

Ops resumed on the night of 7/8 October with two targets. - the Fokker Aircraft factory at Amsterdam for four Nursery crews; and barges and shipping at Lorient for five '1st class operations ' crews, a new term. (Three crews were also detailed for training over the Irish Sea.) Cloud cover over Amsterdam caused two crews to attack nearby docks, with the others attacking the airfield at Schipol. Cloud cover also limited attacks on Lorient – two crews reported releases over the target but were unable to assess results, whilst the others attacked Dieppe, Le Havre and Cherbourg. An Op for 9 October was cancelled and then, on 10/11 October, four crews were detailed to attack the Wesserling synthetic oil plant in Cologne, where three reported bomb releases in the target area. The fourth crew believed they had also hit the target but, on landing, the bomb load was found to be still in place and the debrief mentions 'faulty manipulation.' Two aircraft sustained flak damage, and one of these experienced considerable difficulty in locating the airfield on return.

Weather over planned targets precluded further operations until the night of 14/15 October, when things did not go well. The main target was another synthetic oil plant, in Stettin, for which five crews were selected. All reported attacks in the target area but two failed to receive a message telling crews to divert to Marham on account of fog at Leeming and both crashed, having run out of fuel. Flt Lt Tomlinson's aircraft came down near Thirsk – three crewmembers abandoned the aircraft successfully, but their colleagues, Plt Off Dickinson and Sgt Neville, were found in the aircraft that caught fire on crashing. By contrast, all of Sqn Ldr Ferguson's crew were all able to abandon their aircraft before it crashed near Otterburn. Three Nursery crews had also been sent to Le Havre and two reported having made attacks, with one diverting to Abingdon as conditions at base deteriorated. The third crew – Sgt Wright, Plt Off Cooney, and Sgts Caswell, Henry, and Davies – were all killed during their return leg when their aircraft hit a barrage balloon cable near Weybridge and crashed.

Operations were planned but cancelled, including one with a target at the Skoda works in Pilzen, Czechoslovakia, but the next flown was on 19/20 October, when two crews were sent to attack rail yards at Osnabruck. One reported a successful attack, but hit a mobile airfield Chance light unit on landing at Abingdon. The second had to return soon after coasting out due to severe vibration in the port engine and, after landing, it was discovered that some 6" of the propeller had broken off in flight. The possibility of another raid in Italy was raised for 21/22 October, but the target finally emerged as an aircraft component factory in Stuttgart. Four crews were tasked, with a further two briefed for an attack on an oil refinery at Reisholz. One crew reported attacking the Stuttgart target, another reported attacking an opportunity target, and a third had to turn back after two hours when it lost an engine. The fourth crew went missing over Germany, and was later confirmed through the Red Cross as having crashed and died – Flt Lt Phillips, and Sgts Gordon, Lofthouse, Mapplethorpe and Wills. Only one crew was able to get airborne for Reisholz which was obscured by mist, so it bombed a railway line west of Dusseldorf. Autumnal Yorkshire fog meant that all four returning aircraft that night had to divert.

By late October, with the Battle of Britain won, the government accepted that the immediate threat of invasion had passed. Accordingly, the need for dedicated Standby arrangements introduced earlier in the summer could be called off, and the last listing recorded was for five crews on 25 October 1940.

On the night of 24/25 October, Plt Off Peers' Nursery crew successfully attacked dockyards in Hamburg, as did a force of Wellingtons in what appears to have been an effective raid.. They reported very bad weather out and back, plus severe and accurate AA fire and searchlights in the target area. Having returned intact, and no doubt rather more experienced than when they took off, they may have been charmed to see that they were on that day's final Invasion Standby list and the night flying programme. The main Op target for the Squadron that night was another synthetic oil plant at Magdeburg, with a return visit to the Reisholz refinery for a single Nursery crew. The latter had a short, if eventful, trip – after a delayed takeoff on account of German aircraft in the immediate area, it had to return to base when the rear turret failed and landed first in error at Catterick. None of the four crews sent to Magdeburg could find the target under heavy cloud. Three attacked a variety of opportunity targets, whilst the fourth brought its bombs home.

During the afternoon of 27 October, a fault in the cockpit heating of an aircraft (P 4959) on dispersal caused a fire that proved disastrous once it reached the fuel tanks. For that night, an unspecified target for operational crews was cancelled, but two Nursery crews were sent to attack the submarine base at Lorient. Plt Off Russell's crew reported a successful attack in good visibility, but Plt Off Peers' crew – still not granted Op status - was unable to identify it in poor visibility, perhaps a result of the first attack, and went on to attack Cherbourg through cloud cover.

For 29/30 October, the primary Target for four Op crews was a synthetic oil plant at Magdeburg, with refineries in Hanover listed as the secondary. None was able to see

the primary or secondary targets under thick cloud, and attacks were made on a variety of opportunity targets. One crew reported seeing three German aircraft after coasting out, and these may have been responsible for an attack on Leeming around 1830 hours. Bombs were dropped but no damage was done on the station. Two Nursery crews also set off for Wilhelmshaven, with both reporting successful attacks. On return, one had to land at Topcliffe, having been unable to make radio contact with Leeming. The other – Plt Off Peers' crew - got to overhead the station, was cleared to land, but did not do so and headed off to the north and eventually crashed at an altitude of 2500 feet, near the village of Slaggyford in Northumberland. The aircraft was wrecked but the crew emerged relatively unscathed, with the Captain reporting having received incorrect steers (QDMs).

The record for October closed stating that 55 tons of bombs had been dropped during that month, with a total of 549.7 tons since 1 April 1940.

November opened with a number of flights cooperating with searchlight crews and, on 2 November, five crews flew south to use Bassingbourn as a forward base for an attack on a telecoms factory in Milan. There was a cancellation each day until, finally, forecast weather conditions permitted takeoff on 5 November – but fortune had conspired against them. One crew suffered multiple instrument failures and could not depart. Weather forced a second to turn back after some two hours, and they reported having attacked the docks at Ostend. Weather and engine troubles also forced a third to turn back; its bombs were thought to have been jettisoned "safe" somewhere between Le Havre and Dunkirk, but all bombs were still on board after landing. The fourth crew had to turn back with engine trouble whilst trying to climb out of icing conditions over France. Its bomb load was jettisoned safely. (In such conditions, the instruction in the Order for the mission that crews should "report on the crossing of the Alps with sketch maps of prominent peaks to assist future missions" looks grimly humorous.) Against this background, it is sadly ironic that the fifth crew reported "Ops complete' at 2359 hours, but then failed to return. They ditched in the North Sea, twenty miles off Ramsgate, at around 0610 hours, having run out of fuel. Plt Off Jones and crew died in the crash.

Hamburg featured once again as the target area for newer crews on the night of 6/7 November, when three were detailed to attack a synthetic oil plant. The record book notes: *"1745: All the aircraft took off on operations over Germany in spite of attempts at hindrance on the part of a Heinkel 111."* The thirty minute gap in takeoff times between the first and the two other aircraft no doubt resulted from just this hindrance. Sqn Ldr Sawyer's crew could not find the Primary Target but claimed a successful attack on the Secondary, the inland docks at Duisburg, albeit cloud obscured observation of the outcome. On return, they landed in error at Catterick, but took off quickly once again to return to Leeming. Sgt Towell's crew had also to attack Duisburg, claiming bomb bursts seen in the target area. However, Plt Off Williams' crew was able to report having attacked the designated Primary, after spending an hour in the area trying to identify it!

For 7/8 November, an attack on Italy was cancelled, leaving Fg Off Steyn's crew as the only one on Ops that night, with another oil plant at Wesseling his target. His crew claimed a successful attack, reporting the pinpoints used for identification and, notably, in view of the trials undertaken a few weeks beforehand, taking flash photographs after two of the four sticks dropped. On the next night, the allocated target was an aircraft component factory, near Stuttgart, with six crews detailed for a raid to be mounted from RAF Honington. An attack on the airfield there occurred, but appears not to have caused too much disruption to takeoffs, with a He 111 shot down in the process. Two crews reported having attacked the target; the others were unable to locate it, despite an hour's search by one, and all attacked a variety of opportunity targets.

After that, planned Ops to Italy, Channel Ports and Germany were cancelled until 12/13 November, when two crews flew to Topcliffe to operate from there against the submarine base at Lorient. Fg Off Landale's crew was first off but some two hours later crashed in the Welsh hills, well off the planned outbound route. The aircraft was destroyed and, whilst all 5 crewmembers escaped with injuries, the Second Pilot, Sgt Goldsmith, died later. The second crew reported having attacked the target by on ETA and after a fix, but cloud prevented any proper assessment of the results. On the next night, 13/14 November, two targets were assigned – another oil plant at Merseburg for five Op crews, and a

Naval Armaments Store at Hagen for two Nursery crews. One of the former was unable to get airborne; another had to turn back with engine trouble; two were unable to locate the target under cloud and bombed opportunity targets. The fifth aircraft failed to return - Sqn Ldr Ferguson, OC B Flight, and crew. A faint radio signal had been heard at 0155 hours, by when they should have been on the return leg and, with no known graves, it is likely they came down in the North Sea. Neither Nursery crew was able to locate either primary or secondary targets beneath cloud and mist, and elected to attack opportunity targets instead. A similar target allocation occurred on 15/16 November. Two were assigned: shipyards at Hamburg for five Op crews, and the airfield at Eindhoven for two Nursery crews. One Op crew was unable to takeoff after suffering a number of unserviceabilities; the four remaining crews all reported successful attacks in good weather conditions. Wg Cdr Bufton took one of the Nursery crews and both reported successful attacks.

By mid-November, some months after the Squadron's move, construction of Leeming's concrete runways was not yet complete and, around this time, the records show that soft ground after much rain had begun to be a practical problem affecting taxying and takeoff – echoes of similar difficulties at Dishforth during the previous winter.

For 17/18 November another planned operation over Italy from a forward base was cancelled, and another synthetic oil plant target at Gelsenkirchen was substituted. One aircraft became bogged down whilst taxying and was unable to get airborne; and another had to return to base within minutes with a faulty Airspeed Indicator and a smell of burning that was traced to a large oil leak. None of the three other crews was able to attack the designated target, but all attacked alternatives and all had to divert to Waddington on return as Leeming's weather precluded landings. A single Nursery crew reported having attacked the submarine base at Lorient, albeit that one stick failed to release. That crew had also to divert on return.

Planned operations on the following days were repeatedly cancelled and, whilst a degree of night, bombing, and cross-country training proved possible, there were no further Ops until 26 November, when docks and shipping Antwerp became the target for two Nursery crews. One had to return to base soon after takeoff with a faulty compass, but Plt Off Bridson reported a successful attack after some ninety minutes spent in locating the target! Also on the afternoon of 26 November, five crews deployed to RAF Wyton as an advanced base for a raid that night on the Arsenal at Turin. One aircraft sustained slight damage in a collision with a Whitley of 51 Squadron and was unable to takeoff; another had to turn back en-route as it became apparent that its fuel consumption was excessive. The three other crews reported making successful attacks on the primary target, and it is of note that one crew reported encountering heavy flak <u>over</u> neutral Switzerland and that the lights of Geneva were switched off as the aircraft approached.

Topcliffe was to be used once again by two Nursery crews tasked for another raid on Le Havre on 27/28 November. However, one crew was delayed by getting bogged down before they could taxy at Leeming, and they then had to return once they found that the port undercarriage leg would not retract. Plt Off Steyn's crew reported having pressed home a successful attack in three sticks and, for the first time, mentioned that "the screamers used on bombs caused the searchlights to douse quickly." 29 November was a day of changing plans with, in the end, a single Nursery crew going to Linton-on-Ouse to mount a successful raid on the docks at Bremen. Weather at Leeming precluded the crew's return to base on 30 November; indeed, no flying was possible at all that day.

With new personnel continuing to be posted-in, the unit crew state at the start of December was 15 Total: 9 Operational, 2 Nursery, and 4 Training, and the CO and his Flight commanders are recorded as having regular conferences on Training around this time. In addition, a distinct change in the detail of crew reports is noticeable, with much more being said about how targets had been located and identified.

Four crews were detailed for yet another attack on the submarine base at Lorient on 2/3 December. However, a fracture in the oil supply to the rear turret prevented one takeoff; two were unable to locate either primary or secondary targets under heavy cloud; but the fourth reported having bombed in very difficult conditions. Poor conditions

at Leeming meant that returning aircraft were diverted to Abingdon and Boscombe Down. They could not fly north later that day and weather at base prevented any operational flying until 7 December, when five crews were detailed for a raid on marshalling yards in Dusseldorf, with mixed results. Two reported successful attacks; another was persuaded by cloud en-route that it would not find the designated target so it released a stick plus incendiaries over a factory near Cologne, but went on to attack the target with its remaining bombload. Neither of the other crews could find the primary target, so one attacked the designated secondary in the Dusseldorf area, whilst the other attacked an airfield near Ostend. Severe icing and electric storms were reported by all crews – one had both engines stop, another lost height due to ice accretion, and one captain suffered temporary blindness after a lightning strike. A single Nursery crew (Sgt McHale) left to attack the docks and shipping at Boulogne, and carried out an attack through cloud after an intensive search. They, too, encountered icing and electrical activity

South African, Fg Off James Henry Steyn DFC - a 10 Sqn pilot had completed a full tour of operations at Dishforth and was posted to a training unit at Kinloss. when his Anson crashed on Ben More Assynt in the north-west Highlands. The crew's grave, 700m amsl is the highest in the UK.

on return and, after a day off, they were back on the roster for the night of 10/11 December, when their target was the inland docks at Duisburg. The mission had to be mounted from Linton-on-Ouse and, on reaching the target area, they found partial cloud cover and both sides of the Rhine flooded, making it difficult to locate the port before bombs were finally dropped on two reciprocal headings.

Conditions at Leeming improved sufficiently for the five Op crews detailed for an attack on a power station in Mannheim on 11/12 December to launch from base. However, one was unable to takeoff due to a fuel supply problem, and another had to return within thirty minutes with a faulty Airspeed Indicator and a hydraulic leak. The other three crews all reported successful attacks.

On the night of 16/17 December, Mannheim was once again the objective but, on this occasion and for the first time, it was the city itself that was the target. The War Cabinet had authorised an attack on a German city after the bombing of English cities, notably Coventry and Southampton. Whereas most raids of this period were being carried out by handfuls of aircraft, 134 bombers were detailed for what was the largest attack of the war so far. Seven crews were briefed, but unserviceabilities prevented one crew taking off. Five others reported successful attacks in the target area, ringed by fires caused by earlier aircraft. Plt Off Brent's aircraft lost its starboard engine nearing the target and began to lose height. Bombs were dropped southwest of Cologne, all turret ammunition was fired off leaving Holland, other ammunition guns and moveable objects were jettisoned over the North Sea and the aircraft was able to climb back to 2500 feet and eventually recover safely to Bircham Newton. (In fact, the raid was later shown to have been less successful than first thought. The incendiaries dropped by the first Wellingtons had not been accurate, the largest fires were not in the city centre, and the main force bombs were scattered.)

On 17/18 December, Mannheim was once again the target, but for a much reduced force, including five crews from 10 Squadron. Once again, an aircraft became bogged down and could not get airborne and only one crew reported having attacked the designated target. The others were defeated by the weather – one reported having bombed a flak concentration at Ostend, another went for searchlights around Dunkirk, and the other brought its bombs back. All had to divert to North Coates due to fog at Leeming, and were unable to return to base until the next day.

There was a return to Berlin on 20/21 December, when the targets were an aircraft component factory (Primary) and a power station (Secondary), and six crews were detailed and briefed. Five reported having made attacks in the general target area, with two noting what appeared to be a well-lit dummy town. The other crew failed to locate a Berlin target and bombed an airfield near Rotterdam, having dropped a single 250 lb bomb to test if it might be a dummy target. All lights were then extinguished and an attack was initiated. Plt Off Bridson's crew had to abandon their aircraft over Norfolk – an engine had failed over Germany, the other began to lose power and height could not be maintained - with all crewmembers landing safely. A single Nursery crew, operating from Linton-on-Ouse, attacked the docks at Flushing.

During the evening of 22 December, an aircraft crashed and caught fire near the airfield during night flying. Plt Off Flewelling, a Second Pilot, was killed and 3 others were injured. On the next night, 23/24 December, three crews were detailed to attack a naval armament facility in Mannheim, with a further three Nursery crews briefed to attack Boulogne. One of the Op crews reported having attacked the primary target, another attacked an opportunity target, and the third brought its bombs back. One Nursery crew reported a successful attack, whilst another could not find the port and attacked an airfield at Abbeville. The third had engine problems, jettisoned their bombs in the general target area, and had great difficulty in locating Leeming on return – the record states "an extensive tour of England was carried out" and the aircraft landed some 90 minutes after the others.

A crew conference was held at 0930 on Christmas Day and crews were then permitted to leave the station as there would be no Ops that day. Weather then prevented Ops on Boxing day but on 27/28 December, and after two changes of target, six Nursery crews were finally briefed for an attack on Lorient. One got away an hour after the others once their aircraft was moved onto firmer ground. Five reported attacks on the target but the intense searchlight activity that all reported prevented the sixth crew from identifying it and an attack on Cherbourg was carried out. One crew also reported AA fire over Portsmouth as being more intense and accurate than any enemy fire encountered! Then, on 29/30 December, the primary target was the industrial port at Frankfurt., with six crews detailed. Two had to return to base having experienced severe icing that affected their airspeed indicators and three crews reported having to drop on ETA in the target area, with AA fire penetrating the cloud layer. The other crew also experienced heavy icing, could not climb clear of cloud, failed to locate the target, and released its bombs into the general target area.

On 31 December the general condition of the airfield, aggravated by more rain then snow, precluded all flying for the day. A new crew reported for duty, giving a total of 20 crews at year's end – 8 Operational, 6 Nursery, and 6 Training.

Howgozit?

The Squadron had now been on operations for some 16 months since the outbreak of war. Whereas the day bomber force had suffered many losses to enemy aircraft and AA fire, like the night bomber force generally, 10 Squadron had been relatively fortunate and could even claim a German aircraft as having been shot down. Encounters with night fighters had been rare for the Luftwaffe had still to create an effective force, but that would come. Nonetheless, crews were displaying considerable bravery in often flying over target areas for lengthy periods whilst being sought out by searchlights and AA fire. In retrospect we can question how effective all of this valiant effort was, given the navigational shortcomings of the time. Early on, Wg Cdr Staton had shown concerns in this regard with his attempts at target marking, and these must have been shared by others at senior levels. Their validity would be brought home forcefully in the Butt Report of August 1941, as an analysis of post-attack photography made clear that very many crew reports of successful attacks were simply mistaken, and that it was unrealistic to expect accurate attacks on this or that factory set amongst extensive built-up cities. That remains a challenge today. Yet, in the aftermath of Dunkirk and with cities under attack by Luftwaffe bombers, RAF Bomber

Command's crews and aircraft were one of the few methods by which offensive action could be taken against Germany at the time. The 4-engined heavy bombers, with better navigational aids, would gradually become available in the coming year but 10 Squadron had another twelve months of Whitley ops ahead of it before it would re-equip with the Halifax. And the year began with a now-familiar mix of misfortune and success.

1941

On 1 January, Bremen was the first target of the new year. Seven crews were briefed but one aircraft was unable to get airborne, and two crews had to return with instrument failures that made night landings decidedly challenging. Four crews reported successful attacks, with weather at Leeming resulting in all diverting to Mildenhall. There had been previous reports of searchlights being used to form a cone, with AA Fire concentrated on the apex, but all of the experienced, returning crews made a particular point of this tactic as experienced on this night. 'Coning' would be a significant part of German defensive tactics for the rest of the war. In this regard, many of the Squadron crew reports since late 1940 had included remarks saying that switching on the IFF (Identification, Friend or Foe) equipment over a target appeared to lead to a light being extinguished. There was no technical reason for this to be so, but it was to become a widely-held belief in Bomber Command squadrons. So much so that in 1942, and in case it did some good, a modification to the kit was installed to provide a so-called jamming switch for use over enemy territory. A later study concluded that it had no appreciable effect but, as crews set such store by it, the switch should be retained on morale grounds.

For 2 January the F540 recorded: *"Fairly heavy snow fell during the day but freezing conditions later hardened the aerodrome and thus improved its condition."* Accordingly, for 3/4 January, an Op was planned, with Bremen once again the target. Three crews were briefed but yet again one crew was unable to get airborne, whilst another had to return with instrument problems. Two crews were able to press home what they reported as successful attacks, with the results obscured by the large number of fires already burning in the area. (As already noted, the navigational accuracy of the night bomber force was highly variable. 71 aircraft in all took off for Bremen but later reports noted that Hamburg, to which no aircraft were dispatched, recorded twelve fires that night, with six classified as large.)

4 January was an eventful day. Two Nursery crews left that afternoon, bound for different targets, one in Dusseldorf and the other in Hamburg, but both were recalled almost immediately after takeoff. A major German naval cruiser was reported to be in Brest harbour, and a mixed force of fifty-three aircraft was mustered to attack them, for which 10 Squadron briefed four crews. One was airborne for just five minutes, abandoning the operation with high oil temperature and low oil pressure in the starboard engine; another was almost lost as an engine cut out at 100 feet, but picked up just in time to avoid a forced landing with a fully loaded aircraft; a third needed an airframe change; and the fourth, one of those recalled, had to be extricated from the position in which it had ended up when a flap failure had forced an overshoot of the flarepath on landing back! Despite all this, three crews were able to press on and all reported making attacks through extensive cloud, with the crew that had experienced the earlier overshoot sustaining light damage after landing when their aircraft taxied into a hole, part of the airfield construction work at that time.

Airframe availability, weather and airfield condition all helped prevent operations on the following days, though cross-countries appear to have been possible at restricted takeoff weights. However, by 10 January Leeming was considered unsuitable for full-load takeoffs, and that night's Op was scheduled from Linton-on-Ouse, with aircraft flown there to be loaded for another attack on a Cruiser in the harbour at Brest. Six crews reported making attacks but were unable to assess results in haze. On 13 January, the target for the evening was Boulogne, with operations once again mounted from Linton. Six crews took part but, thanks to a "blanket of cloud from York to target area,' none could locate the target. Interestingly, three jettisoned bomb loads over the sea, whilst their colleagues brought theirs back home.

For the night of 16/17 January eight crews and aircraft left for Linton, to mount an attack on rail facilities in

Wilhelmshaven. Two were forced to return – one with mechanical problems, and the other after suffering severe icing that temporarily stopped both engines, plus a lightning strike that badly affected the Second Pilot. The others all reported heavy icing and electrical activity, but claimed successful attacks without seeing the results. Fg Off Skyrme's crew had sent a W/T message reporting an attack, but nothing further was heard from them. As they have no known graves, it seems probable that they came down over the North Sea.

Airfield conditions, local weather and forecast weather over the Continent all combined to prevent any other operations during January. The crew state at 27 January was 10 Operational, 7 Nursery, 4 Training - and a first sign of better things to come can be seen in Plt Off Hillary's posting to 35 Squadron, the first Halifax squadron then forming at Linton-on-Ouse. Six aircraft had been positioned at Linton on 31 January in anticipation of an Op that night, but the weather in Yorkshire and over Germany ensured that nothing would happen for some days.

Conditions eventually improved sufficiently for an Op to be attempted on the night of 3/4 February when, once again, the target was to be the *Hipper* Class cruiser in the harbour at Brest. Five crews were taken across to Linton but, in the event, one could not get airborne – aircraft could be bogged down there too, it seems. The others pressed on despite heavy cloud, but none was able to locate the target. Three dropped through cloud in the harbour area, the fourth elected to attack Cherbourg, and none could see enough to assess their results. On the following night two Nursery crews were briefed for an attack on shipping and facilities at Calais, again mounted from Linton. Cloud cover defeated both – one jettisoned its bomb load, whilst the other brought theirs home, with both reporting seeing enemy fighters, though no attacks were made. Weather forced both to divert to Driffield, spreading the squadron aircraft ever more widely.

On 5 February RAF Leeming staged a practice gas attack, with all personnel using gas masks for three hours, so it is no surprise to read that "telephone conversations were handicapped, being almost inaudible." On the positive side, the two aircraft were flown back from Driffield and on 6/7 February, for the first time in a month, it proved possible to mount an Op from Leeming. Six crews took off to attack the docks and shipping at Dunkirk and, when none could find the designated target under total cloud cover, three made attacks in the Dieppe area, two at Calais, and one at Boulogne.

Weather intervened once again and, on 9 February, given the condition of the airfield, it was ruled that ops would have to be mounted from the Squadron's former home, RAF Dishforth. Leeming would be used as a satellite airfield, with all aircrew and ground personnel remaining there, and a bus service was arranged between the bases.

At the same time, Air Marshal Sir Richard Peirse, CinC Bomber Command, was planning to refocus his Command's efforts following receipt of a revised directive instructing him that his primary aim was now to be the destruction of German synthetic oil plants. He planned one major raid on an industrial target, before concentrating on the oil plants. Thus, on 10/11 February, missions from Dishforth were planned for both Nursery and Operational crews. The main objective, for seven crews from a total force of 222 that night, comprised industrial targets in Hannover. One crew had engine trouble and elected to attack an active airfield, De Kooy, in Holland; the other six all claimed successful attacks within the target area, already well-lit with fires. The Nursery operation for three crews had Rotterdam as its target. One crew failed to get airborne; a second lost an engine, jettisoned their bombs in the sea and returned; and the third reported an accurate attack through partial cloud. (Three Stirling aircraft of 7 Squadron also attacked Rotterdam, the first Op for the first of the new 4-engine bomber types.)

After that, weather prevented operational flying until 15/16 February, when two targets were nominated – the Fischer oil plant in Sterkrade for Op crews, and docks etc at Boulogne for Nursery crews. Two crews claimed to have made successful attacks, whereas four others could not locate the plant in haze and went on to attack the docks or other targets at Duisburg. Three of the Nursery crews reported having hit Boulogne; a fourth lost its W/T and was unable to fix its position so attacked an opportunity target, the airfield at Haemstede. Crews on both ops were issued

diversion instructions on their return to Dishforth and ended the night on seven different airfields.

The weather once again intervened to enforce a break in the operational tempo. During the period, it appears that Gp Capt Staton and Wg Cdr Bufton discussed the likely impact of impending pilot postings to OTUs and to the new Halifax unit, No 35 Squadron. After these, a new crew list on 22 February included 10 Operational, 3 Nursery, and 8 Training crews.

Ops resumed on 23/24 February when seven crews were detailed for an attack on the docks at Calais, but still mounting from Dishforth. One had to return with a failed generator, overshot the runway, and seriously damaged the aircraft. Four reported successful attacks on the primary target; one was unable to identify it through haze and attacked the docks at Boulogne, the nominated secondary target; and the sixth could not identify a target over Calais or Boulogne, so brought its bomb load back to Dishforth.

A series of attacks on Cologne followed. For 26/27 February, seven crews were briefed, but a generator failure prevented one from getting airborne. The remaining six all returned (to Leeming) with reports of successful attacks. (100 other crews made similar reports, yet postwar access to city reports show only 10 HE and 90 incendiary bombs as having fallen on the edge of the city, plus others in villages to the west.) On the night of 1/2 March, Cologne was again the target for a mixed force of 131 aircraft. Eight crews were to mount from Dishforth and return to Leeming although, in the event, all had to divert elsewhere. Seven returned claiming successful attacks, and it is known that the attacks overall that night were much more accurate. Sgt Hoare's crew reported a successful attack by W/T, and was later sending SOS messages and was fixed off the Dutch coast at 0048 hours. Nothing further was heard from them and it is likely that they crashed in the North Sea. For a third night, 3/4 March, Cologne was again the target, with seven crews detailed to mount from Dishforth. A fuel leak prevented one takeoff, and a second crew had to abandon the op with engine problems. Three crews reported successful attacks over Cologne, another bombed early over Neuss as it was losing oil pressure, whilst the fifth found Cologne obscured and bombed München-Gladbach. All crews returned to Leeming, save one that landed in error at Topcliffe and bogged down there.

Although some training flying was possible, no further operational missions were flown for a week as Bomber Command repeatedly had to stand the night bombing force down on account of bad weather. Operations would resume in different circumstances.

CHAPTER 5

NEW DIRECTIONS

The War Cabinet had become greatly concerned by German U-Boat successes against Atlantic convoys, and a new directive was issued to CinC Bomber Command on 9 March 1941. On the surface, the large cruisers, *Scharnhorst* and *Gneisenau*, were sinking and capturing merchant vessels and the *Bismarck* was soon to join them. In the air, the long-range Focke-Wulf Condor bombers were also a significant threat to British shipping. So, with the severance of sea trade links with the USA becoming increasingly possible, the new directive quoted Churchill's own words: "We must take the offensive against the U-boat and the Focke-Wulf wherever and whenever we can." For much of the next four months, Bomber Command's main effort was to be directed against targets related to the basing and construction of these threat systems.

As this new phase of operations began, the turnover of personnel was now a well-established feature and the 10 Squadron crew state on 14 March 1941 was 9 Operational, 8 Nursery, and 3 Training. The break in operations that had permitted a training programme to run meant that a number of newer crews were quickly brought from Training to Nursery status, and Dishforth was still in use for full-load takeoffs.

Attacks were made on the shipyards at Hamburg on the nights of 12/13 March and 13/14 March. A mix of 8 Op and Nursery crews was employed on the first attack, with one crew forced to jettison its bombs early after an engine failure. The visibility was good, and the reports suggest that squadron crews were in a tighter group than hitherto. One was able to confirm that two others had bombed accurately, and one of those in turn noted that it had been able to do so because Plt Off Humby's aircraft was the focus of enemy attention. Humby's crew was held by searchlights and AA fire for some 12 minutes whilst making its attack, sustaining a number of hits – a piece of shrapnel was reported as having struck the W/T operator's posterior, but without injury! (And as a reminder that the aerial activity was not all in one direction, the records state that a night flying programme at Dishforth was completed "after many interruptions on the part of enemy aircraft.") Another four crews flew on the following night, all claiming successful attacks despite extensive searchlight activity in support of fierce AA fire. Sqn Ldr Holford's aircraft was hit in the starboard engine "which ceased to be of service" and it also lost its Airspeed Indicator. Nonetheless, the crew nursed it back to Bircham Newton, almost certainly with flap or other damage, for a dramatic landing that took it well beyond the runway to the edge of a quarry, passing through hedges and a Nissen hut. The W/T operator sustained a leg injury, but his colleagues emerged unscathed. Two of the other aircraft were also badly damaged, but got home safely.

Six crews launched from Dishforth on the night of 14 March to attack oil storage tanks in Rotterdam. One had to return about 40 minutes out with engine oil pressure falling and temperature rising. Subsequent examination revealed a radiator blocked by a piece of paper, traceable to an airman from July 1940! The other five reported successful attacks with fires and exploding tanks. Nonetheless, Sgt Watson's aircraft was badly damaged by AA fire that required extensive internal firefighting. The crew got it back to Bircham Newton where a wheels-up landing was made, with no injury to the crew. Two returning crews made a point of

reporting 'tiresome' searchlight crews on the English coast that had ignored all standard recognition procedures – possibly explained by German air intruder activity that night that delayed the landings of returning aircraft at Leeming.

Weather forecasts for Leeming and potential target areas precluded further ops on subsequent nights, although a crew was detailed for a drogue-towing test of the station's AA defences. Others were tasked with special Marconi W/T tests with Coastal Command, finally carried out on 21 March. The target for a mixed force of 99 aircraft on the night of 18/19 March was the port at Kiel. 10 Squadron detailed seven crews with varying results. One was unable to get airborne; four reported attacks although cloud prevented observation of results; two were unable to make a positive identification and made opportunity attacks. One of these had an engine seizure overhead base on return and caught fire. Sgt Watson, whose crew had survived an internal fire just four nights earlier, died in the subsequent crash after his colleagues had safely bailed-out.

After the recent losses, three replacement aircraft were flown in from three quite separate locations – Brize Norton, Aston Down and Kinloss – and an op for a single Nursery crew was mounted on the night of 20/21 March from and back to Topcliffe. The target was familiar - the naval facilities at Lorient – and a successful attack was claimed but with no results observable through haze. After this, weather once again interrupted ops but, notably, news came on 26 March that six crews would shortly go to 35 Squadron for Halifax conversion.

On 27 March, seven aircraft were sent to Topcliffe for an attack on Dusseldorf that night. Two crews were unable to takeoff – one being bogged down, whilst the tail-gunner of the other became ill – and another two were forced to return with mechanical problems. Three other crews pressed on – one was confident that it had attacked successfully, and another felt that it had done so through cloud. The third could not identify the primary target through cloud and attacked an unidentified airfield north of Dusseldorf. On return, the crew had to abandon the aircraft after engine failure, probably as a result of fuel starvation, and all parachuted safely near RAF Cottesmore. (The debrief for this crew also mentions that they had dropped packets of tea over Breda in the Netherlands on

A sample tea bag, with a message from the free Dutch Indies: "Chins Up, The Netherlands will arise".

its outbound leg. Tea planters in the Dutch East Indies had donated some 4000 lbs of tea to the Dutch Government in exile in London for distribution over the Netherlands, and similar 'tea bombings' were carried out for much of 1941.)

The airfield at Leeming was declared unfit for use on 28 March. (The Luftwaffe would appear not to have been similarly handicapped for Leeming was attacked by single Heinkel 111 bombers, each pursued by Spitfires, on both 1 and 3 April, with minimal damage and no loss of life.) As the days passed, a plan was made to use aircraft already at the Dishforth Maintenance Unit on 7 April but was later cancelled, and no flying at all occurred until the afternoon of 8 April when ten crews left for both Dishforth and Linton to brief for an attack on Kiel that night. With 229 aircraft listed, this was to be the largest attack of the war to date. In the event, three of the 10 Squadron crews had to abandon the mission and return to base after a variety of equipment failures, whilst some of the others suffered malfunctions but were able to press on. Fires already burning in the target area helped locate it – one aircraft then had a substantial hang-up and only one bomb could be dropped, five others reported

successful attacks in poor visibility, and another attacked the docks at Bremerhaven.

A further attack on Berlin was planned for the night of 9/10 April, with six of seven crews successfully departing from Dishforth and Linton one again. Only two claimed attacks in the primary target area. Two experienced high fuel consumption and attacked earlier opportunity targets over Germany; another lost an engine and attacked canal locks at Brunsbuttel; the sixth made a late landfall and decided to attack the shipbuilding yards at Hamburg.

A maximum effort was called for on 11 April, so nine crews flew to Dishforth, only to return as the op was cancelled. On 12 April, five crews left from Dishforth for another attack on facilities at Brest. Two reported having attacked over heavy cloud cover, whilst the others proceeded to Lorient where the visibility was much better. Significant personnel movements were also recorded for 12 April. Wg Cdr V B Bennett took over as CO as Wg Cdr Bufton was posted to command and form 76 Squadron, the second Halifax squadron, and notification came that the move of both air and ground personnel to that unit, via an attachment to 35 Squadron, the first Halifax squadron at Linton, was to begin at once. From a total of 20 crews a month earlier, a new list dated 16 April showed 7 Operational, 4 Nursery and 5 Training crews.

Five crews attacked Brest on the night of 14/15 April, with haze and searchlight glare making observation of the results impossible. Five crews reported attacks on Bremen in similar conditions on the night of 16/17 April, whilst a sixth attacked docks at Cuxhaven as its aircraft began to lose height. The last radio contact with Sgt Salway's crew was made as they were approaching the Dutch coast outbound. It is likely that they came down shortly afterwards as they are buried in the Sage Cemetery at Oldenburg that contains many graves of airmen brought from the Frisians and Northwest Germany.

Conditions at Leeming improved sufficiently for Ops to resume from there, with the first occurring on the night of 20/21 April when three Op crews went to Cologne and four Nursery crews went to Rotterdam. Cologne was covered by cloud and two crews used AA fire detonations as an aiming point, with the third attacking an airfield, thought to be Venlo. Conditions over Rotterdam were better and three crews reported attacks there, with the fourth attacking an airfield. The docks at Brest were the target for eight crews on the next Op on the night of 23/24 April. Three had to abandon the mission early, and the others reported attacks, albeit that cloud and mist made positive identification and observations impossible.

On the night of 25/26 April, Bremerhaven was the target for two Nursery crews. Unable to identify the target, one crew opted to attack an active airfield at Terschelling. The other reported a successful attack, later proved to be on Tonning in Denmark. This crew was detailed for a raid on Rotterdam on the night of 29/30 April and reported a successful attack in the target area. On that night too, seven Op crews joined an attack on Mannheim. Six reported dropping over the target area but, as so often, there was a cover of thick mist making positive identifications and observations very difficult. The seventh crew had an overheating engine outbound and elected to cut short its sortie by attacking Ostend.

The first raid in May 1941 was on the night of 3/4 May, when the target for nine crews was once again Cologne, with the Oporder instructing that "Captains are to see that all watches and clocks in aircraft are advanced one hour at 0200 hours." (The seeming implication that crews, and particularly navigators, would not fly using GMT looks very odd. The UK had changed to permanent 1-hour Daylight Saving Time in February 1940, and reverted to GMT in October 1945. Double Summer Time, as here, was used each summer from 1941 to 1945.) Instrument failures caused two to abandon and return early. An electrical fault meant that one of them was unable to jettison its bombs and, on landing back, the aircraft overshot the runway and swung sharply into an uneven area of new runway construction, collapsing the undercarriage. The aircraft was wrecked but, as on other occasions, the crew escaped intact. Six of the remaining seven crews reported attacks on the designated target – but, as so often, with thick cloud and haze causing difficulties. The seventh crew could see nothing at all over Cologne and went on to attack Dusseldorf where clearer skies enabled them to report results. A search party was formed after daybreak to scour the airfield for unexploded incendiaries released from one aircraft on return!

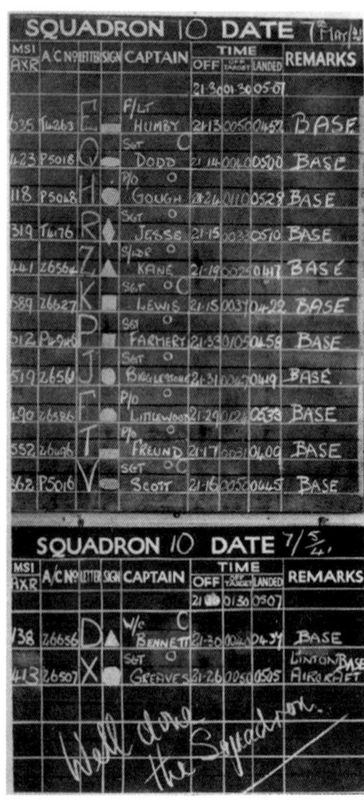

The Squadron Ops Board dated 7 May 1941 applauding the crews for a job well done after a raid on Brest on the Scharnhorst and Gneisnau the previous night.

Two targets were briefed for the night of 5/6 May – Mannheim for six Op crews, and Boulogne for four Nursery crews. One of the latter had to return early after a generator failed; two reported successful attacks, although one had a partial hang-up; and the fourth jettisoned its bombs after failing to find a valid target. Five crews reported attacks on Mannheim through cloud, with the sixth opting to attack Frankfurt.

Crew strength had been growing since the postings-out to the new Halifax force, and a list published on 7 May showed 22 in all, 12 Op, 2 Nursery and 8 Training. That night, a mix of 12 Op and Nursery crews was briefed for a raid on Brest where both the *Scharnhorst* and *Gneisenau* were known to be lying, with all crews reporting bombs dropped around the dock area. A few enemy fighters were seen but none attacked. A large raid on Bremen was planned for the night of 8 May, with many crews listed for a second night running as a special effort was called for. The Squadron briefed fourteen crews in all, with thirteen getting airborne. Of these, one had to abandon the mission due to unserviceability and lost an engine on landing, badly damaging the aircraft after a swing and an undercarriage collapse. Ten returned safely and reported bombing the target area through cloud. Plt Off Guest's crew radioed an 'Ops complete' message, and nothing more was heard from them before they ditched off the Dutch coast and were taken prisoner.

Eleven crews were detailed for Ops on the night of 10/11 May, when the target was Hamburg. Ten got airborne, but two had to abandon the mission with mechanical problems. Seven returned to report successful attacks in clear weather, with fires already burning as they arrived. Plt Off Gough's crew radioed an 'Ops complete' message but nothing more was heard and all five men are buried at Kiel.

Planned Nursery Ops on the night of 12/13 May were cancelled on account of weather, but an attack on Mannheim once again was briefed for eight Op crews. An oil leak prevented one departure, and another crew returned to base after radio failure. Two crews found that they could not climb above 9000 feet, so both attacked Dunkirk and returned to base. Heavy cloud hampered the others – at debrief, two felt they might have released over Cologne, with the other two claiming attacks on the primary target through cloud. Three crews reported having seen a number of fighters near to them, although no attacks were made.

Different targets were chosen for seven Op and four Nursery crews for the night of 15/16 May – Hannover and Dieppe, respectively. The visibility around Dieppe was very poor – one attacked Calais, whilst the others released their bombs in the general target area. One of the Op crews suffered an engine failure just after takeoff, safely jettisoned its bombs in fields and returned. Five crews returned claiming successful attacks through thick haze. Sgt Bigglestone's crew appears to have had severe navigational problems and was attacked by a German fighter, whereupon it released its bomb load at a position estimated to be over the Netherlands. The fighter broke off the attack after they fired a red flare; no serious damage or injuries were sustained.

No Ops were planned for the night of 17 May so, after a Crew Conference, the CO decided that his aircrew needed exercise and arranged transport to take one and all for "a tramp in the hills." So, duly enlivened, ten crews were detailed for Ops on the night of 18/19 May – seven to Kiel, and three to Emden. Weather for the route over the North Sea to

Kiel was clear, but haze and cloud were found in the target area. With observation difficult, six crews reported accurate attacks. The seventh attacked Hamburg after "a presumed navigational error" prevented it reaching Kiel. Two of the Emden crews reported successful attacks, but without verification, and the third had to attack an alternative target. The last four aircraft to return all had to Divert to Driffield.

Bad weather prevented any further Ops until the night of 27/28 May, when ten crews were detailed for a raid on Cologne. Two crews attacked airfields in the Netherlands – one did so to permit itself evasive action once coned by searchlights and heavy AA fire, and the other on account of weather. The other eight all reported attacks in the target area, but with the combination of weather, searchlight glare and accurate AA fire limiting observation of results.

The Butt Report: By this point in 1941, postflight debriefs indicate that night attacks over Germany had become more risky. Whilst daily instructions still allowed anything up to 30 minutes over target areas, it is notable that crews were no longer searching at length over them as had often been done some months earlier. Bombs were now mainly dropped in a single stick, and on ETA if positive identification was impossible. Searchlights and flak had become a greater threat, now also en-route from/to the coast over Belgium or Holland, with aircraft sustaining damage, and the first skirmishes with German night fighters had been reported. The rate of losses to enemy action for 10 Squadron and the night bomber force as a whole was on the increase. At the same time, doubts were growing in the minds of some about the real effectiveness of the bombing campaign to date. (Those minds were not primarily amongst the Air Staff of the day, though we might recall that Wg Cdr Staton had experimented with target marking as early as April 1940.) One of Churchill's Scientific Advisers, Professor Frederick Lindemann, later Lord Cherwell, was sufficiently concerned to commission a report, and he asked a member of the Cabinet Office staff, Mr D M B Butt, to carry out a study of the accuracy of recent attacks. This was done using around 650 photographs taken by bombers over targets in June and July 1941, with the results correlated to post-flight debriefs. (It is likely that some of those photographs came from 10 Squadron sorties as occasional remarks about the fitting of cameras can be found in the unit records.) The outcome is detailed in many of the later histories of the bomber campaign but, in general, it showed that only one in three aircraft making claims had bombed in a target area, itself defined as being within five miles of the aiming point and thus stretching any definition of 'precision' as regards point targets very considerably. When extended to sorties flown, the proportion was no better than one in five. Other factors - weather, moonlight, target difficulty - might reduce the proportion to one in ten. Not surprisingly, the 'Butt Report' released on 18 August was not well received at HQ Bomber Command, but Lindemann had access to the PM who, in turn, took it up with MRAF Portal, Chief of Air Staff. The Air Staff response looked for an enormous increase in the size of the bomber force to 4000, so that more bombs could be dropped. Debate went on for the rest of the year, as did the bomber offensive, with little improvement in its results. It remained, however, the only weapon available for taking the war to Germany, and the number of better-equipped four-engine bombers was increasing.

With no Ops scheduled on 31 May or 1 June, crews were sent on homing exercises on each day. On the first occasion only one of five crews was able to locate the target aircraft; on the next day, with twelve crews involved, the exercise was carried out "fairly successfully," but with jamming of the target aircraft's signal reported as a difficulty.

Fourteen crews of all standards were detailed for the night of 2/3 June, when the target for some 150 aircraft was Dusseldorf. One crew, unable to climb out of cloud, jettisoned its bombs over the sea and returned. The others pressed on and released bombs over the target area, covered in cloud and, with no sign of the 'Fire Raisers' that were supposed to have preceded them, flak and searchlight beams were the target indicators once again. The Bomber Command Diaries note that 107 aircraft claimed to have bombed the city, whereas local records subsequently showed only light and scattered damage with 5 citizens killed and 13 injured – and this is only one example of the many such disparities noted over this period of the bombing offensive. That said, it ought also to be remembered that just four years earlier the biplane and open cockpit Night Bomber force would have been incapable of even proving a nuisance over

Germany from UK bases.

Ops were planned for most subsequent nights and later cancelled on account of weather until the night of 8/9 June. It also appears that, whilst daily crew states were still being declared as Operational, Nursery and Emergency, no distinction was being applied when it came to allocation to targets. (The hitherto regular publication of crew lists also appears to have ceased.) Thus, fourteen crews were detailed and departed for a raid on Dortmund, with one having to return early with engine problems. All experienced poor weather en-route and sought for the target through the now customary haze and low cloud. All reported successful attacks, with several detailing the pinpoints used to establish position. A number of aircraft suffered survivable flak damage, and most had to divert on return, with conditions at Leeming preventing recoveries until the afternoon of 10 June.

The next feasible Op was on the night of 12/13 June, with sixteen crews briefed for an attack on marshalling yards at Schwerte, south of Dortmund. Two were unable to takeoff owing to unserviceabilities, and a third had to land back fully loaded when the aircraft's undercarriage would not retract. Twelve crews continued to attack in the target area, some reporting more difficulty with target identification than others. Plt Off Littlewood's crew had engine trouble from early-on, turned back just after the Dutch coast and jettisoned their bombs in the sea. The port engine then failed, they lost height and finally had to ditch. Their dinghy was deployed but landed in the sea upside down; the observer had been concussed and had to be carried out of the aircraft by the Wireless Operator, Sgt Wilkinson, before he recovered; and the aircraft sank about three minutes from impact with the crew clambering into the still upturned dinghy. After some six hours in the sea, the crew was finally rescued by a RAF launch, guided to the scene by one of a pair of Heinkel 111s, and taken ashore at Yarmouth. As twelve crews were required, some of those who had just returned had to be detailed for another raid on Schwerte on the next night, 13/14 June. The CO was unable to get away for, taxying to takeoff, his aircraft passed onto a soft patch of ground and its undercarriage collapsed. Despite weather and haze, seven crews claimed attacks on the primary target, three attacked Dortmund, and another bombed Hagen.

Thirteen crews were detailed for a raid on Cologne by 105 aircraft on the night of 16/17 June. Twelve reported bombs released in the target area – positive identification appears to have been well-nigh impossible due to haze and intense darkness. The other crew jettisoned its bombs as an engine lost oil pressure, with a shutdown likely. In the event, the crew nursed the aircraft safely back home. One of the crews reported an attack by a night fighter but, after a brief exchange of fire, the aircraft flew off. The subsequent Op on the night of 18/19 June required eleven crews for an attack on Bremen. All got away but, once again, weather conditions in the target area were difficult and made precise indentification near-impossible, and flak and searchlights were very active. Bombs were dropped on ETA and one crew flew back to Bremerhaven and attacked there. Sgt Bradford's crew was last heard at 0250 hours on its return leg and is believed to have gone into the sea shortly afterwards.

Dusseldorf was the target on the night of 21/22 June. Ten crews were briefed and all took off. Haze was again a problem in the target area, as were flak and searchlights. Bombs were dropped but no clear results could be reported and city records show only two bomb loads from some 56 aircraft in the area. The plan for the night of 24/25 June was for a return to Cologne, with the main central station as the aiming point. Eleven crews were briefed, and one had an early return as its aircraft lost oil pressure. Another bombed an airfield near Courtrai after running into intense flak south of Brussels and discovering that the rear turret door had jammed. For the remaining nine crews, industrial haze over Cologne was once again a problem but all reported attacks in the area. 54 Whitleys and Wellingtons were involved that night but the city records report only 11 bombs, 5 houses damaged, and no casualties.

Twelve crews were detailed for a raid on Bremen involving 108 Wellingtons and Whitleys on the night of 27/28 June. All got away but one returned early with radio failure. Two attacked opportunity targets – one had lost some 10,000 feet in altitude due to icing and could not regain height. Cloud and haze covered the target area, and four crews reported dropping mainly on ETA using DR information. One knew it had overflown and reported bombing Hamburg. It cannot have been alone for 76 bombing incidents were recorded there that night. Four crews failed to return, the highest number of Squadron crews lost on any raid to date. The crews of Sgt Rickcord and Sgt Shaw were all killed. FS Lewis, a WOP/AG, was the sole survivor from Sgt Knape's crew. He was captured but escaped in March 1944. He was recaptured, is understood to have been handed over to the SS and shot on 1 August 1944, and is buried in Poland. Plt Off Watson's crew abandoned their aircraft over the Baltic, near Kiel, after being hit by flak. Fortunately they were picked up from the water by a German minesweeper, but Plt Off Watson was found to have died. Sgt Nabarro, the Second Pilot, escaped in November 1941 and eventually returned to the UK in October 1942. The story of the multiple escapes that finally took him to the unoccupied south of France and then by sea to Gibraltar was recognised by the award of a Distinguished Conduct Medal.

Despite the losses, the pace of Operations did not slacken and there would be more. Ten crews were detailed for an attack on Duisburg, one of three targets in the Ruhr on the night of 30 June/1 July. Unserviceability prevented two takeoffs and one crew returned with an engine problem. None of the others was able to identify a coasting-in point on account of heavy cloud cover over the North Sea and subsequent cloud and haze made precise target identification impossible. So, as ever, the crews did what they could and dropped their bombs on Estimated Time of Arrival (ETA) with, inevitably, an uncertain outcome. Plt Off Barrett's crew was shot down over Germany, with three crewmen surviving as POWs. A further loss was incurred over the UK, when Sgt Beveridge's aircraft was attacked by a Luftwaffe intruder and lost both engines. His Observer, Sgt Bassett, and WOP, Sgt Lawson, were able to bale out safely, but Sgt Beveridge and his Tail Gunner, Sgt Alcock, were killed as the aircraft crashed close to Thetford, in Norfolk. (Four-man crews, without a Second Pilot, began to be seen quite regularly from around this time.)

The following week amply demonstrated how difficult matters had become. On 3/4 July, eight crews joined a raid on the Krupps works in Essen. A gunner becoming ill in-flight and a mix of technical problems forced two to abandon the mission, whilst the other six pressed on. Cloud cover meant that none made a reliable landfall pinpoint and as they continued to find the Ruhr area under haze, bombing on ETA and DR, it is no surprise that later reports show that bombs fell in many towns that night. Searchlights were effective, fighters were in evidence, and Flt Lt Landale's crew reported attacks by a Me 110, resulting in considerable damage around the aircraft nose but no injuries. Two nights later, on 5/6 July, nine crews were part of a raid on rail facilities in Munster. The weather was much better than at Essen, the town could be seen and crews believed they had made accurate attacks – but once again there were fighters around. Flt Lt Webster had to take violent evasive action during one encounter, and Plt Off Goulding's aircraft was shot down over Holland with the loss of all four crew. Osnabruck was the target for a raid on the night of 7/8 July, for which nine crews were detailed. The engines of one aircraft would not start and the fault could not be rectified in time, so eight got away. The conditions were favourable once again and seven reported accurate attacks, although later reports suggest that damage was fairly widespread. Sgt Black's aircraft was last heard of at a position off Flamborough Head and is believed to have crashed in the North Sea with the loss of its crew of four. Ten crews were detailed for a raid on Hamm on 8/9 July. All got away, but three had to abandon the mission for a variety of reasons. Three made it to the target, one attacked the airfield at Schipol, two bombed Munster, and again there were fighter encounters – Sgt Lewis' crew of four was lost without trace that night. Finally, eight crews got airborne on the night of 10 July for a target coded as 'GO 1097,' only to be recalled within minutes for weather reasons.

CinC Bomber Command, ACM Sir Richard Peirse, received a new directive on 9 July. This stated that he should now direct the effort of his main force towards the dislocation of the German transportation system and the destruction of the morale of the civil population as a whole, and of industrial workers in particular. At Leeming, crews were made ready each day and, on 17 July, were briefed for a raid on Cologne, but it was not until the night of 20/21 July that Ops were resumed with two distinct targets – Cologne for five experienced crews and the port of Rotterdam for four newer crews. Cologne was protected by active flak batteries and a heavy cloud layer and, with crews forced to bomb on

ETA with very little visual reference, it is not surprising that the outcome was scattered. The crews sent to Rotterdam fared better and fires in the dock area were started.

The two German battleships in the harbour at Brest were still considered to pose a considerable threat and, as a plan to send a force of 150 bombers against them was about to be implemented, it was discovered on 23 July that the *Scharnhorst* had been moved well south to La Pallice and a force of thirty Whitleys was sent to attack the port that night. The Squadron fielded nine crews and all bombed the dock area, with bursts seen in the basins. A number of raids followed, with some damage done to the ship that was then moved back to Brest, shipping water.

Reverting to Germany, six crews were detailed as part of a small Whitley and Hampden force sent to attack Hanover on the night of 25/26 July. The crews made landfalls along the northern coast of Holland and continued to find the city covered by haze, but with very active searchlights and flak batteries. Bombs were aimed as accurately as circumstances permitted. Plt Off Littlewood's crew had met thunderstorms over the North Sea, were initially unable to get above 7000 feet and, caught in searchlights near Osnabruck, they had released a 1000 lb bomb that caused the lights to go out. At ETA for the target they were unable to see anything and released the remainder of their bombs. On the return leg, Sgt Holder's crew did battle with a Ju 88 over the North Sea, sustaining slight damage but no injuries, before landing at Bircham Newton to refuel. However, two crews were lost that night, most probably as a result of similar fighter encounters, with no survivors. Sqn Ldr Landale's aircraft came down in the North Sea: three bodies were eventually washed ashore, and one was buried in Holland and the others in Germany. Plt Off Spiers' aircraft crashed near Hasselt in Belgium.

The next two Ops had Channel ports as their targets. Three crews were part of a small force sent to Dunkirk on 27/28 July, with all bombing on ETA through extensive cloud cover. On 30/31 July, six Nursery crews joined others to attack Boulogne. One had to return early when the aircraft intercom failed, but none of the others could see the target through total cloud cover and bombs were variously jettisoned or brought back to base.

Early August saw a series of attacks on Frankfurt, beginning with seven of eight crews successfully departing on the night of 3/4 August to find heavy cloud in the target area. Four decided to bomb on ETA, whilst the others attacked Aachen and Gerolstein. A larger raid was planned for the night of 5/6 August, for which the Squadron detailed twelve crews. One had to return almost immediately with an unserviceable intercom system whilst the others continued to find much better weather conditions. Ten claimed to have bombed as briefed, and the remaining crew reported attacking Cologne on deciding they would be too late over Frankfurt. Finally, on 6/7 August, two crews set off to join a group of others when, once again, the weather was particularly adverse with icing a particular problem. As a result, Sgt Robertson's controls jammed, his engines lost power and his aircraft went into a spin. Bombs were jettisoned and control regained at 4500 feet before a return to base. Despite the weather, Sgt Lager's crew managed to get to Frankfurt but they could see no result from their bombing. In parallel with these raids, newer Nursery crews were sent to attack Calais on two nights with reasonable success. The focus changed to the submarine yards at Kiel on 8/9 August, when twelve crews were detailed for that night's raid. One had to return early when loss of power made maintaining height impossible, but the others pressed on through heavy cloud to attack in the face of heavy flak over the target area. On the return, Flt Lt Thompson's aircraft iced-up and went into a spin, repeating Sgt Robertson's experience of a few nights previously. He managed to regain control around 4000 feet, when it was discovered that Sgt R Myers, his Tail Gunner, had baled-out during the spinning dive. His fate is unknown and he is commemorated on the Runnymede Memorial. The navigational weakness of the Whitley was demonstrated once again when Sgt Robertson's aircraft had to land well beyond Leeming at Fleming Kirkpatrick, in Dumfriesshire, short of fuel, and with the crew uncertain of their position. Plt Off Littlewood's aircraft was shot down in the target area, with no survivors. (Plt Off Littlewood had been the Captain of a different crew that had ditched on 13 June.)

Crews were alerted to the likelihood of Ops on most of the following nights, and briefed for Hamm on one occasion, before the next Op that was flown - a return to Cologne on

the night of 16/17 August. Twelve crews got airborne. One turned back on losing the aircraft W/T radio and recorded that "owing to faulty navigation most of the time was spent in wandering over England" before finally landing at Linton on Ouse. Another crew was unable to locate the primary target and went on to attack Essen. Seven crews reported having made attacks over Cologne in poor visibility and with intense opposition and the CO, Wg Cdr Bennett, had a particularly eventful flight. He spent some thirty minutes over the city, gradually descending from 12,500 feet to 5000 in an attempt to find a reliable aiming point. His WOP was then hit in the hand by a flak splinter; the intercom, IFF and radio were damaged; and there was no communication possible with the Tail Gunner as he did battle with a fighter that fell away, apparently damaged. On leaving the searchlight belt, he gained height again and finally landed at Church Fenton, short of fuel. In fact, weather at Leeming caused all returning aircraft to be diverted but there were three that failed to return. Plt Off Pearson's crew was lost over Belgium; Sgt Lager and his WOP died over Germany, whilst their colleagues became POWs; and Sgt Craske's crew all became POWs. Sgt Calvert, his Observer, later escaped from Stalag Luft 3 in May and was subsequently shot by civil police in Dresden on 20 May 1942.

Cologne was once again listed as the target for 18/19 August, seven crews were detailed to join that night's force and, generally, things did not go well. One aircraft had to be withdrawn at engine start and, on arrival over Cologne, conditions prevented crews from identifying ground features. Four returned to report having bombed, but later assessments suggest that a decoy fire had attracted most bombs. Two Squadron crews did not come home. Sqn Ldr Kane's aircraft came down at Lanaye on the Belgian/Dutch border, just south of Maastricht, with only himself and his WOP, Sgt Mourant, surviving as POWs. Plt Off Evill's aircraft crashed a little further north, with no survivors. Also, over the target, Sgt O'Driscoll's aircraft was coned by searchlights just after bomb release. Evasive manoeuvering was initiated, his right arm was injured by a flak splinter and the Second Pilot took control in a dive. Despite a compass malfunction, control was finally regained and an emergency landing made at Bircham Newton with no tyres, where the aircraft was written off. This incident, added to the six RAF aircraft lost in action, suggests that the attacking force had much the worst of the night. Two crews had also been sent to Dunkirk – one managed to spot the target and bomb through a break in the cloud, whilst the other reported attacking what they believed to have been Ostend. (Later on 19 August, a night flying training accident resulted in a badly damaged aircraft, but with no injuries. Unfazed, "the night flying programme was continued on another aircraft.")

The end of the month saw a continuation of much that had gone before. Two Nursery crews were sent to bomb the docks at Le Havre on the night of 22/23 August. One had problems in the aircraft's engine control system and could not make the aircraft climb, so the mission was abandoned and the bombs jettisoned. The second crew reported having bombed the target but on the return, flying at just 2000 feet, the aircraft hit high ground near Dent in the Yorkshire dales. The captain, Plt Off Liebeck, and his second pilot, Sgt Fletcher, both died in the crash but the three other crewmen escaped with superficial injuries and shock. After a cancellation on the previous night, a relatively large mixed force for the time – 99 aircraft – was sent to attack Cologne once again on 26/27 August. Eight crews were detailed, but intercom failure prevented one from departing and rising oil temperature forced another to return early. The city had near-complete cloud cover and bombs were released in what was reported as a successful attack. Later that night, a single crew was sent to join a small force in an attack on Le Havre. Cloud obscured the port and an opportunity attack was made on a factory some miles away. Frankfurt was the target once again for over 100 aircraft on 28/29 August. Nine Squadron crews were briefed and, once again, an unserviceability reduced the number despatched to eight. Seven bombed the primary target in bad weather conditions – the results that night were scattered - and Flt Lt Clapperton elected to attack Coblenz. To end the month, eight crews were detailed for a raid on Essen and, yet again, only seven took off. The city had complete cloud cover, results of bombing could not be seen and later reports showed that very little damage had been done.

Meanwhile, Plt Off Neil Ralph Blunden RNZAF had

Plt Off Neil Blunden.

trained to Wings standard in New Zealand and with a group of RNZAF personnel had travelled across the Pacific, Canada and the Atlantic to fly with the RAF. He kept a diary of his travels and service until the night before he died in one of the *Tirpitz* raids of March 1942, and permission has been given by his family to use extracts here. He wrote in a straightforward style, remarking often on aspects of life in England as seen by a visitor. He completed training at 10 OTU, RAF Abingdon, was posted to 10 Squadron, and arrived at RAF Leeming on Friday 29 August 1941:

> "Got transport to RAF Station, approx 2 mls – we are mls from nowhere..... Marvellous Mess and tucker!!! ..... Unfortunately the flying personnel are not to be billeted off Ops stations. Rooms in Mess are very good and spacious. Whole station is very well kept and clean – very efficient CO, he is Group Captain and was in film 'Target for Tonight.' May go on trip tomorrow night! As second pilot. Get Battle Dress on Monday."

(In fact, he did fly as second pilot to Sgt Robertson on the Essen raid, and acquiring his Battle Dress turned out to be a minor saga, only satisfactorily completed on 16 September.)

The next change of command occurred when Wg Cdr Bennett was posted out on promotion to be Station Commander at the Squadron's former home, RAF Dishforth, on 3 September. His successor, Wg Cdr JAH Tuck DFC, arrived from No 19 OTU on 8 September.

On the morning of 2 September the customary crew return was sent to Station Operations and a little later word came that crews would be needed for Ops that night. Crews were briefed at 1600 hours for three targets – Le Havre for nursery crews, and Berlin and Mannheim for Op crews – only for the latter to be re-briefed for Frankfurt two hours later. Heavy cloud and haze covered the city once again and two crews elected to attack alternate targets at Mainz and Koblenz. The five others attacked through cloud. Neil Blunden, "*stooging for Plt Off Hacking,*" wrote:

> "…. Saw two enemy fighters and a Whitley over target! Bags of searchlights round and about target and surrounding area… ground haze made target ident very difficult."

Of the two Nursery crews, one attacked the briefed target at Le Havre, whilst the other attacked Ostend after making a landfall much further north along the Channel coast.

6 September was an unhappy day. As nine crews were briefing for an attack on Hüls, word came that a Squadron aircraft had crashed near RAF Acklington, near Newcastle. The crew had landed there after becoming lost in poor visibility during an Air Test and, on taking off again, had collided with some High Tension power lines. Sgt Stuart RCAF, Plt Off Austin and Sgt Bryant were killed, and Sgt Whitlock was injured. Two crews were then lost on that night's Op – Sgt Poupard's crew was shot down by a fighter over Holland with the loss of all aboard, but Sgt Holder's crew survived to become POWs. Intercom failure forced one of the other crews to abandon the mission, five reported drops on or around the primary target, and one bombed a rail junction at Haltern.

Although a mixed force of 197 aircraft was mustered for an attack on Berlin on 7/8 September, the Squadron was asked to provide just two crews, with an additional four to be sent to Boulogne. One of the Berlin crews reported having bombed as briefed, but Sgt Robertson's crew elected to cut things short and attacked the port of Rostock in view of the high fuel consumption they were experiencing. All four crews on the Boulogne raid reported accurate attacks in clear conditions. On the next night, nursery Ops were cancelled and just four crews were called for an attack on Kassel. An engine refused to start on one aircraft and the undercarriage of another collapsed on takeoff. The airframe was effectively a write-off but there were no injuries to the crew and, reasonably enough, "the sortie for the crew was abandoned." Sgt Baker's crew reported having attacked the target, whilst Plt Off Purvis' crew was unable to get a positive identification and attacked a town to the south. There were more crew and aircraft losses on 11/12 September when eight crews were detailed for a raid on the docks at Warnemünde. All got away, but one was back within thirty minutes with a failed intercom system. Six reported bombing on ETA over cloud and most bombs were later reported as having landed in the town. On the return, Plt Off Hacking's aircraft ran out of fuel and had to be ditched off Flamborough Head. His Tail Gunner fractured an arm due to the crash impact, but all five men were safely aboard the Destroyer, HMS Wolsey, within an hour for recovery to port. (They were released from hospital in Grimsby on 14 September.) Plt Off Purvis' crew had probably also completed the mission but were last heard from at 0448 hours, descending with engine trouble towards the sea some eighty miles from the coast. Searches were put in hand but, in this case, nothing was found. On the following night, shortly after three crews had set off for Cherbourg, another aircraft was badly damaged during night flying. Sgt Webb was unable to control a swing on takeoff and the undercarriage collapsed, but with no injuries to the crew. Cherbourg was well covered by cloud and all three crews bombed, although one reported having seen their bombs overshoot into the sea.

Ops were cancelled on a number of the intervening days before the next positive call on 20 September, when six crews were detailed to join a force in an attack on Berlin. An unserviceability meant that only five took off but, around the north German coast, all were recalled on account of worsening weather over the target. In turning back, all were able to attack alternate targets – two attacked the port of Rostock, and the others bombed Warnemunde, a seaplane base at Wismar, and flak and searchlight batteries at Sylt. The attack at Wismar awoke a heavy flak response and it seems likely that splinters punctured the fuel tanks on Sgt Rochford's aircraft. It ran out of fuel and he had to ditch around 0500 hours off the Yorkshire coast. All five men got safely into their dinghy and were soon picked up by a rescue launch and taken to Grimsby. Another aircraft was lost in a training exercise on the morning of 23 September, when an undercarriage collapsed once again after another vicious swing on takeoff. Sgt Wieland and his crew emerged uninjured.

There were further Op cancellations until 26 September, and five crews were briefed for an attack on Mannheim that night. All duly got away, but a recall message, together with instructions to land at Waterbeach, was issued as they were close to the French coast. However, before returning, Sgt Tripp chose to attack Dunkirk and Plt Off Hacking bombed an unidentified airfield in northern France before going on to land at Bircham Newton. Eight crews were detailed for a raid on Stettin on 29/30 September. There was one early return and high fuel consumption caused three crews to attack nearer targets at Warnemünde and Rostock. The other four found Stettin and attacked in reasonable conditions. (Neil Blunden was Second Pilot to Sgt Whyte that night and noted that he had been precisely one year in the RNZAF on that day. He appears to have been given the landing, after virtually ten hours in the air:

> "Heavy landing and bent prop tips on ground! Ugh.") Three crews were briefed to join an attack on Hamburg on the last night of September. They found the city covered by cloud so all bombed on ETA, with two of them taking minor flak hits.

Stuttgart was the target for main force crews on 1 October, and five crews were detailed for that Op. There was one early return and the others pressed on to bomb on ETA over complete cloud cover. However, on the return leg, Plt Off Godfrey's Observer missed a significant wind change and the aircraft overflew England, above cloud, to head out over the Irish Sea. DF fixing finally caused the crew to turn north till fuel starvation led to a ditching twenty-two miles off Pembroke. As the crew left the fuselage, the dinghy failed to open and had to be cut free before the aircraft sank in under three minutes. They were picked-up a few hours later by a naval trawler and landed at Pembroke Dock. One of the newer crews had also been sent to Boulogne that night and managed to bomb through a gap in the cloud cover, but with no indication of results. The upper wind over the UK that night appears to have affected this crew too till, finally, uncertain of their position, the Captain landed at Hunsdon in Hertfordshire before recovering to Leeming later in the day.

The next call was for six Nursery crews to carry out another attack on the port area at Dunkirk on the night of 3/4 October and all did so in fairly clear conditions. (Neil Blunden had been allotted a crew and this was his first

sortie as Captain. We may see experience talking in his comment: *"Second pilot far too damn chatty"!*) Weather on the Continent closed down bomber operations for the next week or so and crews did not get as far as briefing until 12 October, when nine crews were detailed – six for a raid by 152 aircraft on Nuremberg and three for another comprising 99 aircraft for Bremen. Four of the first group reported reaching Nuremberg and all claimed accurate attacks in reasonable conditions – but, in truth, most bombs that night fell on two towns sixty-five and ninety-five miles away. Plt Off Hacking's crew had attacked Wurzburg as an alternate target and, although the intercom system on Sgt Peterson's aircraft had failed about an hour after takeoff, he elected to press on and attack the docks at Dunkirk. The Bremen group had mixed fortunes – one had to abandon the mission for technical reasons and the others reported bombing using DR over cloud. On returning to a misty Leeming, Plt Off Joyce landed and struck a crashed 102 Squadron aircraft, seriously damaging his starboard wing, but with no injury to the crew. (Neil Blunden, now fully operational, had to make a precautionary landing at Coningsby as his fuel was running low:

> "Interrogated and had breakfast, and then went to sleep in an armchair in front of fire in ante-room.")

On the afternoon of 13 October, Sgt Dean's crew left on a cross-country training flight and must have become a little lost as they were two hours overdue when they landed that evening at Topcliffe. Then, on the next morning, an aircraft that was delivering a ferry crew to Topcliffe overshot on landing and went through a hedge, damaging the undercarriage. On 16 October, four crews joined a small raid on Ostend that was invisible under cloud and mist. None could see a target so their bomb loads were jettisoned in the sea. After that, weather forced further cancellations of planned Ops until 20 October, when nine crews were called for a raid on Wilhelmshaven. There were two early returns after system failures, and another missed the target and attacked Wesenmünde. The other six claimed attacks on the primary target but city reports state that few bombs fell there that night. Four crews briefed for Le Havre two nights later, but only three got away. Conditions on the French coast were difficult and only one crew managed to complete a first-run attack. By contrast, Sgt Lloyd flew over the area for almost an hour with only slight opposition before a gap in the cloud permitted an attack. The third crew could see nothing at all and eventually jettisoned their bombs in the sea.

A raid on the Kiel harbour area by over 100 aircraft was planned for the night of 23/24 October and ten crews were detailed to join. All took off but one was back in a little over an hour with an oil leak. The others found the target area covered by cloud and eight dropped using their best DR position whilst the ninth dropped in a built-up area thought to be Neustadt.

The Squadron's first Halifax II aircraft was delivered on 24 October, the real hint of a more capable future, but the Whitley would remain in use until mid-December. Accordingly, eleven crews were initially detailed for a relatively large raid on Hamburg on the night of 26/27 October. The Acting OC then withdrew two of the less experienced crews on hearing the weather forecast; an unserviceability later reduced the number to eight; and a failed exactor control system immediately after takeoff, so often the cause of early returns, reduced the number still further. Plt Off Kenny found his fuel gauge had failed as he approached Sylt so he bombed the installations there before returning to base. Flt Lt Clapperton's crew could see so little at ETA that they elected to attack Bremerhaven, with no results seen. The five others managed to attack Hamburg, but not without incident and flak and searchlight activity in the general area was intense. Sgt Tripp's aircraft was picked up by lights on the way in and held for forty minutes whilst he carried out evasive action, during which they released their bombs from just 600 feet. They were then attacked by a Me 109 that they also successfully evaded. His crew remained uninjured though his Tail Gunner had to be relieved for a period on account of cold. Neil Blunden's aircraft was also caught by lights just after "Bombs Gone", with possibly forty to sixty reported, and evasive action finally lost them at 5000 feet. In the meantime, the port wing had been damaged and his Tail Gunner had been badly wounded in the chest and stomach by flak that had also caused fires that he extinguished with his feet and gloved hands. Two colleagues managed to move him forward from the badly damaged turret into the fuselage. With his engine power dropping, he finally landed at RAF Docking, a satellite of Bircham Newton, in Norfolk, so that FS Clements could be got quickly to hospital. (Not surprisingly, an extended diary entry describes that sortie, ending:

> "Plane a mass of holes, all over and under – and believe both engines hit to account for loss of power. Landed with approx ½ hour's petrol left perhaps?")

The next call was for another Channel port raid, with Cherbourg the target for the night of 28/29 October. Five Nursery crews were briefed, but only four were able to take off. The area was covered by cloud and despite lengthy searching, three were unable to identify a target so duly jettisoned their bomb loads at sea. Sgt Williamson's crew managed to find a gap and believed they saw their bombs fall into the dock area. Crews were readied on the next two days but Ops were cancelled until fourteen crews were detailed for further raids on Hamburg and Dunkirk. An unserviceability reduced the Nursery group to four, and a swing on takeoff causing Sgt White's aircraft to crash into a Nissen hut reduced it further to three. Two reported hitting the dock area through gaps in cloud, whilst the third tried for almost thirty minutes before jettisoning its load at sea. Five of the nine crews destined for Hamburg reported having attacked there over thick cloud, whilst the others decided to find alternates and attacked Auvich, Flensburg, and targets thought to be around Minden and in the general Schleswig area.

There was very little operational flying possible during November, thanks to weather conditions on the Continent and at home and local training was also affected. However, for the night of 7/8 November, ten crews were detailed – five for Berlin, where a major effort was being made involving 169 aircraft, plus five for a much smaller attack on Essen. Only two of the Squadron's Berlin contingent reached the city, bombing on DR Information through cloud. One crew cut the mission short on finding extensive cloud over the German coast and attacked the installations at Sylt; another attacked Rostock; with a third hitting Lubeck after concluding it could not reach Berlin. With only seventy-three crews claiming to have reached Berlin that night, this outcome is comparable with that for the force as a whole. The Essen group fared little better. After an initial engine problem was thought to have been rectified, Sgt Wisher made two attempts to take off, experiencing violent swings on each occasion, and the CO ordered that he abandon the mission. Sgt Owen hit severe icing conditions followed by a mix of system failures and, with his airspeed dropping alarmingly, he jettisoned his bomb load over the North Sea and was able to recover to base. Two crews reported attacking Essen through cloud, with another bombing what was believed to be the Hamm area after no landfall fix could be made. Three Nursery crews completed attacks at Ostend later on 9 November, after which no Ops were flown till 15/16 November when another two Nursery crews were sent to Emden. One reported bombing on ETA over heavy cloud, with no results seen, whilst the other was forced to jettison its bombs after the aircraft lost power in severe icing conditions and went into a spin. Nothing more was possible until four Nursery crews left for the docks at Emden again on the night of 30 November. Two reported successful attacks, a third went to Hamburg on finding Emden wreathed in mist, and the fourth did not return. Plt Off Nelson's crew was last heard from at 1900 hours on their way to the target and must be presumed lost over the sea.

With Halifax conversion proceeding at a pace, 10 Squadron's Whitley era was now near its end. The call for crews on the night of 7/8 December was representative of much that had gone before – two Nursery crews to Dunkirk once again, eight crews for a target in Germany, with outcomes that mixed the endeavour and mishap that had been so much in evidence since September 1939. Sgt Tait's crew got to Dunkirk at 1845 hours with a good deal of cloud around but bombed successfully, with impacts noted within the docks. However, Sgt Barber's crew had left Leeming five minutes earlier but could see nothing once over the target area. Bombs were jettisoned and the crew headed back to Leeming where, on approach to land, he found a 77 Squadron aircraft also landing and blocking his approach. He initiated an overshoot and, whilst trying to regain height, the aircraft struck the roof of a house and crashed and burned near Londonderry, just outside the airfield, with only minor injuries amongst the

crew. The target for 130 main force aircraft that night was the Nazi Party HQ in Aachen, for which eight crews were briefed. One crew had engine problems on startup and cancelled and, like most of that night's force, none of the others reported an attack on the primary target. Bombs were reported by five crews as having been released on alternative targets in the area; the other two decided to attack Channel ports but, in the event, neither did so.

The final Whitley Op was flown by just two crews that joined 58 other aircraft in a raid on Cologne on the night of 11/12 December. Bombs were dropped in hope, using DR navigation and releasing on ETA over thick cloud – no results could be seen and the city has no record of being bombed that night. Worse still, on return both aircraft had serious accidents. One crew undershot the airfield in cloud and high wind and crashed outside the perimeter, but with nobody injured. The other crew fared much less well. When passing over a hill between Pateley Bridge and Ripon, near home, their aircraft was drawn down by a strong air current and crashed. All of the crew were injured, and Sgt Hoskins, the Second Pilot, died from his injuries. So, with no result and two crashes, this was a cruelly ironic end to the unit's time with the Whitley, an aircraft that, with the benefit of hindsight, we can say was ill suited to the precision bombing task it was so often given but that did achieve modest results thanks to the courage and skill of those who flew it.

CHAPTER 6

HALIFAX AND MIDDLE EAST

The Squadron's first long-awaited Halifax II aircraft was delivered to Leeming on 24 October 1941. On the next day, the ferry pilot took Wg Cdr Tuck and a crew to Linton-on-Ouse to begin a short type conversion course and a new era in 10 Squadron's history began. (The Halifax II underwent continual improvements, mainly visible in changes to the nose and tailplane. The unit records do not differentiate between the versions held at different times.) Other crew conversions followed (some were also carried out at RAF Leconfield) until, with 18 Halifaxes on station by late November and a progressive reduction in Whitley ops being flown, the record for 14 December 1941 stated simply:

> "The Squadron would operate with Halifax aircraft only from today onwards. Weather – cloudy with some rain, westerly wind."

(Before continuing, a brief note on significant events in the war around this time is worthwhile. In North Africa, a British offensive had started well but Rommel's Afrika Corps counter-attacked very successfully – 10 Squadron would feel the effects of this in mid-1942. In the Pacific, Japan attacked the US Navy at Pearl Harbor on 7 December and launched invasions of South-East Asia. Within days Hitler declared war on the USA, an act that would bring the Eighth Air Force to England, with huge implications for the bomber offensive in Europe.)

The first operation was flown on 18 December 1941 - a daylight attack by six crews, each carrying four 2000 lb AP bombs, on familiar targets, the German cruisers *Scharnhorst* and *Gneisenau* lying in the harbour at Brest. Wg Cdr Tuck was first off, and things did not go well. He was unable to retract the starboard undercarriage and was obliged to abandon the sortie and jettison his bombs in Filey Bay before returning to base. When it was clear that the CO had abandoned the mission, Sqn Ldr Webster took over and his debrief gives much more detail than seen previously on raid patterns. He led a vic of three with the CO's remaining pair

Halifax Mk 2 passes by some building cumulus.

Being cold and wet was normal for ground crews at Leeming in the winter of 1941/42.

astern to join other 4 Group aircraft over Linton, before flying south to join a group of Stirlings and Manchester bombers over Lundy Island. The 3 Group Stirlings led the attack and the force of some 47 aircraft moved into line astern at 15,000 feet to bomb the cruisers that were clearly visible. Flak was intense and four of the Squadron aircraft took hits before returning to Boscombe Down. On returning to Leeming the following day, Flt Lt Miller's aircraft had a different undercarriage problem – his would not come down and he had to land with it retracted at another airfield.

Wg Cdr Tuck's ill luck with the new aircraft continued for, on 29 December when, having been cleared for takeoff, he collided with another Halifax on a crossing runway. The pilot of the other aircraft, FS Tripp, and his WOP/AG, Sgt Green, were both killed and six others in both crews were injured with both aircraft written-off. Former Flt Lt 'Dusty' Miller saw the incident and, in 1988, recalled it for an Association Newsletter:

> "I was nearby and immediately saw that they would meet at the intersection .... Tuck must have seen Tripp out of the corner of his eye, for his starboard wing lifted and went right through the cockpit over Tripp's head.... I got there just in time to help the WingCo out of his aircraft and, as the port inner motor was still running on about six cylinders and petrol was pouring out of the wing, we ran around the back of the aircraft to be met by a van and driven off the field. The aircraft did not catch fire as we feared."

Meanwhile, six crews had stood by for another daylight Op on 24 December, but it was not until 30 December that a further, smaller attack on the cruisers at Brest was mounted. The Squadron detailed five crews, but one aircraft went unserviceable on start-up and another had to return early with engine trouble. The flight once again joined other 4 Group aircraft over Linton and the three crews that reached Brest reported bombs released successfully into the harbour. Flak was again intense, with all aircraft hit. FS Whyte's aircraft lost an engine to shrapnel and, dropping behind the others, was attacked by a Me 109 from the rear. His tail gunner (Flt Lt Roach) was killed and both inner engines put out of action. The aircraft made a controlled ditching about 80 miles from Lizard Point and, once it proved impossible to extract the gunner's body from the rear turret, the crew took to their dinghy only minutes before the aircraft sank. They were picked up by an Air Sea Rescue launch about five hours later and returned to Leeming on 3 January. The same Me 109 had gone on to attack Sqn Ldr Webster's aircraft and was itself shot down. Sgt Porritt, tail gunner with Plt Off Hacking's crew, excelled himself by shooting down an attacking Me 109, with another as a 'possible,' whilst incurring some injuries. Two more 109s approached but both were shot down by covering Spitfires, and he had yet another exchange of fire with a further fighter that then broke away. The record ends:

> "No further actions took place and the captain made a safe landing at St Eval. This aircraft was holed in the fuselage and wings, both rear turrets were damaged and the starboard tyre was punctured. None of the crew were injured with the exception of the slight wounds of Sgt Porritt."

1942

The turbulent experience of those first two weeks of Halifax operation had provided reasonable indications of what the next two and a half years might bring. However, the initial emphasis would be on targets other than in the German heartland. In November, Churchill had instructed that there was to be a period of restraint in bomber operations after heavy losses, and Air Marshal Sir Richard Peirse, CinC Bomber Command, was removed from his post in early January. It was a critical time for Bomber Command and its place as a prime strategic weapon was debated at length in London, with the Air Staff still arguing for a force of 4000 aircraft. Against this wider background, and with the Squadron still building its capability with a new aircraft, the targets in the first weeks of 1942 were predominantly maritime in nature.

On 6/7 January 1942, four crews were detailed to join another early morning attack on the docks at Brest. All found the area almost completely obscured by cloud, a factor that reduced the defence's AA opportunities, and all claimed bomb releases into the designated target area, although no results could be observed. After that, four crews were briefed to join a raid on the port at St Nazaire later on 7 January. One had to cancel before takeoff with an electrical fault, but the

The contribution made by the Squadron's ground crews can never be overestimated.

Preparing the night's load

others reported successful attacks. There was then a break until 15 January, when three crews were detailed to join a night raid on the port at Hamburg, but engine problems reduced the number to two. Despite near total cloud cover, one crew believed it had bombed accurately by reference to a ground fix. Sgt Schneider's crew were all killed on return when his aircraft crashed near Northallerton, and his injuries were such that no report on the sortie was possible.

Crews were briefed for raids on Bremen on both 17 and 21 January. Two crews flew on the first raid, and both bombed over cloud. Flt Lt Miller returned on the second night and again bombed through cloud. Then, on 26/27 January, three crews briefed for a raid on Hannover. Two claimed accurate attacks in the target area, whilst the third attacked a road/rail junction at Friedrechstadt after a series of navigational difficulties.

The record for 28 January states:

> "Six aircraft and crews .... were detached to Lossiemouth to carry out an operation to be detailed by HQ Bomber Command."

The operation was to sink the *Tirpitz*, a German battleship then well protected by the steep cliffs surrounding it on three sides in the Faettenfjord, an offshoot of the larger Aasenfjord, moving inland from Trondheim, where it was also well camouflaged and protected by anti-submarine nets and booms. The vessel was considered a most significant threat to convoys to Russia and to RN operations generally and a series of attempts to cripple the ship would be made in the weeks that followed, although that would not occur before 1944. Wg Cdr Tuck with three other Squadron crews took off from Lossiemouth early on 30 January as part of a larger force of 16 Halifaxes and Stirlings. In cloud and icing conditions, all four experienced unexpectedly high fuel consumption and had to turn back. Other aircraft in the force reached the Norwegian coast but none were able to carry out attacks. The crews remained at Lossiemouth for several days, presumably hoping for more favourable weather, before returning to Leeming on 7 February.

Further south, whilst the repeated attacks on Brest had failed to sink either the *Scharnhorst* or *Gneisenau*, the ships

had been damaged and Hitler directed that they should be recovered to home ports in Germany, together with the cruiser *Prinz Eugen*. A comprehensive plan was drawn up to maximise their protection whilst at sea and their passage through the Channel began on the night of 11 February with a forecast of extensive cloud cover. However, both ships were spotted by two Spitfires during the morning of 12 February, and Operation FULLER was initiated in response. Unfortunately, most of Bomber Command had been stood down due to the poor weather so a frantic effort was required to get aircraft ready to search for and attack them. Seven Squadron crews were briefed and took off just after 1600 hours to patrol an area off the Dutch coast, each armed with fifteen 500 lb SAP bombs. The Kriegsmarine had chosen its time well and the crews found heavy cloud cover and poor visibility as darkness fell. Sqn Ldr Thompson's crew got a brief glimpse of a large ship and, at 9000 feet, dropped its bombs in a single stick at 1812 hours, but they were the only 10 Squadron crew to do so. The others jettisoned their bombs and returned to base, having done no worse than the many others who flew that day. Two of the ships hit mines but all were safely in port by early on 13 February – the "Channel Dash" had succeeded, albeit at some cost to the German ability to threaten Atlantic convoys.

Six crews briefed on the previous evening for a small raid on Cologne on 14 February, but the lack of hardstandings came into play once again as aircraft tried to taxy. An aircraft became bogged down and another trying to negotiate a path around it suffered the same fate. The four others all reported attacks over heavy cloud, with no results seen. Nonetheless, 14 February would be a significant day for Bomber Command as it was on that day that a new Directive was issued to the CinC. This took account of the navigational weakness that had been revealed in the Butt Report and effectively abandoned the targeting of individual industrial targets in large built-up areas. For the most part, targets would become the cities themselves and the primary objective should be *"the morale of the enemy civil population and in particular the industrial workers."* A week later Air Marshal Sir Arthur Harris was appointed as CinC and he resolutely took forward the 'area bombing' instructions that he had been given, using the growing force of more potent and better equipped aircraft then coming on line. In addition, he sought a much greater concentration of aircraft in time and space over targets, a need that would create the 'bomber stream.' That said, the changes would take a little time.

Although the surface threat to the Atlantic lifeline had been reduced by the move of *Scharnhorst* and *Gneisenau*, the U-Boat threat remained and another raid on St Nazaire was planned for the night of 15/16 February, for which six Squadron crews were to provide the Halifax element. One crew had to abandon the mission when the pilot found it impossible to close his cockpit hatch, left open in error – an oversight that no doubt involved a subsequent chat with his Flight Commander. The other five continued and bombed over thick haze. FS Lloyd's crew appear to have been given erroneous bearings on recovery and found themselves over hills with their fuel running out. They abandoned the aircraft over Cumberland, all without injury. The campaign against major German surface vessels continued and, on 22 February, three crews were detailed to bomb Mandal airfield in Norway, as part of a diversion plan to cover a Fleet Air Arm attack on the damaged *Prinz Eugen* taking shelter in a fjord. One claimed a successful attack despite cloud cover, and the others bombed the airfield at Lista. Following that, for three nights from 25 February, attacks were made on the docks at Kiel, where the *Gneisenau* was at anchor. The Squadron provided six crews on the night of 26/27 February, when their target was the ship in a floating dock, if it could be identified. If not, the Deutsche Werke shipyard where the battleship had been built was to be attacked. Five crews reported bombing the latter accurately against active defences. Notably, Sgt Gribben's debrief mentions that he was detailed to drop two of the new 4000 lb high-charge 'Cookie' bombs specifically on the shipyard, the first mention of these weapons in the unit record – and it is an interesting speculation whether it was one of these that penetrated the *Gneisenau's* forward hull, causing damage that ended its seagoing career. The sixth crew (Sgt Weiland) unfortunately failed to return.

In an early demonstration of the potential power of the forces at his disposal, the new CinC ordered his Command's largest raid to date on 3 March, when 235 aircraft were detailed for an attack on the Renault factory at Billancourt, west of Paris, that was making many thousands of trucks for the German forces. 10 Squadron provided seven crews, three armed with two 4000 lb bombs, and six reported accurate attacks. The seventh arrived outside the permitted time period and, on its return, attacked the airfield at Beauvais.

On 9 March, Wg Cdr Tuck took another detachment of seven aircraft and crews back to Lossiemouth for a fresh attempt on the *Tirpitz*. FS Whyte's crew had an interesting flight north. To begin with, when routing up the North Sea they flew over a coastal convoy and were fired upon by its escort vessels. One scored a direct hit on the port inner engine and damaged its propeller. Inside the aircraft this caused some consternation as one of the groundcrew travelling in the rest area felt it was time to leave. He pulled the cord on a parachute that billowed about whilst he tried to open the escape hatch before being restrained by colleagues. The aircraft then crashed on landing, thanks to defective brakes, and a replacement had to be sent from Leeming. In the event, no operation was flown and the crews returned to base on 13 March.

The next event of note was a formal station visit by HM The King and Queen Elizabeth on 25 March before Wg Cdr Tuck took another detachment to Lossiemouth, this time eleven strong, on 27 March. A twelfth crew followed the next day and an operation against *Tirpitz* was flown on the night of 30 March by a force of 34 Halifaxes. 76 Squadron provided a first wave, with 10 Squadron providing a second wave of ten together with twelve aircraft from 35 Squadron. On this occasion each was carrying a load of four modified 1000 lb naval mines but, in the event, cloud covered the ship, no attacks were possible and most of the mines were jettisoned. Six aircraft were lost that night, two of them from 10 Squadron (the crews of Sqn Ldr Webster and Plt Off Blunden RNZAF), and both are thought to have been shot down by AA fire in the Hemnefjord area, west of Trondheim. Three bodies from Plt Off Blunden's crew – those of Plt Off Day, second pilot; Sgt Richards, engineer; and Sgt May, WOP/AG - were recovered from the water in subsequent days and were buried in the churchyard at Heim on 4 April. Sqn Ldr Webster's body was also found on 4 April, at the Terningen Lighthouse, 55 miles west of Trondheim, and he was buried in the Stavne cemetery nearby. And on 4 April, the depleted detachment returned to Leeming, where it would be ten days before the next Op was flown. Neil Blunden's diary had closed with what was his final entry on 29 March:

> "Warmer day and some sunshine. Met, no joy for this evening – decided on at briefing at 1530 hrs. Quiet evening in Mess and went to bed at 2130 hrs. After briefing cancelled, I got crew together and showed and explained to them the target and my plan of action etc."

(Of interest, the first mention of No 10 Conversion Flight appears in the March record when Sqn Ldr Thompson and Sgt Mortimer (WOP/AG) were posted there. Later in 1942 it would merge with other Halifax Conversion Flights to form No 1658 Heavy Conversion Unit at RAF Riccall, through which some thousands of men would pass on their way to 4 Group Halifax squadrons before the war ended.)

After the break, an Op was scheduled for the night of 14/15 April. Eight crews were detailed as the Halifax element of a total force of 208 aircraft targeting the docks at Dortmund, and five claimed accurate attacks. High fuel consumption proved a significant factor that night, with three crews forced to attack nearer alternate targets and only two making it back to Leeming without landing at an airfield in the South, as four of their colleagues had to. Two aircraft failed to get home at all. WO O'Driscoll's crew had to ditch as their fuel ran out - all were picked up uninjured - and Plt Off Hughes' crew had to abandon their aircraft over Surrey as their engines stopped. His six colleagues landed safely but he was killed as he tried to make a dead-stick landing near Hindhead.

On 15 April, the Squadron saw another change of command as Wg Cdr Tuck was posted out. The new CO was an Australian, Wg Cdr D C T Bennett, posted in from command of 77 Squadron, another Leeming squadron still using the Whitley. Within days, on 23 April, he led the next detachment of eleven crews plus aircraft back to Lossiemouth for another crack at the *Tirpitz*.

Whilst the bulk of the Squadron was away, Ops still came up for some of the remaining crews and, for the first time, the record refers to some of them as 'freshman' crews.

Newly appointed OC 10, Wg Cdr D.C.T. (Don) Bennett was shot down over Norway on 27 April 1942 in a raid on the battleship Tirpitz, on his first Squadron mission. He escaped through Sweden rejoining the Squadron a month later only to leave again on 1 July to form the Pathfinder Force.

Two of these were detailed to join a mixed force for an attack on the docks at Dunkirk on the night of 24/25 April. The weather was good, and accurate attacks were claimed. (Both also dropped packets of Nickels south of Dunkirk, the first occasion this had been explicitly recorded for many months.) On the night of 26/27 April, two crews were detailed to join around 100 others for a raid on Rostock, but neither was successful. One had to jettison its bombs and return after losing its fuel gauges and rear gun turret; the other was experiencing oil leaks and oxygen supply problems and elected to bomb Hornum, an earlier target. On the same night, Sgt Allen's 'freshman' crew had an eventful sortie to Dunkirk. They were coned by twelve searchlights and sustained moderate flak damage that damaged an engine. The need for constant evasive action prevented dropping with accuracy and so forced the pilot to jettison his bombs early. His crew joined a raid on Ostend on the night of 29/30 April, when all went off much more smoothly!

In the meantime, a force of over 40 Halifaxes and Lancasters had gathered at both RAF Lossiemouth and RAF Kinloss. The *Tirpitz* was still in Faettenfjord but it was not until the afternoon of 27 April that the weather was suitable for a raid. An attack in two waves was planned for the night of 27/28 April, with the first carrying 4000 lb bombs to begin the attack from 6000 feet, and 500 lb bombs for use against flak and searchlights. The second wave consisted of the Halifaxes of 35 and 10 Squadrons, tasked to fly at around 250 feet to drop their mines between the ship and the rising shoreline, down which it was hoped that the mines would roll to explode alongside. 10 Squadron supplied nine crews that night, led by Wg Cdr Bennett. Visibility was good but, as they ran in and closed to the target at heights later reported as from 250 to 1500 feet, it became evident that the ship was obscured by a smokescreen and they were subjected to heavy and accurate AA fire. Mines were released as accurately as was feasible, although four crews reported one or two mines hanging-up. Years later, Wg Cdr Bennett, whose starboard wing had been on fire, described the scene in his autobiography "Pathfinder:

> "To my horror, the bomb aimer could not see the ship, and with good reason. Below us was a white haze.... a man-made camouflage of which our Intelligence had reported nothing.... A split second later the ship's superstructure passed beneath us, but it was too late to let go, and I vaguely hoped that I would be able to hold the aircraft in the air long enough to turn back for a second run. I completed the turn .... I pointed towards the ship's position and released the mines. I often wonder where they went!"

The CO then headed towards Sweden but could not gain height to clear the mountains, so he turned back to the west and ordered the crew to bale-out. All were able to do so, including the injured tail gunner. On the ground, about an hour later, Wg Cdr Bennett's book goes on to describe his meeting up with his WOP, Sgt Forbes, and their journey into Sweden from where he was returned to the UK a month later. Two others in the crew escaped, and the remaining three became POWs. Flt Lt Miller's crew was also shot down that night after completing their attack. With the starboard wing in flames, he ditched in Trondheimfjord, when Sgts Stott and Annable were killed, with the other five becoming POWs. Flt Lt Hacking's aircraft was also badly damaged by flak, but he was able to nurse it back to Lossiemouth.

With no obvious damage done to *Tirpitz*, another attack was ordered for the night of 28/29 April, and 10 Squadron managed to field seven aircraft and crews, five of whom had flown on the previous night. It was decided that the two phases of this raid should be flown much more closely together, and that the Halifaxes would carry five mines each. Visibility was again very good, the flak was just as intense

Plt Off L.G.A. Whyte & crew April 1942: one of the crews detached to Lossiemouth for attacks on the battleship Tirpitz.

and the smoke cans were again ignited to form a covering screen. Runs were made at heights from 200 to 1000 feet, with crews reporting better drops than on the previous night. Four aircraft were damaged by flak, with crewmen sustaining injuries, and two of them had to land at Sumburgh on the Shetlands. It was later established that the *Tirpitz* had suffered no significant damage, and on 30 April the crews flew back to Leeming. Air Marshal Harris then issued this message to Station Commanders:

> "The courage and determination shown by your crews in the attacks on Tirpitz was indeed worthy of immediate and outstanding success. Moreover, undismayed by their first experience of the full fury of the defences, they returned with undiminished ardour to the charge. Never was more asked, and never was more given of outstanding devotion to duty. We shall, I hope, yet find that their efforts have not been in vain but, be that as it may, your crews have set an example unsurpassed in the annals of British Arms."

With Wg Cdr Bennett's fate unknown, he was replaced as CO on 4 May by Wg Cdr J B Tait DSO, DFC, who had first joined 10 Squadron as a new pilot in August 1936. On the following day, word was received that Wg Cdr Bennett was in Sweden, as were two others of his crew, FS Colgan and Sgt Walmsley. He returned to Leeming to resume command on 4 June for what was to be another foreshortened period. (Wg Cdr Tait moved on and, perhaps fittingly, was OC 617 Squadron when Bomber Command finally put *Tirpitz* out of action with Barnes Wallis' 'Tallboy' bombs in November 1944.)

The Squadron rejoined the main offensive on the night of 3/4 May, when ten crews were briefed for a raid on Hamburg, with one crew later withdrawn before takeoff by HQ 4 Group. Seven crews reported dropping on ETA over cloud. Another could not close its bomb doors and the resultant drag and speed reduction forced it to bomb an alternative target at Gluckstadt, also over cloud. The remaining crew had engine problems, jettisoned its bombs early and, on its return, had two separate engagements with night fighters – the second, a Ju 88, was shot down.

Squadron crews next flew on raids over Stuttgart – six on 5/6 May and two on 6/7 May. (A 'freshman' crew also attacked Nantes on 5/6 May.) On the earlier raid, one crew experienced abnormal fuel consumption and had to bomb

an airfield alternate, with another releasing over Saarbrucken for the same reason. The four others reported bombing the primary target, but later records show that no bombs from over 70 aircraft hit the city that night and it is probable that a decoy target at Lauffen, 15 miles to the north, was the aiming point used. On the next night, high fuel consumption led to another crew attacking Saarbrucken, whilst they claimed a fighter shot down about an hour earlier. The other crew encountered glycol leaks on two engines and had to abandon the mission. On the night of 8/9 May, five crews briefed to join a medium-sized raid on the town and Heinkel factory at Warnemünde on the Baltic coast. Sqn Ldr Guthrie's crew was detailed to carry out low-level attacks on searchlights, but all were killed as the aircraft was shot down. The raid was set a particularly narrow 5-minute timeframe and, as the other crews were all slightly late, none was able to attack the primary target but they did bomb others along the Baltic coastal strip. WO O'Driscoll's aircraft was attacked by fighters on both outbound and inbound routes – the crew claimed a 'possible' against a Ju 88 but only after sustaining substantial damage themselves. There were then no further Ops till 19 May, but the Prime Minister visited the Station on 15 May.

A force of 197 aircraft was mustered for a raid on Mannheim on the night of 19/20 May, for which the Squadron detailed nine crews. All took off, but one had to abandon the mission when its intercom failed and another decided to bomb Trier when the Observer took ill and passed out. Six returned reporting attacks on the target area, albeit that the local record shows most bombs fell away from the town. One crew (Plt Off Baker) failed to return and has no known graves. On the same night, two freshman crews were detailed to join 63 others in a raid on the docks at St Nazaire. One had to return after losing an engine, but the other reported seeing bursts from its bombs amongst the docks. Another, smaller raid was mounted on the night of 22/23 May, when two crews were provided. The target was covered by heavy cloud and, like most others that night, neither crew was able to complete an attack. The next Op came on the night of 29/30 May, when the Squadron detailed five crews to join some seventy others for a raid on the Goodrich Rubber factory and the Gnome Rhone Aircraft Engine Works at Gennevilliers, near Paris. Three returned claiming successful attacks; one had to shut down an engine and jettison its bombs; and the fifth could not see the primary target but claimed an accurate attack on an airfield as an alternate target.

Air Marshal Harris mounted the first 'Thousand Bomber Raid' on Cologne on the night of 30 May/1 June 1942. There can be little doubt that he was motivated by a need to prove

Prime Minister Winston Churchill visited 10 Sqn at Leeming on 15 May 1942. Escorted here by the Stn Cdr Gp Cpt Strang Graham and OC 10 Wg Cdr J.B. (Willie) Tait. Graham had been a previous OC10 from April 1937 – April 1938, when the Squadron was equipped with the Heyford bi-plane bomber.

that Bomber Command was a potent strategic weapon for, then as now, competition for budget allocations was an issue in Whitehall. Planning for this remarkable demonstration of capability had been in hand for some time and would need aircraft and crews from beyond the 400 or so then available from front-line squadrons. 10 Squadron together with No 10 Conversion Flight would detail twenty-two crews in all, and an idea of the pressure to maximise participation can be seen in a 1990 Association Newsletter contribution from former pilot Ben Gibbons:

> "I was a Sgt Pilot at the time and had just completed conversion to Halifaxes and was without a crew. For some days we had been aware that something big was in the offing – there was great activity in the maintenance hangars and at the dispersals. During the morning Sqn Ldr Ennis asked me to become part of a scratch crew in the role of Observer/Bomb Aimer. The only other member of his crew who I knew was FS Fred Simpson, who was waiting to re-muster to aircrew as a Flight Engineer.
>
> A senior officer from Group attended the briefing in an atmosphere that could only be described as 'electric.' Throughout the briefing the number of aircraft on the target kept increasing – I think the final number we were given was 1054."
>
> (BC War Diaries gives a final figure of 1047.)

The number of aircraft involved brought about the creation of the 'bomber stream', a tactic to concentrate force effectiveness and minimise the time spent penetrating German night fighter interception areas en-route that would remain in use afterwards. (With one aircraft U/S, the other twenty one got away over 35 minutes, whereas in the early days of the war a group of four Whitleys might have departed in a similar timeframe.) Five crews had technical problems (including one instance in which an electrical fault released the bomb load) and either jettisoned or made attacks on alternate targets. For the others, their debriefs make clear that the target area was already covered by fires by their arrival overhead and, with bomb loads that carried a high proportion of incendiaries, their attacks would have added to the effect. Official figures for aircraft bombing over the city range from 868 to 898, and very considerable damage resulted. Forty-one bomber aircraft were lost, including Sgt Moore's crew from 10 Squadron over Holland. As it happened, Sqn Ldr Ennis' scratch crew had the most eventful time over Cologne, as Ben Gibbons remembered:

> "We reached the target without incident but, on the run in, we were picked up by a so-called master searchlight and eventually coned by a group of about six and became the target for an intense barrage of flak. After "Bombs Gone" the skipper headed for the deck, finally levelling at about 50 feet. One engine was out, the radio was out, the rear turret had been demolished and the gunner seriously wounded. The skipper told us to prepare to abandon aircraft.
>
> This required me to place all documents in the chart destructor to ensure that these, in particular those relating to the on-board equipment called 'Gee' which was quite new and, as far as we knew, had not fallen into the wrong hands. However, the skipper managed to get control of the aircraft and decided to head for home. Unfortunately, I was unable to retrieve the documents from the destructor, so it was homeward bound on my memory of the flight plan.
>
> Sgt Groves, the rear gunner, was brought to the rest position, his injuries attended to, given an injection of morphia, and connected to the oxygen system. Visibility was good. We hedge-hopped our way back to the coast to be welcomed by searchlights that eventually led us to Manston ... We were lucky to land when we did – just after we cleared the runway a Whitley came in to land with wheels up, completely blocking the runway."

All in all, the first 1000 Bomber Raid had gone well and another was planned for the following night, 1/2 June, when the target was Essen. The Squadron called on the Conversion Flight once again and twenty crews were detailed. (In the end, the force that night was 956, no doubt largely reflecting the losses and badly damaged aircraft from maximum effort on the previous night.) This raid was considerably less successful – the target was covered by haze and low cloud and it appears that bombs were scattered around towns in the Ruhr. For 10 Squadron, two crews had unserviceabilities that prevented takeoffs, twelve reported attacks in the area, two attacked alternates, and one was forced to jettison its bombs owing to engine trouble and fighter attacks. Eight

Halifaxes were lost that night, three of them from 10 Squadron – the crews of Plt Offs Senior, Joyce and Clothier. (Some days later news came that Plt Off Senior's crew had been safely picked up off the Dutch coast, apart from Tail Gunner, Sgt Whitfield, who had been killed.) More attacks on Essen would follow in the immediate days ahead, mixed with others on Bremen and Emden.

So, on the next night, 2/3 June, Essen was again the target for 195 aircraft, with 10 Squadron detailing nine crews. One had to abandon the mission, one bombed Aachen, and seven reported attacks on the town over heavy cloud but, as so often, local records report little of significance as having occurred. No doubt once recce photography showed this to be so, another raid was mounted on the night of 5/6 June for which the Squadron detailed fourteen crews. One failed to get away, one had to turn back when a crewman became ill, and three jettisoned bombs over the sea thanks to technical problems. Low haze cover was again an issue but those who returned reported attacks in the target area. WO Peterson's aircraft had all four engines cut out over the target and lost 3500 feet in altitude before they could be restarted. It was later attacked by a Me 110 that damaged the port wing and fuel tanks before the tail gunner got in a burst that brought it down. FS Rochford's crew was brought down over Germany, with three killed and four taken as POWs. Essen featured again on the night of 8/9 June, for which nine crews were detailed. Three had to jettison and turn back for a mix of timing restrictions and engine problems. The remaining six reported the customary haze cover as preventing any assessment of results. The final night of the 'campaign' against Essen was that of 16/17 June, with eleven crews detailed. All encountered heavy cloud, and the eight who reached the Ruhr all elected to attack Bonn as an alternate target. Engine failures en-route had forced the other three to jettison bombs and abandon the mission. (Alongside high explosives and incendiaries, the records show that leaflet dropping was still a part of many sorties. The Order for the Essen raid on 5/6 June indicates that 42 packets of five different Nickels were to be loaded on each aircraft for dispersal over Germany.)

Fitted between raids on Essen was one on Bremen on the night of 3/4 June, with five crews detailed to join a force of 170 in all. One crew was unable to get away and another attacked an early alternate target as engine problems developed. The three others all reported attacks on the target in reasonable visibility. Following the Essen raids, there were two on Emden on successive nights. Seven crews were detailed for 19/20 June. Sgt Ball's aircraft failed to gain height after takeoff and crashed at North Otterington, but he and his crew were able to escape before it burst into flames. With the target heavily obscured by cloud, one crew elected to attack Osnabruck whilst the others bombed on DR and radio fixes. The town's records mention no damage being done. Six crews were detailed to join the follow-up raid on the night of 20/21 June. One crew had to abandon the mission early on with a mix of technical problems. The others reached the target area and reported accurate attacks, and town records indicate this having been a more successful raid than on the previous night. Another 1000 Bomber Raid was planned for Bremen on the night of 25/26 June, although the Command total for the night fell a little short at 960. 10 Squadron and the conversion Flight briefed fifteen crews, of which two had to jettison bombs en-route after encountering technical problems. The target was well-covered by cloud, but nine reported bomb releases over the area whilst four attacked suitable alternates. Another large raid on Bremen was mounted on the night of 2/3 July but, as the Squadron then had other things on hand, only two crews were detailed to join it. Plt off Fegan's crew reported an accurate attack, but Sgt Lawer's crew failed to return.

For most practical purposes, 10 Squadron was then taken out of the offensive on Germany. Things had gone badly in North Africa, where there was a conflicting and equally pressing need for heavy bombers, and Ben Gibbon's recollection of the 1000 Bomber Raids sets the scene:

> "So ended my days as an Observer, but perhaps it should be mentioned that at that time Second Pilots were not being carried – the new mustering Bomb-aimer was to assist the Pilot during takeoff and landing. But when we were detached to the Middle East for 16 days, we had neither Second Pilots nor Bomb-aimers. The dedication and versatility of the Flight engineers in the right hand seat was invaluable, whilst

the Observer continued to drop the bombs. We stayed in the Middle East for 8 months, but that is another story ......"

THE MIDDLE EAST – 10/227 SQUADRON

By mid-1942, the British military situation in North Africa was fraught. General Rommel's advance through the Western Desert had left Tobruk in Axis hands and although General Auchinleck had finally managed to halt the German advance some 90 miles west of Alexandria, the risk was high that Egypt might fall and, with it, the Suez Canal and the fastest sea route to India and the oilfields of the Middle East. With Tobruk occupied, the Germans posed a considerable threat to British supply routes in the Mediterranean and, in late June, it was decided that a heavy bomber reinforcement of theatre air forces was required.

10 Squadron and 76 Squadron were directed to detach a number of aircraft and crews to the area to carry out attacks on the Italian fleet. As far as 10 Squadron was concerned, Wg Cdr Bennett did not share London's strategic view, as he made clear in his postwar autobiography 'Pathfinder,' where he stated that this was a waste of 'a perfectly good heavy bomber squadron capable of hurting Germany.' He was therefore delighted to be told just before departure that he was to hand over command and report to HQ Bomber Command for the creation of the Pathfinder Force. Sqn Ldr Seymour-Price (now promoted to Wg Cdr) took command of the detachment and his crew, with WO Peterson's, left Leeming on 29 June as an advance party. Seven crews then left on 5 July, and a further seven on 6 July, and each of the sixteen aircraft detached took three groundcrew. These forty-eight NCOs and airmen were insufficient to provide the depth of support needed for sustained operations, and they were to be reinforced by airmen of No 227 (Naval Cooperation) Squadron, a Beaufighter unit in-theatre that had sustained heavy losses and was temporarily disbanded. Former armourer Sgt Fred Brinton was one of those airmen and wrote for the Squadron Association Newsletter in 1985:

> "Flying Bristol Beaufighter aircraft we operated from El Dabbha, about 100 miles west of Alexandria. We had always been given to understand that our main base was in Malta, so it came as no surprise when all our aircraft and aircrews flew to Malta towards the end of June 1942. What did surprise the remaining members of the squadron was our movement with all our tentage and maintenance equipment by road to Aqir in Palestine, to receive and operate a detachment of 16 Halifax aircraft from 10 Squadron, flown out from the UK by way of Gibraltar, at the beginning of July 1942. With these 16 aircraft came a nucleus of specialist ground crews in various ranks and trades......
>
> .... As part of Operation Bareface all the RAF personnel concerned with the detachment had been told that they were to fly out to the Middle East 'for one week on a special operation to bomb the Italian fleet.' Everyone believed that. Wives, families and friends were all left behind on the understanding that everyone would be back to their UK base at the end of that week. No bills were paid, no accounts settled, and cars and cycles were left beside homes, hangars and squadron offices 'until next week.' Some airmen who flew out with that 10 Squadron detachment were medically unfit for service overseas while others were to be married in the coming week.
>
> Once arrived at Aqir, it rapidly became clear to the whole detachment that:
>
> a. There were no plans for a special operation against the Italian fleet and
> b. Everyone was now posted to Middle East Command – full stop.
>
> As can be imagined, nobody was best pleased.... There was understandably anger and bitterness over such treatment, but firm leadership soon established the fact that 'if you can't take a joke, you shouldn't have joined' and 10/227 Squadron, as it was officially known, settled down to operations against Tobruk."

Fred Brinton's recollection is amply borne out by the 10/227 Sqn ORB, in somewhat exasperated terms that could suggest that HQ Bomber Command may have been misled about the duration of the intended detachments:

> "Both No 10 and No 76 Squadrons had come out for Operation Barefaced, which was supposed to be completed in 16 days. Nothing was known of this in the Middle East

Servicing at Aqir, Palestine in September 1942 where the Squadron detachment was renamed 10/227 Sqn to change, on its completion to No 462 Sqn.

> and both squadrons were put on ordinary night bombing operations. At the same time, the personnel were warned to be prepared for an indefinite stay in the Middle East. Consternation and dissatisfaction were caused by this discovery as nobody had made any private arrangements at home because they had been assured that the Squadron would be returning at an early date. More than one member of the squadron had left England having made arrangements to get married on his return. Others had left wives and other dependants totally unprovided for, and others still had come for the supposed 16 days although they were, in actual fact, unfit for Overseas Service."

And just getting to this point had been far from uneventful. Of the sixteen aircraft and crews that had left Leeming, only fourteen finally arrived at Aqir. (Aqir was in what was then Palestine and is today an Israeli airbase.) The two advance crews arrived on 5 July. The CO flew on to Ismailia to meet and confer with the AOC and staff of 205 Group, to give an indication of the Halifax's capabilities (inferring that the staff were not entirely clear about the assets being given them), and to attend a meeting at which the creation of 10/227 Squadron was agreed. He then appears to have taken a Wellington to Kasfareet to collect twelve bomb winches but had to force land in the Sinai desert on his return, with the journey then completed by train and another Wellington. Meanwhile, eleven of the other fourteen aircraft left Gibraltar for Egypt – the other three were unserviceable – but one had to turn back. The crew spent some time over the Rock jettisoning fuel but, with visibility deteriorating, Flt Lt Hacking was forced to attempt an overweight landing and the undercarriage collapsed on his third attempt. The crew emerged unhurt, but the aircraft was written off. Five made it to Aqir on 9 July, but a sixth ditched off Alexandria, having failed to find a landing ground and run out of fuel. (The crew and groundcrew got ashore by dinghy and continued on to Aqir.) Four aircraft landed at Luqa, on Malta – three had been unable to climb to height at full load, and one had suffered an engine failure over Tunis. They were then fortunate to survive unscathed five raids on the airfield by both night and day, before flying on to Aqir on 10 July. One of the delayed aircraft at Gibraltar reached Aqir on 11 July, a second made it on 27 July, bringing news that one aircraft and crew (Plt Off Hillier) had been sent back to the UK 'on instructions from the Transit Authorities at Gibraltar.' (The unit ORB at Leeming shows them as having arrived back there on 14 July, 'surplus to requirements,' though there is no indication of what level of operational command, at whatever location, made this decision.) The last aircraft and crew stuck at Gibraltar needed a new wheel and finally arrived in-theatre on 13 August.

By then, as Fred Brinton said, the unit had settled down to ops against Tobruk. However, it was quickly established that the Halifax had insufficient range to operate from Aqir. That location was retained as the Squadron base, with all major servicing carried out there, but operations were mounted from forward airfields in the Canal Zone - either Fayid to the north or Shaluffa (LG 224) to the south – with crews

10/227 Sqn aircrew at a pre-flight briefing in September 1942

landing back there or at Aqir. (The record shows Sgt Brinton and 24 airmen being sent to LG 224 on 12 July to provide a forward maintenance party.)

The first bombing operation was flown by WO Peterson's crew on the night of 11/12 July against slight opposition, with poor visibility preventing any assessment of accuracy. Four aircraft followed on the next night, with a similar outcome. By the night of 13/14 July, the intensity and accuracy of the flak awaiting the four aircraft tasked had increased – Plt Off Drake's aircraft was hit, he was unable to jettison his bombs and, on making an emergency landing at Almaza, he crashed and the bombs exploded. He and one of his gunners were badly injured and several Egyptian firemen were killed.

As operations continued, the record shows the targets listed as either Tobruk itself or the jetties and shipping in the harbour. An inability to observe bombing results persisted on many nights, and opposition was varied from slight to heavy and accurate. Subsequent operations were flown on 11 nights in July and on 17 nights in August, in numbers varying from one to seven. (In the midst of all this, the CO and a group of unit officers and airmen went to Abu Sueir to join a parade of theatre personnel to be inspected by Winston Churchill, returning to the UK after a visit to Russia to meet Marshal Stalin.) The smaller numbers may have been the result of maintenance difficulties – Fred Brinton remembers many hydraulic faults and a colleague expressing an opinion that the number of engine changes required was probably keeping a Spitfire squadron permanently grounded.

Operations continued into the following month, and on 2 September an Administrative Instruction was issued stating that personnel from 10 and 76 Squadrons would be posted to No 462 Squadron, and those of 227 Squadron would be held pending a decision on further posting.

Before that took place, an operation was launched out of the established pattern. On the afternoon of 5 September, five crews joined another four from 76 Squadron for a daylight attack on the occupied airfield at Heraklion on the northern coast of Crete. One of the 10 Squadron aircraft returned early after an engine failure. The others met heavy opposition and, for the first time from Aqir, enemy Me 109 fighters. Three were able to bomb the airfield, whilst the fourth suffered a hangup. All four were damaged and one, flown by Flt Lt Hacking, had fire spread from an engine into the starboard wing. Some parachutes were seen to leave the aircraft, which crashed killing the pilot and both gunners (Sgts Carson and Porritt). Sgt Bradley (WOP/AG) and Sgt McFarlane (FE) baled out, were picked up by Cretan partisans, and lived in the hills for six months, helping to pass information by radio to HQ Middle East. They were eventually picked up by a RN motor launch and taken to Egypt. The navigator, Plt Off

The 10/227 Sqn detachment later moved from Palestine to Fayid in Egypt. Tobruk, Libya, and Iraklion, Crete were two of their frequent targets.

Turner RCAF, was captured by the Italians on Crete. The Germans wanted to take him but the Italians shipped him to Rhodes by a fishing boat intercepted by a RN submarine that took him off.

On 7 September 1942, the amalgamation of Nos 10/227 and 76/462 Squadrons took place to form No 462 Squadron RAAF at Fayid, albeit that, at the outset, and given that they had arrived in-theatre from the UK with 10 and 76 Squadrons, very few personnel were Australian. The eight remaining Halifax aircraft of 10/227 and those of 74/462 Squadrons were also allocated to the new squadron, and towards the end of the month it was agreed that 462 Squadron would be designated a RAAF squadron within the Empire Air Training Agreement, with its RAF personnel replaced by RAAF personnel as these became available. Wg Cdr D O Young DSO DFC AFC was initially appointed as CO but fell ill and was replaced by Wg Cdr Seymour-Price after a few days.

(The various mergers around this time have caused some confusion, in that the loss of Flt Lt Hacking's crew is listed as a 462 Squadron RAAF loss on the official Australian War Memorial website. However, despite the Admin Instruction having been dated 2 September, the record books of both 10/227 and 462 Squadrons agree that the RAAF squadron did not form until 7 September, after the attack on Heraklion, and its first operation, back to Tobruk, is recorded for the night of 8/9 September.)

CHAPTER 7

MELBOURNE, A NEW HOME

The detachment of the greater part of the Squadron to the Middle East in early July 1942 for what turned out to be a considerably longer period than expected did not mean that operations from the UK came to an end. However, as RAF Leeming was not fully useable, three crews were sent to RAF Topcliffe to operate from there and, with Wg Cdr Bennett posted out, a new CO, Wg Cdr R K Wildey DFC, was appointed on 26 July and confirmed in post on 1 August.

The crews who moved across to Topcliffe on 19 July were those of Sqn Ldr Griffiths, Plt Off Fegan and Plt Off Hillier, and the latter two went into action that night, joining a raid on the U-Boat yard at Vegesack, near Bremen. The target was completely covered by medium level cloud, and Plt Off Hillier returned to report an attack fixed in the target area. Nothing was heard from Plt Off Fegan's crew after departure. Two nights later, both remaining crews at Topcliffe joined 290 others in a raid on Duisburg. Once again cloud prevented visual aiming and later reports showed that bombing had been largely wide of the target. Plt Off Hillier's crew joined return raids on Duisburg on the nights of 23/24 and 25/26 July.

A new crew (Plt Off Black) was sent to Topcliffe to bring the detached number back to three, and all joined a raid on Hamburg docks on the night of 26/27 July, claiming accurate attacks in good visibility. Another crew (Sgt Hampton) was one of three participating in the first large raid on Saarbrucken on 29/30 July. They had to abandon the mission and jettison their bombs when the rear turret guns proved unserviceable, but the others reported completing attacks over the target area. A further crew (FS Saunders) joined the others at Topcliffe on 31 July. They got airborne that night for Dusseldorf – the only Squadron crew involved - but had to turn back as defective superchargers prevented their climbing above 9000 feet. The next raid joined was to Duisburg on the night of 6/7 August when two crews reported accurate attacks in good visibility.

On 11/12 August, two crews, still operating from Topcliffe, were detailed for a raid on Mainz. Plt Off Black's crew lost their airspeed indicator (ASI) after coasting out. They stalled, spun, lost 3000 feet, recovered and pressed on. However, once under flak attack in the searchlight belt on the other side, the lack of an ASI left them unable to take adequate evasive action and they finally decided to jettison their bombload and come home. Sgt Hampton's crew was more fortunate and was able to report a successful release over Mainz through partial cloud cover. On the same night, another three crews took part in an attack on Le Havre, reporting successful attacks in the target area.

The move to Melbourne: On 17 August instructions were received that the Squadron was to move as quickly as possible to RAF Melbourne, a satellite station of RAF Pocklington, some twelve miles east of York, and the move was completed over the following two days. A temporary airfield had been constructed there in 1940, it would eventually have three concrete runways and three hangars, and 10 Squadron would be its sole occupant for the rest of the war. It would eventually attain a strength of some 1500 in all - aircrew, groundcrew, ops and signals support, and general admin staff, all commanded by a Group Captain. However, to begin with, Ops were flown from Topcliffe and Pocklington until late October. Harry Wordsworth had been overseas for over four years and was posted to Leeming prior to the move and supplied his memories of that time for a 1986 Association Newsletter:

"I was surplus to establishment in the MT Sections and got all sorts of gash jobs like training young airmen to handle and fire Sten guns and teaching personnel from other Sections to drive as B Class drivers. I also collected the first 4000 lb bomb racks from HP Cricklewood. In May of 1942 I finally left Leeming for Melbourne, whilst the runways were still under construction, along with most of the buildings. Myself and a Sgt Clements, a storekeeper, were the only RAF personnel on the camp and our job was to handle any stores and equipment that had arrived. Each night we left Melbourne for Pocklington where we were billeted.

Gradually personnel were posted in as the billets became available, but conditions were primitive and generally rough but, as far as I am concerned, far more acceptable than the Western Desert and Greece where we were under canvas. I remember how lots of the blokes thought Melbourne was the back of beyond at first, but they soon settled down to very little time off and plenty of work. Lots of us were issued with cycles as the billets were often half a mile away from the Messes and places of work....The Squadron at first operated in small numbers from Pocklington until the runways at Melbourne were completed, but soon they arrived and operations started in earnest."

Tail Gunner Chas Harrison, who had survived a Whitley ditching on the upturned liferaft in June 1941, had rejoined the Squadron in early 1942 to start a new tour. He was on the CO's crew as it moved from Leeming and later recalled:

> "It was like leaving the Ritz to live in a doss house initially, although we settled in after a short while......There were only about seven crews operational, the rest were converting and the Melbourne runways were not ready yet. We could use the airfield except for operations. We would air test the plane down to Pocklington, it was bombed up and we operated from there."

A note on the formation of the Pathfinder Force may be appropriate at this point given that, as stated in Middlebrook and Everitt's 'Bomber Command War Diaries,' *"10 Squadron can be said to have been the cradle of the Pathfinder Force."* The claim rests not only on Wg Cdr Bennett's posting out to form and command it but, more significantly, on the prior work done within the Air Ministry by Gp Capt S O Bufton, a previous Squadron CO, and then Director of Bomber Operations, to make the case for the establishment of a specialist Target Finding Force in the face of opposition from Air Marshal Harris at HQ Bomber Command. (And it

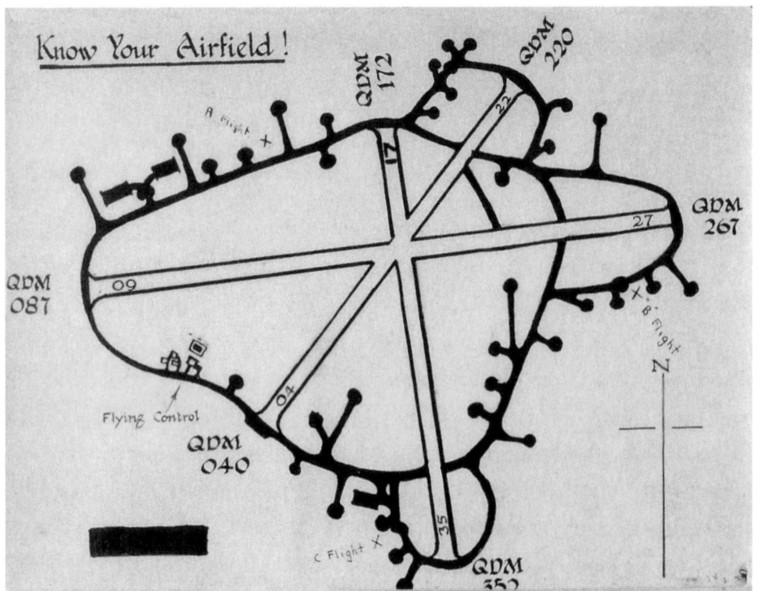

Melbourne Airfield Diagram: Many airfields in Yorkshire and Lincolnshire were situated close to others. It was not unheard of for aircraft to land at the wrong airfield, particularly in poor visibility weather conditions.

The airfield construction workers showed their humour when laying Slab No 186 on the northerly runway at Melbourne in early 1942. Still visible today, Hitler's face is drawn in the concrete of the 6 but the 8's image is unclear as to its identity.

should also be recalled that by the beginning of June 1940 the Squadron had been testing marking with flares, and it seems improbable that the then CO, Wg Cdr Staton, did not later discuss the issue when he became Station Commander with Bufton as Squadron CO.) No 35 Squadron was transferred to be the Halifax component of the new force, perhaps only because 10 Squadron had been depleted by the deployment of its Middle East detachment. The new Force went into action on the night of 18/19 August. It had variable success at the outset, but went on to be a substantial contributor to the effectiveness of the bomber offensive.

Inevitably, with the Squadron having to be built up to normal strength, its contribution to operations in the second half of 1942 was limited, albeit that it established a second Flight once again on 16 September. Ops were flown from Pocklington to Saarbrucken by three crews on the nights of 28/29 August and 1/2 September. All reached the target area on the first night, two bombing through considerable cloud and the third finding that an electrical fault precluded bomb release. Sgt Cobb's aircraft had to evade and fight off two fighter attacks. One crew had to return early on the second occasion with a glycol leak affecting a starboard engine. The remaining two both reported attacks on the primary target, with visibility sufficient to identify bomb bursts.

Two crews bombed Karlsruhe on 2/3 September, and three went to Bremen on 4/5 September, when all reported accurate attacks. Sgt Cobb's aircraft was attacked once again, this time by a Ju 88, shot down by a long burst from the tail gunner, Sgt Williams. On the night of 6 September, Plt Off Morgan's crew set off alone to join some 200 others for Duisburg but was shot down over Germany. Two crews joined an attack on Frankfurt on the night of 8/9 September – one reported bombing over haze in the target area and Sgt Cobb, fighter-free that night, reported releasing over Mainz.

Six crews were detailed to join some 470 others for a raid on Dusseldorf on the night of 10/11 September, and all six reported bomb releases in the target area. Five were detailed for a return to Bremen on 13/14 September – one had to abandon the mission with engine problems, and the others reported bombing in the target area. One aircraft was lost as it crashed on a diversionary landing at Great Massington, with no injuries to the crew. On the next night, four crews were detailed to join an attack on Wilhelmshaven. All reported accurate attacks in what was later assessed to have been a very successful raid.

For the night of 16/17 September, four crews were briefed to join a raid on Essen. Two were able to attack the target, but engine problems forced one to return early. Sqn Ldr Carter's crew had also to return after experiencing high fuel consumption and loss of power - unfortunately, their brakes failed on landing and they overshot onto the Hull-York road. The crew escaped unharmed, but the aircraft was written-off. Another aircraft was lost on return from the next Op, to Saarbrucken, on 19/20 September. Five crews had briefed, one had to abandon the mission, and the others reported bomb bursts in the target area. On return, Sgt Wilmott's aircraft hit a tree in the Pennines and the damage caused made the aircraft almost uncontrollable. He gained height and ordered his crew to abandon the aircraft. All landed safely apart from the Navigator, Sgt McDougal, who was killed.

A number of raids on shipbuilding yards at Flensburg in Schleswig-Holstein now followed. A single crew was involved and reported an accurate attack on the night 23/24 September; and three crews set off on the night of 26 September but were recalled when over Denmark, with one claiming a Me 110 shot down over the sea on the return. However, five crews took off on the night of 1/2 October, but only one returned and the loss of four crews was the Squadron's worst night up to that point in the war. FS Hayes' crew became POWs, but all twenty-one members of Plt Off Jones, Sgt Campbell and Sgt Moller's crews were killed. These were all very new crews – Sgt Campbell had flown two abandoned missions and the others were on their first Op. Indeed, by a cruel irony, the official posting date to the Squadron of Sgt Moller and Hayes' crews from the Conversion Flight is recorded as 4 October, after they went down. And it is doubly ironic that it is on that night that the first record is to be found of a new pilot flying as Second Pilot with another crew to gain experience before taking his own crew on a raid – something that became standard practice from then on.

Another new crew was lost on the next night when Plt Off Baxter's crew went off on their own to join a raid on Krefeld

and did not return. All seven became POWs. Small groups of crews were sent on the next three Ops without losses – four to Aachen on 5/6 October, two to Osnabruck the following night and, for the first time in the Halifax era, five crews went on a Gardening mission, laying mines around the Frisian Islands, on the night of 10/11 October. (NB: These mines were not round, with horns. Aerial mines were cylindrical in shape, some nine feet long, with a parachute to give a vertical entry into the sea, where they sank to the surface awaiting acoustic or magnetic detonation.) The greatest number of crews for some time – eleven in all – was mustered for a raid on Kiel on the night of 13/14 October. One had to abandon the mission when the rear turret proved faulty, nine reported having attacked the target area and one (Plt Off Lindsay) failed to return. Crews reported heavy flak, but later town reports state that little AA fire was used and it is probable that they (like the majority of others) had attacked a decoy fire site away from the city.

The CO, Wg Cdr Wildey, and his crew went down over Germany on the next Op, to Cologne on the night of 15/16 October. He and his two gunners were killed, with the rest of the crew becoming POWs. (He and Maj Ward, who died on 21 September 1917, are the only two Squadron Commanders to have been lost in action.) The seven other crews returned reporting successful attacks, but it is likely that these were made on a decoy site. Sqn Ldr Carter was given acting rank and appointed as the new CO.

Work at Melbourne had proceeded to a point at which the first Op to be mounted from there took place on the night of 23/24 October. On that same night, in North Africa, the barrage that launched the Battle of Alamein began and, to put additional pressure on the Italians at home, a bombing campaign against targets in Northern Italy had been initiated on the previous night. Nine crews were detailed for an attack on Genoa involving some 120 aircraft, but one aircraft swung off the runway on takeoff. The others reached the area and, despite heavy cloud and haze, seven reported having attacked the primary target. The eighth attacked Savona, some thirty miles away, as an alternate and later analysis indicated that most others had, in fact, done likewise, thanks to the cloud cover. Sgt Somerscales reported an odd incident. On return, near Lake Geneva, another Halifax that had followed his aircraft for 150 miles overtook and, once ahead, fired on him from around 200 yards, slightly wounding his second pilot! The last Op for the month was flown by Plt Off Margett's crew, operating from Pocklington to join a raid on Milan on the next night, and bombing over cloud on ETA.

With Squadron aircraft now able to take off at full weights from Melbourne, November and December saw a mixture of Ops flown with, of course, a very similar pattern of cancelled takeoffs, enforced returns to base, encounters with enemy fighters and crew losses that had been seen up to that point. The Italian campaign continued, with Genoa and Turin the chosen targets, and these missions were interspersed with further attacks on German industrial towns and Gardening sorties. Genoa was attacked on the nights of 7/8 and 15/16 November. For the first mission, it would appear that there were particularly adverse winds at height on the return as most crews elected to land at airfields to the south. Indeed, Plt Off Munro's crew felt that they were about to run out of fuel, so he ordered the crew to bale out and then found that he could achieve a landing at Downham Market in Norfolk. Happily, all of his crew also came down safely. There were five raids to Turin before the end of the year, with mixed results but no losses. Indeed, losses were relatively few in the last two months. Sgt Hale's crew went down on a Gardening sortie into the Frisians on the night of 8 November, and Plt Off Collett's crew was brought down over France whilst on a raid to Stuttgart on the night of 22/23 November. He and five colleagues became POWs but Sgt Jenson, his WOP, was killed. Not all of the losses were on Ops. On 30 November, shortly after takeoff on a formation training sortie, FS Willmot's aircraft crashed just beyond the airfield, killing all eight crewmen. Completing the picture for 1942 were three further Gardening missions and raids on Hamburg, Stuttgart, Frankfurt, Mannheim and Duisburg.

1943

The first Op of the year saw eleven crews participate in a large minelaying mission by some 120 aircraft into the Frisian Islands during the evening of 9 January. Eight returned to report mines laid in allotted areas, with no opposition. However, two of the last three away completed

their drops but one suffered flak damage, whilst the other had to fight off a Ju 88. Sgt Fish's crew was brought down over Holland, with all seven crewmen killed. 'Gardening' missions would once again become a standard part of the Squadron's repertoire and would often be used to give new crews a taste of operational flying before being exposed to attacks into Germany.

By the start of 1943 the U-Boat threat to Atlantic shipping was again causing great concern in Whitehall and, in mid-January, a new Directive was issued to Bomber Command requiring a sustained assault on the U-boat bases and the towns nearby on the French Atlantic coast. By this stage, the U-boat pens themselves were so heavily protected under concrete that it was the ports and towns that took the brunt of what followed, and Lorient and St Nazaire were soon almost completely deserted by their local populations. Lorient had the highest priority as a target, and a raid with 122 aircraft was immediately mounted to attack it on the night of 14/15 January. 10 Squadron briefed ten crews – six reported successful attacks, one had to return early with engine trouble, and three sorties were cancelled as a result of deteriorating weather. Three more raids on Lorient followed during January, with two additional minelaying missions.

A change of command occurred on 3 February when Wg Cdr Carter moved on and was replaced by Sqn Ldr D W Edmonds from 102 Squadron, with the rank of Acting Wg Cdr. As a related matter, a significant administrative change would be initiated within Bomber Command from March 1943, with the introduction of an intermediate level of command - the Base - between Group HQs and Stations.

__Above:__ Leo Groak (centre) warms his hands with the rest of the ground crew who serviced Fg Off G. Hewlett & crew's Halifax HR691. Life-long friendships frequently resulted between the air and ground crews: in Leo's case with Tom Thackray, the flt eng on Hewlett's crew.

__Left:__ Melbourne, East Yorks could be a cold place to be when servicing a Halifax outside in the open. Here Mk 2 Halifax BB 194 (ZA-E) in January 1943 is being serviced. Standing at the bench on the left is engine fitter LAC George Tait.

The aim was to relieve those responsible for operations of many of the support functions that absorbed commanders' time, and this was generally effected by grouping two satellite stations under one main Base station, commanded by an Air Commodore. So, from 24 March 1943, RAF Melbourne was part of 'Pocklington Base,' commanded by Air Cdre G A 'Gus' Walker, with Elvington completing the trio. In September, the terminology changed to '42 Base,' and in April 1944 the station moved to a new '44 Base' centred on RAF Holme-on-Spalding Moor.

This Merlin-engined Halifax Mk2 DT 788 failed to return from a Cologne raid on 14 February 1943. Pilot Sgt J.D. Illingworth and five of his crew were taken prisoner and WOp Sgt H. Kay was killed; there was no mid-upper gunner on this aircraft.

Ops for the Squadron in February involved a return to the main offensive against targets in Germany on six nights, a raid on Turin, four on the U-boat bases at Lorient and St Nazaire, and a single minelaying sortie. One crew was lost over Holland, on the night of 14/15 February, during a raid on Cologne - Sgt Kay, the WOP, was killed whilst his colleagues on Sgt Illingworth's crew became POWs. Continuing the offensive more deeply, a return to Berlin as a target by some 300 aircraft was mounted on the night of 1/2 March. The Squadron detailed ten crews and all returned reporting bomb releases on the Pathfinders' target markers, with fires visible from up to 100 miles distance. However, subsequent reporting revealed that the Pathfinders had difficulty in achieving concentrated marking – the early H2S radar was unsuited to breaking out individual parts of a large built-up area – and that bombs had fallen over 100 square miles. Nonetheless, with the more potent bombers now available, even such a spread was capable of causing significant damage and this effect would become a feature of later attacks. A new crew carried out minelaying sorties on each of the following nights whilst, on 3/4 March, colleagues in eleven crews were detailed to join some 400 others for another raid on Hamburg. One had to return early as an engine overheated, but the others reported attacking using both markers and visual identifications.

With Bomber Command continually growing in strength, what become known as 'The Battle of the Ruhr' was about to begin. There would be occasional targets chosen elsewhere, but for the coming months the concentration would be on the industrial cities comfortably within range that permitted maximum bomb loads to be carried, and that could now be very accurately marked by the Pathfinders' *Oboe* equipped Mosquitoes. The campaign opened with an attack by 442 aircraft on Essen on the night of 5/6 March for which 10 Squadron detailed fourteen crews. Two had to jettison bomb loads and return early with technical problems, but all of the others reported making attacks over the target markers against a background of fairly heavy flak and considerable searchlight activity.

A new crew carried out a successful minelaying sortie on the evening of 7 March. That was followed on the next night by another substantial, but more distant, raid on Nuremberg. Fourteen crews were provided for this, with all crews attacking on the markers against only moderate opposition. On the night of 9/10 March the target was Munich, with nine crews briefed for the Op. One was forced to jettison and return early when a crewmember became ill, but the others all reported attacking in built-up areas. To permit the start of evasive action, Sgt Brunton's crew also reported having dropped their bombs after being coned by searchlights on entering the target area.

Losses were sustained over the next days, starting with a crash in training near the airfield on 10 March in which all six crewmen died. On the following night, when ten crews set off for Stuttgart, all returned safely. However, on the night of 12/13 March, when twelve crews joined over 400 others in a raid on Essen,two did not return. Sgt Dickenson's crew came down over Germany and Sgt Barker's over Holland, with no survivors. Flak over the target was heavy, with some aircraft damage. Sgt Vinish's tail gunner was also badly injured, and Fg Off Dawes reported spending some twenty minutes over the target as he sought to evade two sets of searchlight cones.

Local weather prevented all but a few local flights over the next nine days until, on the night of 22/23 March, ten crews joined a large and accurate raid on the base at St Nazaire. The Ruhr campaign continued on the next night, with thirteen crews joining a large raid on Duisburg and attacking on both ETA and markers over cloud. Apart from another new crew going to St Nazaire on 28 March, the month ended with two attacks deeper into Germany – Berlin once again. Although all twelve crews who participated on the night of 27/28 March reported dropping on PFF markers, the large raid that night was later assessed as not very successful. Ten crews participated in a return raid on 29/30 March when weather conditions were difficult and four had to abandon the mission in conditions of severe icing and static. The other six reached the city and reported releasing on the markers, although it would later emerge that most bombs fell in open country to the southeast.

The next month began with three crews joining a raid on Lorient on 2/3 April for what would be the Squadron's last visit to the U-boat bases for some time as the Ruhr campaign gathered momentum. On the next night, thirteen crews were detailed for a raid on Essen. The town's defences were very active and four aircraft sustained light damage in what appears to have been a relatively successful raid. On the next night, some of those crews went out again as the Squadron sent fourteen to join a large raid of over 500 aircraft on Kiel. For twelve crews, bomb release had to be on markers over complete cloud cover. Sgt Williams' crew was late off and eventually accepted that they could not achieve their target time. They jettisoned their bomb load to return, were almost immediately attacked by a Ju 88, lost intercom, the bomb doors fell open and about forty minutes later, an engine stopped. An approach to land at Thornaby was initiated but, at 500 feet, the remaining engines cut out and the aircraft crashed in a nearby field, with the crew relatively uninjured. Fg Off Wann's crew failed to return and presumably went down over the North Sea.

It can be seen from the records that, by April 1943, the Squadron had grown in size and was operating in three Flights. Bad weather prevented flying Ops until the night of 8/9 April, when eight crews were detailed for a raid on Duisburg. Plt Off Hellis' crew had to return early as they lost their Air Speed Indicator, but the other seven were able to reach the cloud-covered target and release on PFF markers. The next target was Frankfurt, for 500 aircraft on the night of 10/11 April, when eleven crews returned reporting another drop over cloud. There is no mention of prior marking and later analysis suggests this was an unsuccessful raid overall. A large raid on Stuttgart was planned for the night of 14/15 April. Twelve crews were briefed, but illness and unserviceability forced two to return early and FS Hancock's crew was brought down over France. The others reported attacks in relatively good weather.

The next two raids were on targets well away from the Ruhr: the Skoda works in Pilsner, Czechoslovakia, on 16/17 April, and the Baltic port Stettin on 20/21 April. Eleven crews were amongst the 327 mustered for Pilsner and all reached the target area where one crew found that their bomb release mechanism failed. The others reported dropping using a combination of markers and visual identification, as they had been instructed to do. (In the event, it appears that only a handful of crews on the raid returned with photographs indicating that they had been within three miles of the factory.) The return flights were eventful for some. Near Reims, FS Virgo's aircraft was attacked by a Me 110 that FS Hill, his tail gunner, was able to shoot down. Flt Lt Wood lost his port outer engine to flak over the target and his port inner to flak on the French coast. Fortunately, he had sufficient height to cross the Channel for a crash landing near Lewes, where all of the crew required treatment for shock or injuries. Fg Off Dawes hit a tree and crashed whilst attempting a landing at Harwell, his crew faring better than their colleagues at Lewes. Thirteen crews were detailed to join over 300 others for the raid on Stettin, where marking and visual identification proved much more effective, although another bomb release failure prevented one crew from completing its attack. Sgt Glover's crew did not return and became POWs.

Following a break from Ops, nine crews were detailed to join another raid on Duisburg with over 500 others on 26/27 April, and this was followed by participation in Bomber Command's two biggest minelaying Ops to date – in the Frisians once again (six crews) on 27/28 April, and in the Kattegat (nine crews) on the next night. The month ended with a return to Essen, when twelve crews were detailed to join some 290 others. Engine and oxygen system problems made two crews turn back. The other ten all reported having to drop on PFF markers over heavy cloud cover and later photography suggested that this had been a reasonably successful raid.

The weather in much of May was unseasonably unsuitable for flying Ops and, adding in avoidance of the Full Moon phase around 19 May, this meant Ops being flown on just seven nights in the month. The first major attack on Dortmund was mounted on 4/5 May using 596 aircraft, with 10 Squadron contributing thirteen of them. One crew was behind its target slot and dropped its bombs elsewhere, but the others reported attacks using markers and visual identifications in good conditions. With poor conditions back at base, diversions to Leeming were initiated and, on the way, FS Weddes' aircraft hit high ground and crashed. He survived, with his Flight Engineer and Tail Gunner, but the five other crew members were killed. This raid was followed by a seven day break, during which it appears that a good deal of training was undertaken on the days when flying was possible – formation, night firing and X-countries, fighter affiliation and bombing exercises. After that, thirteen crews were detailed to join a large raid on Duisburg once again on 12/13 May. Two aircraft were unserviceable at start-up; three crews were unable to maintain height in icing and were forced to return early; and Sgt Beveridge had to jettison his bombs near the Dutch/German border on coming under attack by a Me 110 during which hits were taken in the tail, bomb doors and fuel tanks. Sgt Compton, the Tail Gunner returned fire, saw hits and claimed the aircraft as probably destroyed – and about thirty minutes later had to call for more evasive flying to escape a Me 109 that, in the event, did not continue the engagement. The other seven crews reached Duisburg and all reported attacks on the markers despite very active defences. This was later assessed as a highly successful raid and Duisburg was not attacked again during this campaign. There was one more raid before another break, when thirteen crews were detailed for Bochum on 13/14 May. FS Mills crew was lost over Holland, but ten made it to the target and another bombed an alterative in the Dortmund area. Sgt Beveridge's crew had an infinitely more eventful sortie than that on the previous night. At 18,000 feet between Dusseldorf and Koln, the aircraft was caught in searchlights and during violent evasive action a nearby flak burst caused the rudder to overbalance. The aircraft then turned on its back, the nose dropped and height was lost. Once level again at 7,000 feet, it was discovered that Sgt McCoy, the Mid Upper Gunner, was missing from his turret. He had baled out without instructions and did not survive. Shortly afterwards, the aircraft was caught by lights once again, its bombs were dropped and a turn for home made. Over Holland about an hour later, the aircraft was engaged by two Ju 88s making repeated passes. Fire was returned by Sgt Compton, the Tail Gunner, and after one burst, one of the JU 88s was seen to go down in flames and explode. He then experienced a temporary stoppage as the other 88 returned and evasive manoeuvres continued. The crew debrief concluded:

> "Aircraft then returned safely to base. Damage sustained was port tailplane, elevator, rudder and port outer tank holed, and one gun unserviceable. None of the crew were injured, the only casualty being the Mid Gunner who baled out."

(The crew's next mission was altogether a tamer affair as they had only a bomb hangup to report – and it is no surprise to see Sgt Compton's name amongst those awarded the DFM in May.)

A nine-day break in Ops followed, but the Squadron kept busy as a sample F540 record shows:

> 16 May. No operations. Nineteen crews of A and B Flights carried out formation practice. Two A Flight crews did Beam practice, three crews of C Flight carried out Bombing practice, two carried out Fighter Affiliation, five formation practice and six crews carried out Bullseye exercises. Weather at base was fine and clear, little wind."

However, when Ops resumed, they did so at a pace – twenty-one crews for Dortmund on 23/24 May, eighteen for Dusseldorf on 25/26 May and for Essen on 27/28 May, and seventeen for Wuppertal on 29/30 May. The first, to Dortmund, involved over 820 aircraft and was highly successful. One Squadron crew had to return when their intercom failed, but seventeen made it to the target. On the way home, Sgt Watson's crew had a 45-minute encounter with three Ju 88s and a Me 109, during which their aircraft sustained damage in several places. They lost an engine but claimed a 'Probable' on one of the 88s, and all made it safely back to base. However, three crews did not return – those of FS Denton, lost over Germany; Sgt Hine, down over Holland; and Sgt Rees, probably down over the Baltic. A cancellation and an early return reduced the crews over Dusseldorf two nights later to sixteen, for what was later assessed as a very scattered raid involving some 750 aircraft. Eighteen crews were detailed with 500 others for a return to Essen on 27/28 May. One had to turn back as their aircraft lost power and could not climb, and two did not return. Both were lost over Germany – two of Fg Off Rawlinson's crew became POWs, whilst their colleagues were killed, as were all members of WO Price's crew. The other fifteen crews reported being over the target, using Gee fixing and PFF markers. The final Op of the month saw seventeen crews detailed to join 700 others in a raid on Wuppertal. Three had technical problems that caused early turnbacks and one, FS Clarke's crew, was lost over Germany. The remaining thirteen crossed the target for what was assessed as a most successful raid.

It need hardly be said that the human cost of the WW2 losses in Bomber Command at home was very considerable – children lost fathers, wives lost husbands, and engagements were cut short. For example, Squadron captain Flight Sergeant Jack Denton had met Freda Smith at a dance in York in February 1943; they quickly fell in love and were soon engaged to be married – until his crew was shot down in the 23/24 May Dusseldorf raid. Freda was a Quaker, working with the Friends Ambulance Unit in London, whose beliefs might normally have precluded a relationship with a bomber pilot. Years later, her niece Dr Nicola Carter found the many letters that passed between the couple and, as Esdaile Carter, used them as the basis for "Pilot and Pacifist: a World War Two History," published in 2013. There must be dozens of similar tributes that could be written arising from 10 Squadron's losses alone.

After another Ops break during which all three Flights kept busy at base, the next Op was mounted on 11/12 June, when twenty crews were detailed to join some 763 others for another raid on Dusseldorf. One cancelled due to unserviceability and two returned after encountering severe icing. The remaining seventeen reached the city for what would be assessed as the most damaging raid of the war there. On the next night, seventeen crews briefed to join a 503-aircraft raid on Bochum. Technical failures forced two to return to base and Sgt Innes' crew was lost over Germany. The others all reported bombing on the markers in what was seen as a generally successful raid. The next Op was away from the Ruhr, on 19/20 June, when the target was the Schneider armaments factory at Le Creusot in Eastern France. 290 aircraft were involved, with sixteen 10 Squadron crews detailed. Sqn Ldr Debenham's crew had to turn back with an oil leak and had an active encounter with a Me 109 that was claimed as destroyed, whilst Sgt Watson's crew was lost over France. The other fourteen reported attacks on the target factory, with the position identified visually. On 21/22 June, seventeen crews were detailed to join a raid by 705 aircraft on Krefeld. There were two early returns but the others all reported attacking the markers in what was later assessed as a highly damaging raid. For the next night, with Mulheim the target, nineteen crews were briefed. Again there were two returns with mechanical problems and sixteen reports of bombs released accurately over the target area. Sgt Pinkerton's crew did not return. On 24/25 June there was a return to Wuppertal by over 600 aircraft, with fifteen Squadron crews briefed for the raid. On this occasion three crews returned with technical trouble whilst the remaining twelve reported releasing on the PFF markers in another highly effective raid. The pressure continued with eighteen crews detailed for an attack on Gelsenkirchen on the next night. One had to return when it became apparent that, having got well off track, it could not make its concentration time. The other crews reported attacking the markers that were seen, but the *Oboe*

Mosquitoes had a night of variable success with a knock-on effect on the damage assessment. The final Op in June was on the night of 28/29, when a new attack on Cologne was planned using 608 aircraft. 10 Squadron detailed eighteen, of which one returned with an overheating engine that restricted the aircraft's climb capability. Two crews did not return – FS Geddes' crew and Plt Off Peate's were both lost over Holland.

Another substantial attack on Cologne with 653 aircraft was mounted on 3/4 July, with the Squadron briefing nineteen crews. Two had to abandon the mission with technical problems, whilst Plt Off Cox elected to jettison his bombs and return early after losing a port engine in an encounter with a Ju 88. Their colleagues all reported attacking the markers, and this is known to have been a successful raid. FS Morley's crew was lost over Belgium – he and his Mid-Upper Gunner, Sgt Sadler, were killed, with their colleagues taken as POWs. After a short break, twenty crews were detailed for a 418-aircraft raid on Gelsenkirchen on 9/10 July. Engine-related problems caused two to abandon the mission, and the others reported bombing over cloud. However, the PFF Mosquitoes had an unfortunate night and the raid was largely ineffective. Next came a return to Aachen on 13/14 July, with nineteen crews detailed to join 355 others. All of the Squadron crews reached the target, which was badly damaged, and all returned safely. The final Op, on 15/16 July before another break, was to a destination well away from the Ruhr and, unusually, would use only Halifaxes, 165 of them in all. The target was the Peugeot motor factory in Montbéliard, Eastern France, close to the Swiss border. One crew lost a starboard engine and abandoned the mission, with seventeen returning to report bombs dropped on Target Markers. However, it appears that the centre of these was beyond the factory, and most bombs fell in the town. Bombs were also dropped in Besancon, some miles back along the route to the target. There was a local witness to a battle between a Halifax and a Do 217 over the town that ended in both aircraft crashing into it, with the Halifax hitting the main rail station. There was a huge explosion that lit up the town and, despite still having some 45 miles to run to the planned target, this appears to have been taken as target marking by twelve crews later in the stream. It was always likely that the Halifax was JD211, flown by FS Pyle's crew, but identification of bodies found in the wreckage and debris in Besancon proved inconclusive. The crew is commemorated at Runnymede, and the Besancon town cemetery has a grave inscribed to unknown RAF airmen. However, recent investigations have offered proof that the Halifax was, indeed, that flown by FS Pyle's crew. No further identification of bodies is possible, but plans are in hand for a new headstone to be placed in Besancon that will make specific reference to JD211. A second Squadron crew (Sgt Mellor) was also lost over France that night, at Recey-Sur-Ource.

A decided change in focus for Bomber Command now took place as Hamburg became its primary target over a period of some ten days. This new campaign – Operation GOMORRAH - began on the night of 24/25 July, when 791 aircraft were mustered for a raid, with twenty-one contributed from 10 Squadron. A crew had to return fairly soon after takeoff after losing an engine and another had a bomb release failure over the target. All of the others reported attacking over the PFF markers as part of a relatively concentrated raid. The size of the city also meant that the 'creepback' effect that had become evident since the introduction of marking did not prevent serious damage being done. (It had emerged from study of aircraft and recce photography that a tendency for a proportion of main force aircraft to bomb

Single-engine trials for the Halifax were carried out by 1658 HCU based at Riccall, near Melbourne in July 1943. Here Sqn Ldr P. Dobson determines the height loss for emergency flight on one engine.

Sqn Ldr F.J. Hartnell-Beavis & his crew. Shot down during a raid on Essen on 26 July 1942, H-B became a PoW in Stalag Luft 3 of earlier 'Great Escape' fame and Wop Sgt R.A. Smith evaded capture. The rest of the crew were killed.

targets short of visible markers had developed, with each new set of fires moving the aiming point for later crews further away.) Three more raids on Hamburg would follow over the next week but, before that, there was another on Essen on 25/26 July. The Squadron detailed twenty of the planned 705 crews. All took off, but two had to return - one was too late at the concentration point, and the other had engine trouble. Those that reached the target reported drops on the markers in what was a highly effective raid that left the Krupps factory badly damaged. However, Sqn Ldr Hartnell-Beavis' crew was shot down over Holland – he was taken prisoner, but his crew were all killed, apart from WOP Sgt Smith who escaped back to England. The attack on Hamburg resumed on 27/28 July, 29/30 July and 2/3 August, with well over 700 aircraft involved on each night, and with the first night going down in history as that on which a number of conditions combined to cause a Firestorm that only stopped once all flammable material in the affected zone was consumed. However, there is no obvious difference in the crew debriefs for that night regarding what they saw beneath them as they crossed over the city. Twenty, twenty-one then twenty-two crews were detailed for these nights, with only three abandoned sorties in all. Notably, there were no losses, very possibly because the Hamburg campaign saw the introduction of *Window*, aluminium foil strips released to mislead German radars and the tactic worked well until the defences learned how to work around it. However, the final raid did not go well. The weather was not good, debriefs make clear that PFF markers were expected but, with few reported exceptions, these did not materialise, and crews generally bombed on ETA or best estimates of position, whilst a handful attacked alternate targets. Fg Off Jenkins' aircraft was attacked by a Ju 88 before reaching the target and he jettisoned his bombs to help his evasive flying. Despite this and seeing the enemy aircraft spiral downwards on fire, his aircraft sustained severe damage but was able to recover to base. Sqn Ldr Debenham's aircraft was hit by flak after leaving the target, suffered control surface damage and was reduced to flying on two engines. He was unable to maintain height, despite having all gun ammunition jettisoned but was finally able to get an engine re-started over the sea and make it back to Melbourne.

After a week's break the main offensive restarted - Mannheim on 9/10 August with sixteen crews detailed, and Nuremberg with nineteen on the next night. One crew abandoned the first raid due to a navigation error and redeemed itself on the second, whilst another lost an engine prior to the target but was able to carry on. Sgt Dibben's crew went down just off the French coast on the second night and became POWs. There were then two reminders that training sorties were not without risk. FS Plant's crew survived a crash after takeoff on a night cross-country on 11 August and Flt Lt Smith's aircraft crashed near Rugby whilst on another on 20 August, but this time with only his Navigator and Bomb Aimer surviving.

After political direction and with the invasion of Italy planned, a sequence of raids on targets in Northern Italy was initiated, aimed at persuading Italy to break with Germany and make peace. Italy did surrender on 8 September, some days after the successful Allied landings in the south, but remained under German occupation. 10 Squadron took part in only one of these raids, to Milan on 12/13 August when all seventeen crews detailed returned after a successful mission. As an indication of the development of marking tactics, this is the first occasion on which unit debriefs mention a 'Master of Ceremonies' directing attacks over the target area, something that would be common from then onwards. (Later on the MC became the 'Master Bomber.') The next Op was also specially directed – an attack on the research

facility at Peenemunde on the Baltic coast where the German V-weapons were under development. The plan called for 596 aircraft, with 10 Squadron contributing eighteen of them. Fourteen reported completing attacks, two had to abandon the mission, and one bombed an island north of Rostock, short of the primary target, after losing two engines to flak damage. FS Long's crew was lost that night, probably over the sea and probably after completing an attack as a feint attack towards Berlin had diverted German fighters for the first part of the night.

The offensive rolled on with seventeen crews detailed and all completing a raid on Leverkusen on 22/23 August, during which Sgt Goodall's crew claimed a Ju 88 shot down. Marking did not go well that night and bombs fell on a number of towns in and around the Ruhr. On the next night, 727 aircraft were scheduled for another attack on Berlin for which 10 Squadron detailed seventeen crews. Icing forced two to abandon the mission, but the others bombed on the markers in what was later assessed as a partly successful raid. A number of crews reported fighter engagements, with Sgt Lucas' gunners claiming one as shot down. Fifteen crews left for another raid on Nuremberg on 27/28 August, but two had to abandon. On the return, Sgt Baker's aircraft was shot down by fighter over Belgium. Sgt Darvill, the Mid-Upper Gunner, became a POW and Sgt Warren, the tail Gunner, was killed but all six others – there was a Second Pilot aboard that night - managed to evade and return to the UK. Other crews also had fighter encounters and, on the return, near Frankfurt, FS Clarke's aircraft was coned by searchlights and targeted by flak batteries. Considerable damage was sustained and after an evading dive it was discovered that his Bomb Aimer, Sgt Mabbs, had baled out. The aircraft recovered to base, and Sgt Mabbs' fate is not recorded. Before another raid on Berlin there was one on 30/31 August on Mönchengladbach for which sixteen crews were detailed. Fifteen reported bombing the primary target and one attacked Rheydt nearby. On the next night, another sixteen crews were detailed to join over 600 others in a return to Berlin. Technical problems prevented three from getting airborne and another three had to return with engine-related difficulties. The others all dropped on the markers, but these were placed well off the centre. The Squadron was fortunate to have no losses that night as overall losses were particularly high at 7.6%, meaning 47 aircraft, including 20 other Halifaxes.

September began with two crews sent on Air Sea Rescue Sweeps, something that became fairly common in later months, and its first Op was once again to Mannheim on the night 5/6. 605 aircraft were scheduled that night, sixteen of them from 10 Squadron. One crew returned after problems with both inner engines, and fourteen returned to report attacks on the PFF flares in good visibility that resulted in a highly effective raid. Plt Off D'Eath's crew was lost over Germany. Fifteen crews were detailed for Munich on the following night. Two had technical problems leading to a return, whilst eleven reported releases on markers over broken cloud. FS Lindsey's crew also reported an encounter with a Me 210 using a searchlight in its nose. Both gunners fired on it as evasion began and it was seen to fall in flames and explode. Two other crews were less fortunate and did not return. Flt Lt Douglas and Sgt O'Kill, his Tail Gunner, died as their aircraft went down over Germany, with their colleagues becoming POWs and the fate of FS Davies' crew is unknown. A break from Ops followed, with the weather on some days preventing any flying at all until the night of 15/16 September when the target for a force of 369 aircraft was the Dunlop rubber factory at Montluçon, in central France. The Squadron detailed sixteen crews – one had an early return, and fourteen bombed on accurately placed markers. Sgt Dunlop's crew went down over France - two men survived to be POWs and WOP Sgt Bilton successfully made his way back to England. Another French target was scheduled on the next night, the rail yards at Modane on the main line from France to Italy. Eleven crews were detailed, of which three had to return with a mix of icing and technical problems. Plt Off Heppell had also to return. Trying to reach 22,000 feet and get above the cloud, icing caused him to stall and lose height three times. As he reached 18,000 feet again on his fourth attempt, the port wing dipped sharply and, with his control column locked, the aircraft began to turn and spin. He recovered with difficulty, only for another spin to develop. His bombs were jettisoned in the spiral and the crew was instructed to prepare to bale out. Control was again regained at 10, 500 feet, when it was realised that two of the crew had baled out. (After free-falling for some 11,000 feet before he could open his parachute, the Navigator, FS Booth, managed to evade and return to the UK via Spain some three months later, but the fate of WOP Sgt Varley is not recorded.) The aircraft made it back to Tangmere with the Bomb Aimer navigating and several instruments useless. The other seven crews all reached the target area and, as Modane now has a square named 'Place du 17 Septembre 1943,' the raid is clearly remembered as a significant event.

Various return visits to targets were made before the month was out. The first was on 22/23 September, when sixteen crews were briefed for a raid on Hanover involving over 700 aircraft. Equipment problems forced two to return early whilst most others went on to bomb on markers that were misplaced. The exception was Flt Lt Jenkins whose aircraft had sustained considerable damage in an encounter with a Me 110, resulting in an inability to release his bombs. Every effort was made to jettison these without success on the way back to Melbourne where the crew was eventually directed to fly out to sea and bale out. The crew did so over Patrington near the coast and, with the autopilot set on an easterly heading, Jenkins followed. Local Army units were alerted and found the airmen, when it transpired that Sgt Hurst, the tail gunner, had stumbled through a coastal minefield but without coming to harm! (As fate would have it, their aircraft had been specially fitted with a colour cine camera in the flare chute to record the scene over a target – understandably, nobody remembered to extract the film before leaving the aircraft.) The following night's target was Mannheim once again, when eleven crews were detailed to join over 600 others. Three crews had technical faults that compelled early returns whilst the others attacked the markers in good visibility in what was an effective raid. FS Wardman's aircraft suffered considerable damage from Ju 88 cannon fire but was able to recover safely to base. Matters went much less well on the night of 27/28 when Hanover was again the target, with eighteen crews briefed for another very large raid. With two early returns, thirteen returned to report bombing on markers that had been placed away from the city centre. The remaining three crews did not return. Having survived the previous Op, FS Wardman's crew went down over Germany, as did Plt Off Cockrem's crew, with three men in all surviving as POWs. There were no survivors amongst Sgt Rostron's crew, also down over Germany. The final Op of the month was flown on the night of 29/30, when nine crews were detailed for a medium-sized raid on Bochum. An engine fault prevented one from taking off and two thers had to make early returns. Six went on to participate in what was an accurate raid, with Flt Lt Cox's crew claiming a Me 109 that had attacked some ten minutes prior to the target.

Around this time, 10 Squadron's aircraft began to be equipped with a significant new navigational aid known as H2S, a mapping radar, with its scanner mounted in a cupola under the rear fuselage. This gave navigators an ability to pick out the reflections from built-up areas, and to compare the picture with town shapes on their charts, particularly when a land/water contrast was available. Blind marking or bombing had become a reality. (As the system was developed, using higher frequencies and shorter wavelengths, more precision became possible and a version of H2S would still be at the heart of the Navigation & Bombing System of the Victor bomber flown by 10 Squadron some twenty years later.)

October opened with five crews despatched on a Gardening mission, with four reporting their 'vegetables planted' – the first use of this term – in their appointed areas, and the guarded wording of the debriefs suggest that H2S fixing from coastal features was used to establish accurate dropping runs. However, Flt Lt Jenkins crew was unable to drop as the equipment in their aircraft was unserviceable and sea mist prevented getting an accurate pinpoint. Three crews were attacked by night fighters but all came home safely. On the next night, sixteen crews were mustered for participation with over 500 others in a raid on Kassel. Four were forced to return, two because they could not climb as required, one with an unserviceable oxygen system, and one when its Gee navigation equipment caught fire. The others all returned to report dropping on the target markers and, despite these having been dropped long, the raid damaged some significant factories in the area. On the night of 4/5 October, ten crews were briefed for a 400-strong raid on Frankfurt. Sgt Evans' crew had to turn back as their navigator was losing consciousness. – in sharpening a pencil, he had inadvertently punctured his oxygen supply line! The remaining crews reported attacking

over accurately placed markers that led to an effective raid. Sqn Ldr Dean's aircraft was hit by flak over the city and by a 4lb incendiary bomb falling into the fuselage from an aircraft above. Luckily, it did not ignite and was promptly ejected through the flare chute. Further Gardening off the Danish coast took place on 7/8 October – one crew had to return, unable to get an accurate starting fix, but four others were successful. Sixteen crews briefed next day for a large raid on Hanover, but one had to cancel after the Engineer walked into a propeller when preparing to guide his pilot onto the perimeter track! Once airborne, two crews had early returns – one with engine failure and the other due to an inability to climb above 14,000 feet. Thirteen reached the target area that suffered a concentrated attack after accurate marking in good visibility. Plt Off Cameron's aircraft was hit by flak that damaged the undercarriage and flaps, leading to a crash landing at Melbourne that all aboard survived. Sgt Crothers aircraft also sustained flak damage over the target, losing the starboard outer engine. The Starboard inner then cut out some forty minutes later and all wing containers and bomb racks were jettisoned to slow down the rate at which height was being lost. After losing 3000 feet, the inner engine restarted and the crew landed safely at RAF Hardwick in Norfolk.

There was then a break from Ops for fully two weeks, a period around the full moon often beset by bad weather at base, with useful crew training exercises flown whenever possible. These now included 'Y' training cross-countries, a code understood to refer to the new H2S radar. The break ended on 22 October when nineteen crews were detailed for another large raid on Kassel. A mix of icing, technical faults and poor climb capability forced five crews to abandon the mission and return. Marking that night was accurate and eleven crews returned from what was assessed as a devastating raid. However, three crews did not return – those of Plt Off Plant, Flt Lt Wilkinson and Plt Off Heppell all came down near to Kassel, with only Fg Off Pyne, the Bomb Aimer of the first crew, surviving to become a POW. Back at Melbourne, mist and fog set in once again and that was the last Op for the month. (At some point in the month, the records are not specific, one of the Flight Commanders, Sqn Ldr J F Sutton AFC, was promoted and took over command from Wg Cdr Edmonds, who was posted out.)

The return to Ops on the afternoon and evening of 3 November was to prove eventful. The target was Dusseldorf, with eighteen Squadron crews detailed to join another raid of almost 600 aircraft. Two crews made early returns, one experiencing that now recurring inability to climb above 15,000 feet, and two were forced to jettison bombs when under attack. Flt Lt Trobe's aircraft was subjected to a series of attacks by four fighters, was repeatedly hit and three crewmen were injured. Considerable damage was inflicted – the rear turret, hydraulics, intercom, radio and other equipment were put out of action. Two engines failed but Sgt Bridge, the Flight Engineer, managed to restart one and then had to attend to an internal fire, assisted by Sgt Mowatt, the Mid Upper Gunner. Meanwhile, Sgt Bisby, the badly injured WOP, was working to repair the radio and was finally able to obtain a DF fix to help set course for home. The radio then failed again and Sgt Bisby, near to collapse, made another repair before, finally, a three-engine landing was made back at base. (All four were recommended for immediate awards, gazetted just a month later - a DFC for Flt Lt Trobe, the CGM(Flying) for Sgt Bisby, and the DFM for both Sgt Mowatt and Sgt Bridge.) At around the same time, Plt Off Dixon's aircraft had also been attacked by a Ju 88, and he too had to return on three engines. Flt Lt Harden's aircraft was brought down in the target area. He and three colleagues were killed, with a further three becoming POWs. Eleven crews were able to return with reports of attacks on markers. It is likely that the remaining crew, that of Plt Off Cameron, had also completed an attack but their aircraft crashed in Norfolk on its return. The Tail Gunner, Sgt Winstanley, was the only survivor but he too died of his injuries shortly afterwards.

On 11 November, nine crews were detailed for two different tasks – four were to join a raid on rail facilities at Cannes, on the main line into Italy, with five to go Gardening off the Frisians. At Cannes, the markers were attacked but the railway yards were not hit. Further north, three crews laid mines as required, whilst technical problems prevented the others from doing so. All in all, these last months had been hard but there were harder ones to come.

'The 'Ol Ram'. Apart from the bombs denoting the raids carried out by this aircraft 3 swastikas denote the fact that this aircraft's gunners shot down 3 enemy aircraft.

Crew of "The 'Ol Ram". Flt Lt J.T. Hewitt (Can) & Crew.

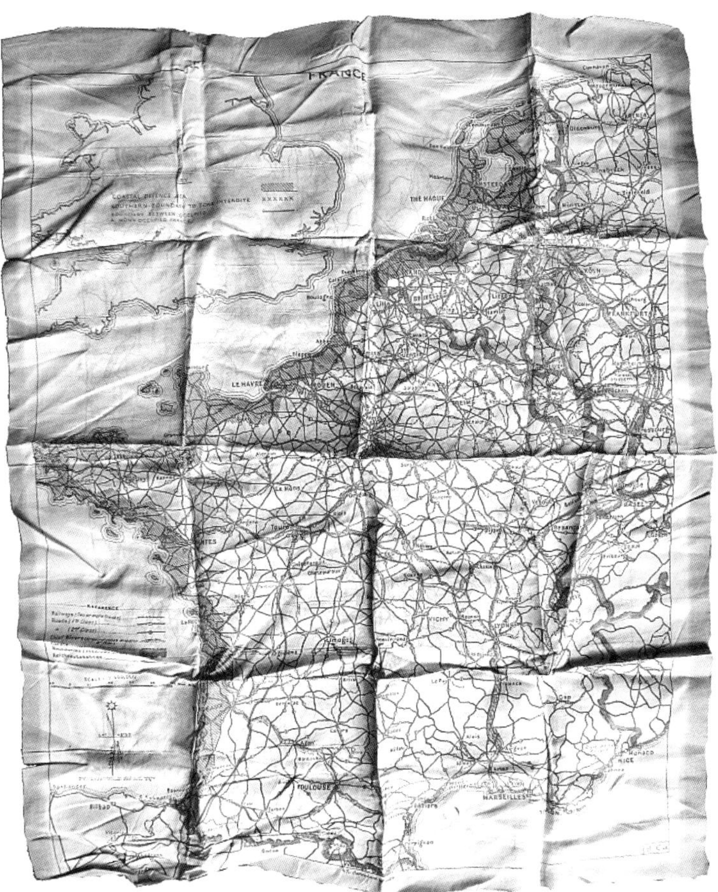

The silk 'escape' map of France issued to all crews. This one belonged to Flt Lt Henry Wood DFC.

Sgt Tom Thackray was a flight engineer on 10 Sqn from March to August 1943. Meeting wife Dorothy whilst undergoing a course at St Athan, they were married on 19 June 1943 and celebrated their platinum anniversary in 2013. Tom was later to edit the 10 Sqn Association Newsletter for 30 years until 2014.

Flt Lt Lawrence Henry Wood's DFC (Distinguished Flying Cross), a Melbourne Halifax pilot from 1944-45, who like so many, was given this award for his bravery.

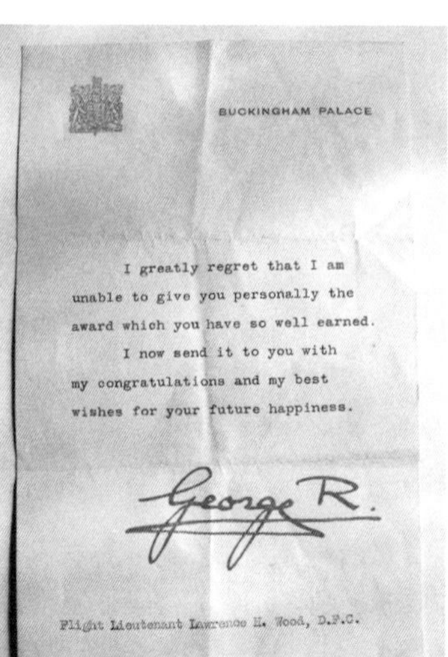

BUCKINGHAM PALACE

I greatly regret that I am unable to give you personally the award which you have so well earned.

I now send it to you with my congratulations and my best wishes for your future happiness.

George R.

Flight Lieutenant Lawrence H. Wood, D.F.C.

Flt Lt L.H. Wood's letter from HM King George VI

Sgt (W) Betty Allan, an MT driver at Melbourne in 1944 –worth coming home for !

The WAAFs of Melbourne – date unknown. One Halifax flight engineer used to find a pair of 'blackout's' (RAF –issue ladies bloomers) resting on the aircraft throttles before each flight. After a safe return they were always there again the next night.

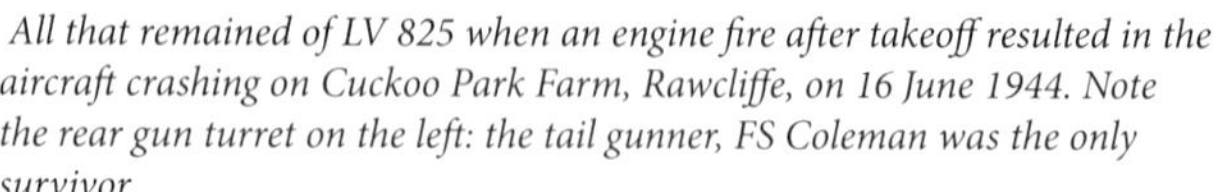

All that remained of LV 825 when an engine fire after takeoff resulted in the aircraft crashing on Cuckoo Park Farm, Rawcliffe, on 16 June 1944. Note the rear gun turret on the left: the tail gunner, FS Coleman was the only survivor.

So young ! Sadly not all crews were lost to enemy action. The crew of LV825, who apart from their tail gunner, all lost their lives in a crash at Rawcliffe near Goole. R: (L-R) Sgt R.A. Crawford - Flt Eng, Sgt C.R.B Lewington - WOP, Sgt C.A.R. Dummer - Mid Up Gnr, FS M.J. Coleman - Rr Gnr F: (L-R) Sgt R.F. Pearce - Nav, Plt Off N.C.C. Leitch RAAF -Pilot, FS W.J. McCarroll - Bmb aimer

Fg Off Fred "Nugget" Worker RNZAF Pilot, Fg Off Ken Stewart Nav, Fg Off Fred Wilkinson RNZAF Bomb-Aimer, FS Fred Tiller Mid-Up Gnr, FS Ian Rowlands RNZAF WOp, Unknown who was only with the crew for a few days, in the absence of their Canadian Flt Eng, and FS Bob Woollard Rear Gnr with their Ground Crew.

The Melbourne Band in which Doug Dent, the Association's founding chairman, played.

As an Air Cdre, 'Gus' Walker, was the much admired and respected commander of 'Pocklington Base' from 24 March 1943. Renamed No 42 Base in September 1943 it comprised the three stations at RAF Pocklington, Elvington and Melbourne.

Back to Melbourne, one of the few airfields to have FIDO which enable landings in foggy conditions. The acronym FIDO stayed in common use after its original 'Fog Investigation Dispersal Operation' trials.

An age-speckled photo of Charles Brignell, the wireless operator (rear left) on this crew. He went on to serve with the Squadron on Dakotas in India before becoming a surveyor after his demob.

CHAPTER 8

THE FINAL PUSH

From mid-November 1943 and over the winter months that followed, Sir Arthur Harris made the execution of a sustained assault on Germany and, in particular, the destruction of Berlin, his primary goal. The task was to prove difficult and costly as the Luftwaffe demonstrated that it could adapt to counter Bomber Command's improved tactics and equipment, with the lesson having to be learned that any aircraft device emitting radio energy enables others to home onto it. The Command's Stirling aircraft were soon withdrawn from the main battle as their loss rate became unacceptable and, in February 1944, the remaining Halifax Mk II and V aircraft, including those of 10 Squadron, were also withdrawn. However, before that and the Squadron rejoining the battle with the improved Halifax III, there was much work to be done.

The Berlin campaign began on the night of 18/19 November with a force of 440 Lancasters targeting the capital, whilst a very similarly sized mixed force executed a diversionary raid on Mannheim. 10 Squadron detailed eighteen crews for this. With a now statistical inevitability, one crew had to turn back when the aircraft's oxygen system failed, but sixteen were able to report bombing the primary target, and the use of H2S for run-in fixing is mentioned in a number of their debriefs. FS Lindsey's aircraft was shot down over France with only the Tail Gunner, Sgt Chambers, surviving as a POW. Sgt Doug Evans and his crew, who had joined the Squadron in August and would go on to complete a tour of Ops in May 1944, had what he believed to have been their worst experience that night. They were coned by searchlights just after 'Bombs Gone,' and he recounted what followed to Naomi Leathey on behalf of the Squadron Association in 2013:

"Only once did I think we had a serious encounter, and it was a very serious encounter indeed. We were suddenly coned. The intensity of the light was just unbelievable and the alertness of our gunners was to be commended because, almost immediately after turning onto course, I was alerted to "Corkscrew Starboard - Go!" Now it was the Air Gunners' aeroplane, not the Pilot's, and I was happy to comply and do as I was told. I throttled back because the aeroplane responds better at lower speed and I combined speed with descent whilst corkscrewing starboard and port and, very shortly after we started this manoeuvre, we were attacked by a FW 190. He was in as much danger as we were, I always thought, because he was also right in the midst of all this flak that was flying around. (I've spoken to a Luftwaffe fighter pilot since and, yes, he agreed that it was a dangerous situation for them but they had to do this as part of their job.) This chap, I think, thanks to our Gunners, did not live to tell the tale. I know that the fighters were very wary not to cross our tail, but this FW 190 came so very close to us from starboard to port that the Tail Gunner said that he could see the Pilot's silhouette in all the light that was surrounding him as he was pouring in the .303s from all his guns. I don't think that Pilot would have survived and the aeroplane appeared to turn over and go down. The tail gunner claimed a 'Probable' on that occasion. Meanwhile, we continued to be coned and we were not aware of a further attack - there may have been a further attack, it's difficult to tell. I was descending the whole time into darkness. In darkness by now, the cone gave up on us, possibly because of their low elevation, and we pulled out of the descent. We were clear of the cone around 7000 feet and my logbook says we levelled at about 5000 feet, having started the manoeuvre at 17,000 feet. We felt very lonely climbing back up all on our

Sgt Doug Evans and crew: Nick Nicholas, WOP; Bert Morley, Flt Eng; Doug Evans, Pilot; Les Duncan, Mid-Upper; Jimmy Manson RCAF, Bomb Aimer; Harry Hull, Tail Gunner; Sims Hamilton, Nav. Seen here around September 1943, the crew completed its Ops tour in May 1944. Doug went on to a long civilian flying career before becoming a founder member of the 10 Sqn Association.

own. Much alertness demanded of us all in case of further attacks, but we got home safely."

A further eleven crews were briefed for the following night's target, the IG Farben chemical factories in Leverkusen, near Cologne. Two crews abandoned the mission with radio, Gee and engine failures as the reasons and a third bombed an early, alternate target when engine trouble developed. There was complete cloud cover over the target, marking did not go well that night, and the outcome was a widely scattered raid. Conditions over England were not good for returning aircraft, with widespread fog, and two Squadron crews crashed. Plt Off Lucas' aircraft had lost an engine after being hit by flak. He attempted a landing at Ford in Sussex (where the Squadron had been wound-up in 1919) but overshot and crashed. He and three colleagues were injured and taken to hospital, with the others treated for shock. FS Holdsworth's aircraft hit a hangar on landing at Tangmere, with all seven airmen killed. Fog then prevented any flying at Melbourne on 20 November and the early hours of that day had seen the first operational use by 35 Squadron aircraft of FIDO (Fog Investigation and Dispersal Operation), the fuel-burning fog dispersal system that had been installed at their base, RAF Graveley in Cambridgeshire. The system would be installed at a number of airfields, including Melbourne.

Sixteen crews were detailed for the next major raid on Berlin – 764 aircraft, on the night of 22/23 November – later assessed as the most effective of the war. Two crews had early returns - one had an escape hatch blow open, and they could not get it to close, whilst the other's Gee navigational equipment failed. Twelve recovered to base to report having completed attacks, and two crews did not return. Fg Off Pont's crew was lost without trace, and Fg Off Hall's crew came down over Germany, with no survivors. Those were the last losses of the month as all crews got back from the next two raids. Seventeen crews left for Frankfurt on 25/26 November, with sixteen making it to the target, and eleven briefed for Stuttgart on the next night, with ten reporting successful attacks.

The next Op for which the Squadron was scheduled was to Leipzig, well into Germany, with 527 aircraft on 3/4 December. Sixteen crews were detailed, with two returning early – one had engine trouble and, as so often before, the second could not carry out a proper climb. It is likely that all of the others reached the target, bombing on markers for what was assessed as a highly effective raid. Most aircraft losses are believed to have occurred after the attack, including Plt Off Walker's crew, shot down by a fighter. Only two of the crew survived, Sgts Brock and McNeil, and both became POWs.

A lengthy break from Ops followed. Training sorties were possible from Melbourne on most days but, notably, no Halifaxes from Pathfinder or Main Force squadrons were used on the next large attack on Berlin on 16/17 December. However, they were finally called upon again on 20/21 December, when the target was Frankfurt once again for a mixed force of 650 aircraft. Twenty-one crews were detailed, with unserviceability preventing two departures. Fifteen crews reported attacking markers, backed up by visual and H2S fixes. However, Sgt O'Connor's crew appears to have made some navigational errors and reported having bombed markers over Mannheim, almost certainly some of those positioned for a diversionary raid. Three crews were lost that night, with only eight of the twenty-two crewmen surviving as POWs. The next Berlin raid involving Main Force Halifaxes came on 29/30 December, when 712 aircraft were mustered. 10 Squadron briefed twenty crews, and nineteen returned reporting successful bombing on flares over cloud, although the night was not without incident. FS Burcher's aircraft lost an engine to flak and, at the same time, was struck by two incendiaries falling from above and causing internal damage - a very similar incident to that experienced in October by Sqn Ldr Dean's crew. FS Hewitt's crew saw a Ju 88 attacking another Halifax above them on approach to the target – both Gunners fired at hit, scoring hits, and the aircraft was seen to fall away by the rest of the crew. A night fighter shot down FS Green's aircraft on its outbound leg over Holland and there were no survivors.

1944

The first Op of the new year did not occur until 6 January, but it is noteworthy that the first mention of *'Fishpond'* appears in the record of training flights on 1 January. *'Fishpond'* was a defensive aid developed from H2S when it was realised that returns from nearby aircraft could be seen on the radar and that, if these kept a steady position, they must come from other bombers in the stream. However, if a return was moving across the screen, positioned at the WOP's station, it was probably an approaching fighter. (In a 1996 letter for the Association Newsletter, ex-WOP Roy Stewart recalled a night when he saw many steady responses although no corresponding aircraft could be seen by the Gunners or anyone else. He was also dropping bundles of *Window* that night, and it was later assessed that it was these blooming as they dropped that he had been seeing as aircraft – a demonstration of *Window's* efficacy in confusing radar systems!)

There were two Gardening missions on 6 and 14 January involving seven and four crews respectively, with no losses, before Main Force Halifaxes were called upon again on 20 January. It seems reasonable to conclude that concern was building over the losses being incurred, in part on account of the Mk II's shortfall in performance when compared to that of the growing Lancaster force. Nonetheless, 264 Halifaxes, seventeen coming from 10 Squadron, were included in a 769-aircraft raid on Berlin on 20/21 January. Sixteen crews reported bombing on markers over total cloud cover, one lost both gyro and artificial horizon outbound and jettisoned its bombs by attacking a flak battery, whilst another got badly off track, jettisoned and came home. German fighters fared well that night and two Squadron crews did not return, both shot down over Germany. Two members of FS Arthur's crew survived as POWs but none of Plt Off Crothers' did so. Another large raid with over 600 aircraft was scheduled on the following night. This was to be the first major attack on Magdeburg, and the Squadron listed twenty crews to participate although, in the event, two crews were unable to take off. Sixteen reported attacking the markers, although it would appear that bombing overall was away from the city, and one attacked an alternate target once it became apparent that navigation errors had been made. It was another excellent night for the Luftwaffe's fighters, with the heaviest bomber losses to date. The Halifax force lost thirty-five aircraft, an unsustainable 15.6%, with one being that of Flt Lt Dixon, brought down near to Magdeburg with only two crewmen surviving as POWs.

The assault on Berlin continued – 515 Lancasters on the night of 27/28 January, and a mixed force of 677 aircraft on the following night. 10 Squadron detailed twenty-two crews for this raid. One crew failed to take off, six had to return after technical failures, and eleven returned to report attacks. Unfortunately, the Squadron incurred four of the twenty-six Halifax losses that night. Sgt O'Connor's aircraft crashed in Denmark, with all aboard killed, and three were lost close

to Berlin – with the crews of Fg Off Large, Sgt Ling and Flt Lt Kilsby, from which ten crewmen survived as POWs. In 1996, for an Association Newsletter, former Plt Off Don Shipley, the Tail Gunner in Flt Lt Kilsby's crew, recalled the graphic circumstances of his baling-out that night:

> "Unless you have ever been over a target area, you have not experienced the Pandemonium that it is, with the searchlights coming in all directions, the bombs exploding, the flak hitting the aircraft, bombers coming in all directions at all levels, enemy fighters coming in, and having to watch the flares laid by the Pathfinders. In all, Hell let loose .... We were going down fast. I could see one or two chutes opening, the flames coming back, the heat searing, the slipstream, being so strong, was frightening. I therefore was faced with it and I threw myself backwards to get out of this burning hulk. To my horror, I was caught up by my right leg, which was trapped, as was lectured to me many times, in the walls of the turret. This was a terrifying experience to me as, at that time, the flames and the slipstream were overpowering. Even though I clawed at the sides of the turret, I was excluded from gaining entrance to free my entrapped leg. I was faced with death .... The next thing I can recall was seeing enormous flames and my first thoughts were that I was in Hell. But, all of a sudden, silhouetted against the flames was a Gerry helmet and a Gerry soldier standing guard beside me.

(Don had come down on Tempelhof airfield, active with Ju 88s and, with other airmen captured that night, was taken to Frankfurt for interrogation and onwards movement to a prison camp.)

Six crews were sent on Air-Sea Rescue sweeps on 29 January, a larger number than usual for this activity and probably not unrelated to the number of aircraft missing from the previous night. There was then a lengthy break from Ops over Germany, during which a wide range of training was undertaken – cross countries by day and night, often dedicated to 'Y' or H2S radar training, more *Fishpond* training, gunnery and bombing practice. However, two minelaying Ops were flown – fourteen crews seeding near Kiel in the early hours of 2 February, and ten off the French coast on 4/5 February. The sorties were generally successful – two crews had problems on the first night and, on the next, one had a brush with a Me 110 that evasive action promptly brought to an end. The value of H2S and its ability to be used for accurate coastal fixing at the start of a minelaying run is clear from the debriefs for these nights, and re-emphasised by noting that the failure of the kit was the reason for one of the crews being unable to drop on 2 February.

On 15 February, eighteen crews were detailed for yet another attack on Berlin. This was to break records in many ways – with 891 aircraft involved, it was the largest force launched against the German capital and the largest non-1000 Bomber raid on any target with, inevitably, the heaviest tonnage of bombs dropped. Four crews had to return with a variety of technical problems, and thirteen reported crossing

FS Jack Walker and crew prior to them being shot down by an Me110 on 20 February 1944. All parachuted to safety and became POWs. Jack rejoined 10 Sqn flying VC10s in 1967.

the target area, where extensive damage was recorded. Forty-three aircraft were lost that night, including the Halifax flown by Fg Off Clark's crew, shot down over Berlin with none of the crew surviving.

Ops were called then cancelled on both 17 and 18 February before the next major raid was mounted on 19/20 February, when Leipzig was the target for 823 aircraft. Eighteen squadron crews were briefed. One failed to take off, one had to return early with engine trouble, and fourteen returned to report dropping on PFF markers over complete cloud cover. Fighters were with the bomber stream from the coast, so there were many sightings. FS Fenny's crew had three separate encounters prior to the target, took hits and survived; Sgt Pearson's crew had just one, and they also emerged uninjured. However, two crews were lost – FS Walker's crew survived as POWs, whilst only the Navigator, Sgt Murray, survived from FS Davenport's crew. He too became a POW. The losses that night amongst the Halifax force, some 15% of those crossing the enemy coast, were such that the Halifax II and Halifax V squadrons were withdrawn from operations over Germany. A full month would pass before 10 Squadron would return, by then equipped with the more capable Halifax III.

In the interim, there was serious Gardening to be done – four missions in five days. Five crews laid mines off St Nazaire on 21/22 February, and ten took off for the Kattegat on 24 February, but all were recalled about thirty minutes out as the forecast showed deteriorating weather in the area. Fifteen crews set off to replenish fields off the west coast of Denmark on the evening of 24 February – one returned with an unserviceable H2S set and another after losing the starboard outer engine. Eleven crews completed the mission, but one found that the bomb doors would not open "presumably due to Nickels jamming the hinges" and brought its mines back to base. The remaining crew was also unable to drop as planned as the marker flares in use had extinguished before they could get over the drop point so they, too, brought their mines home. The fourth mission of this series was on the night of 25/26 February, placing mines in Kiel Bay. Thirteen crews were briefed as part of a force of some 130 aircraft, itself part of the diversionary effort laid on that night in support of a main force attack on Augsburg. Engine and compass failures sent two home early and two were unable to drop, leaving nine to complete the mission.

With Ops over Germany temporarily ruled out, the Mk II aircraft could still be used against less heavily defended targets in France and that is exactly what happened once the airfield became useable after a bout of snow. On 2 March, a raid was planned on the SNCA aircraft factory at Meulan-les-Mureaux, on the northwest edge of Paris. The mission called for 117 aircraft and twenty Squadron crews were briefed. Fg Off Burgess' crew had to abandon the sortie after the pilot's escape hatch ("fastening previously checked") blew off on takeoff. Their recorded flight time was 3 hours 20 minutes, so they clearly pushed on until things on board became unsustainable before returning. A crew had to return with engine problems, and another failed to see any markers over the target. The Pathfinders' *Oboe* Mosquitoes marked accurately and the other seventeen crews bombed accordingly and the factory was seriously damaged. One aircraft had a brief brush with a FW 190 that ended when the Tail Gunner fired three bursts, and there were no losses amongst the force that night.

Eleven crews briefed for a Gardening mission off the French coast on the next night, 3/4 March. Two aircraft were unserviceable at takeoff time, but the other nine laid their mines successfully. The next Op was against railway yards at Trappes, southwest of Paris, on the night of 6/7 March, the first in a series of raids on communication and reinforcement infrastructure in preparation for the Invasion, just three months away. Nineteen crews were called – all got away (three flying aircraft of 77 Squadron) and all reached the well-marked target for what photography proved was a highly successful attack with no losses. On the next night, railway marshalling yards at Le Mans were the target. Thirteen Squadron crews were briefed for this relatively large raid involving 304 aircraft. The target was completely covered by cloud, but was marked. Nine crews bombed on the markers, but four saw no markers and could not drop – there was by now considerable sensitivity about the risk of killing French civilians during such raids, but that could not always be prevented. FS Pearson's aircraft had encounters with two different Ju 88s, and his gunners claimed one as destroyed and the other as damaged.

A significant event on 7 March was the delivery of the Squadron's first Halifax Mk III, and general handling sorties for pilots began on 10 March as other aircraft arrived. The deterioration in performance and handling of the essentially similar Mk II and Mk V Halifaxes had been largely a consequence of increased all-up weight as new equipment – eg H2S radar and engine exhaust shrouds - was fitted and, by this point in the war, they had become relatively under-powered compared to the newer Lancasters flying above them. The Bristol Hercules powered Mk III brought performance improvements in speed, height and rate of climb and its design included enlarged, squared fins to overcome a rudder stall tendency that, in certain circumstances, had been the cause of Halifax II crashes. More Mk IIIs (2127 in all) were built than of any other variant. (It is probably no coincidence that the records start to refer to a training exercise known as a 'loaded climb' during this period.)

By 22 March, the Squadron had sufficient new aircraft and trained crews for it to be included once again in Ops over Germany, and thirteen crews were briefed to join a force of 816 aircraft attacking Frankfurt. All took off but one had to abandon the mission over problems with both fuel and oxygen supply. Despite a diversionary raid drawing off many fighters, WO Aston's crew had to bomb an alternate target to the north of Frankfurt as they came under six repeated attacks from a Ju 88. His gunners responded and, after a ten second burst, their guns failed, possibly due to icing. The aircraft suffered a great deal of damage but nobody in the crew was injured and a safe landing was made back at base. The other eleven crews bombed on markers, checked with H2S, in what was a very successful raid. Berlin was the target once again for the night of 24/25 March, and fourteen Squadron crews were detailed to join almost 800 others for what was to be the last major raid on the city. At start-up, three magneto drops (Mag drops) and a fuel leak prevented four departures. A feature that night was a particularly strong northerly wind at altitude that had not been forecast and this caused the force to become scattered – one Squadron crew drifted some 55 miles south of track on the outbound leg and, having established that and made a correction, the Navigator calculated that they would be forty minutes late over the target, at which point they abandoned the mission and jettisoned their bombs on the way home. Nine crews reported bombing on the markers, and three of them also reported being blown well south of track on the return leg. The strong winds also resulted in markers drifting considerably and bomb results were scattered. Fighters did well that night and although Squadron crews reported many sightings, no significant encounters resulted.

The next Op was on the following night, 25/26 March, when two crews joined 190 others to fly the Squadron's last Mk II mission against marshalling yards at Aulnoye, near the border with Belgium. Both reported bombing on markers, but it appears that marking that night was inaccurate. Both crews were then detailed with sixteen others for a raid on Essen the following night, a return to the Ruhr that caught German fighter controllers off guard. That said, the Squadron lost two of the three Halifaxes lost to enemy action that night – Plt Off Wilson's crew was shot down over Germany with no survivors, and Plt Off Simmons' crew came down over Belgium. Five of the crew died but both Gunners, FS Corby and Sgt Hendry, survived and evaded capture.

The next Op, to Nuremberg, on the night of 30/31 March was controversial then and has remained so. The moon's phase would normally have prevented Ops being planned for that night but a forecast of protective high cloud cover persuaded Air Marshal Harris to go ahead. The route planned was almost straight and occasioned objections from Group Commanders and there was a report from a Met Flight Mosquito that the cloud cover was unlikely to be as forecast. Nonetheless, the raid went ahead and the Squadron briefed sixteen crews from a force of 795. Perhaps fortunately, three could not take off and there were four early returns. One of these – Flt Lt Burgess's crew – had a hydraulic pipe fractured in a night fighter attack and the bomb load was jettisoned. On approach to Melbourne, lowering the undercarriage effectively emptied the damaged hydraulic system, leaving nothing to power the brakes. The aircraft continued off the runway, through the perimeter fence and across a road to end up in a field, with all aboard unscathed. Everything that night worked in favour of the Luftwaffe and those crews able to press on came home to report very considerable fighter activity, with fighter flares being dropped from above to illuminate the bomber stream both into and out of the target

area. The outcome was the heaviest loss incurred by Bomber Command on any raid, 95 aircraft, or virtually 12% of the force despatched. In the circumstances, it can be no surprise that the Squadron lost a crew when FS Regan's aircraft was shot down on the way to the target, although three crewmen survived as POWs – Navigator, WO Norris RCAF; Bomb Aimer, Sgt Wilmott; and Flight Engineer, Sgt Lawes. The eight crews that completed the mission reported dropping on markers over thick cloud but, in truth, little damage was done within the city. The Battle of Berlin was at an end.

The first phase of an archaeological project to recover what remains of LV881, the Halifax III flown by FS Regan's crew that was shot down on the way to Nuremberg, was carried out during August 2014. The work is being overseen by archaeologists of the Federal State of Hesse, assisted by colleagues from the Netherlands and the University of Winchester. The aircraft came down on a hill outside the village of Hungen-Steinheim, where some items including the bomb sight and other items from the forward fuselage were found. Cleaning and conservation of a number of these has begun in Winchester and a second phase of excavation took place in September 2015, when a commemoration ceremony was held, attended by relatives, a 10 Sqn delegation, and German dignitaries.

After the heavy losses of the previous months over Germany, Bomber Command's focus now shifted to support of preparation for the Invasion, with 'strategic direction' being provided by General Eisenhower's planning staff. As a consequence, the targets chosen would often be rail facilities in France and Belgium and the Squadron had already recent experience of these. (Some were also chosen to support the deception plan designed to encourage the expectation that the Invasion would use the shortest channel crossing to the Pas de Calais.) Germany would not be forgotten, but it is unlikely that crews felt much regret over the change of emphasis.

On 1 April 1944, Wg Cdr Sutton was posted to HQ Bomber Command and replaced by Wg Cdr D S Radford DFC, AFC. On the same evening, three crews joined others for a minelaying Op off the Frisians that was carried out successfully in clear conditions, and with minimal opposition. Training on the new aircraft continued during the following days as weather permitted and Ops were called for two nights but later cancelled. Finally, sixteen crews were called for a raid by 239 aircraft on the rail yards at Lille on the night 9/10 April. All reported dropping on the Target Indicators and, whilst a good deal of damage was done to the rail facilities and rolling stock, a substantial number of bombs fell in French civilian areas, with some 450 deaths

Fg Off Tubby Lawrence, the Sigs Leader, presenting the Adjutant, Flt Lt Paddy Keenan, with a 'Flying Adjutant' badge that used a quill pen as the wing.

VE Day - Job Done, but at a cost.

resulting. Despite the new navaids and PFF marking, precision bombing at night was still a real challenge. Nonetheless, a similar raid followed on the next night, when twenty crews briefed for an exclusively Halifax raid on the yards at Tergnier. A crew could not raise the undercarriage after takeoff and had to jettison and return, but eighteen returned to report having dropped successfully on PFF Indicators. Flt Lt Barnes' aircraft was one of ten Halifaxes lost on the raid – he and his tail Gunner died, three became POWs and three evaded successfully. A return to Gardening in the Kattegat followed on the next night, when eleven Squadron crews were tasked. One had to turn back as its H2S failed, and the others completed the mission against mild opposition. A break from Ops followed from 12 to 17 April and, on 18/19 April, it was back to Tergnier for thirteen crews joining 158 others. All of the Squadron's contribution got away and returned safely, reporting bomb releases on target markers in good conditions, under instructions from the Master Bomber. Six crews also went Gardening in Kiel Bay that night – one had a partial hangup and another had a tussle with a Ju 88 on the outbound leg. It scored some hits but was fought off by the Tail Gunner and claimed as 'damaged.' Two nights later, on 20/21 April, the target for almost 200 aircraft was the marshalling yards at Ottignies, southeast of Brussels. Twenty-one crews were briefed for this and all made it out and back to report a successful raid.

The night of 22/23 April saw a return to the Ruhr, when 596 aircraft were mustered for a renewed attack on Dusseldorf. Twenty crews were detailed and all got away. Nineteen returned to report bombs dropped on the markers, a couple of night fighter encounters and many searchlight cones in operation. The aircraft of Australian 'A' Flight Commander, Sqn Ldr Trobe, was shot down but he was able to make a forced landing in the most southerly Dutch province of Limburg, after some crewmembers had baled-out. He and five others were able to evade capture, whilst both Gunners became POWs. On the following night, thirteen crews were briefed for a relatively large (114 aircraft) Gardening mission in five areas of the Baltic. The Squadron's target area was at the western end, near Kiel, and an unusual aspect was that six crews, all H2S equipped, were tasked to drop flares with their mines to provide markers as a datum for a 189 second timed run for the remaining crews not so equipped. The weather in the area was good and, from the debriefs, this procedure appears to have worked well and all thirteen crews completed the mission.

The next two Ops were again back in Germany. On the night of 24/25 April, a force of 637 aircraft was tasked against Karlsruhe, and the Squadron detailed nineteen crews. One crew was unable to take off when its radio was found to be unserviceable, and there were no early returns. The outbound route took crews into an area of very active lightning activity about an hour before the target and strong winds at altitude drove the Pathfinders north so, despite crews using the markers, a very scattered raid was the outcome. WO Murtha reported attacking Mannheim, some 35 miles north of Karlsruhe, due to compass inaccuracies after a lightning strike, and it appears that his crew was but one of one hundred later reported as bombing the town that night. Essen was once again the target on 26/27 April – nineteen crews were briefed to join a 493-strong mixed force, but technical issues stopped two crews getting away. Sixteen returned to report attacking on markers, in what was assessed to be an accurate raid. The only Halifax lost that night was that of Flt Lt Allen's crew. They were brought down on their homebound leg by a combination of flak and fighter fire after crossing into Belgium. Two from the crew survived – FS Cooper, WOP, became a POW but Flight Engineer, Sgt Weare, was able to evade capture.

The next Op, on the night of 27/28 April, targeted the marshalling yards at Aulnoye once again, and eighteen Squadron crews were detailed. Things got off to a bad start when Plt Off Burcher's aircraft swung off the runway on takeoff and broke its back. Although none of the crew were injured, this incident stopped two further takeoffs. Flt Lt Kennedy had a hydraulics failure and was unable to raise his flaps after takeoff but, after going out to sea to jettison his bombs, he was able to land back safely. The other fourteen crews reached the target for what proved to be an accurate, concentrated attack and on the way home, WO Murtha's aircraft had a nine-minute encounter with a Do 217 that attacked from ahead and below with its tail guns, such that his gunners were unable to engage it. Some hits were taken and damage caused, but there were no injuries.

For the next month or so, and with just one exception, Ops for the Squadron became a mix of Gardening and attacks on railyards, often on the same nights, and with the significant addition of coastal gun battery sites as targets. Three crews went minelaying on the night of 30 April/1 May, and six on the following night – all using H2S fixing. At the same time on the night of 1/2 May, ten crews crews were part of a raid on the marshalling yards at Mechelen (Malines), midway between Brussels and Antwerp with the only loss from the force that night being Plt Off Lassey's aircraft. It crashed between Brussels and Mechelen with the loss of all onboard, apart from the WOP, Sgt Burton, who succeeded in evading capture. There was then a break from Ops until 6 May, during which a good deal of operational training is recorded – including more initial handling sorties on the Mk III as seven new crews had been posted-in during the last days of April. (This was entirely normal in that period and it is sobering to reflect that the supply of crews in the training system was such that significant operational losses could be made good within days.)

The mix of targets continued with a medium-sized raid on rail installations at Mantes-La-Jolie on 6/7 May, for which the Squadron briefed fourteen crews. All completed the mission, bombing on markers as controlled by the Master Bomber and with burst sightings confirmed in good visibility. Two aircraft had relatively minor skirmishes with fighters on the return leg. After that, another successful Gardening mission was flown by eight crews on the night 8/9 May. On the following night, an attack on a number of coastal batteries was undertaken – 10 Squadron's thirteen aircraft were allocated the battery at Berneval on the Normandy coast, well east of the eventual Invasion beach area. (The battery was, however, close to Dieppe and, on 19 August 1942, it had been the target of a gallant but unsuccessful assault by men of 3 Commando in support of the main Dieppe raid that day.) All of the crews completed the mission, with many reporting bomb bursts in the target area despite scattered PFF marking.

On the night 10/11 May, five railway targets were selected and the Squadron briefed twelve crews to join the raid on installations at Lens, in northern France. This raid went well, with all aircraft reaching the target. However, after bombing Flt Lt Keltie's aircraft had a collision with another Halifax. It suddenly appeared from below and, despite Flt Lt Keltie's reflex avoiding action, damaged both his fuselage and port wing. Control was maintained and a safe landing made back at Melbourne. Four crews had also been sent on a successful Gardening mission that night. On the next night, eight crews set off as part of a raid on another coastal gun battery at Trouville, close to the area chosen for the Invasion beaches. One crew arrived late after all markers had extinguished and could not attack as its bombs hung up; others were a little less late and continued as best they could. Badly forecast winds appear to have been the main cause ofor the late arrivals. The next Op before a break came on 12/13 May, when eleven crews briefed to join 100 others in an attack on rail facilities at Hasselt, in eastern Belgium. They all reached the target and, despite haze cover, reported bomb bursts in the area but it appears that most bombs fell away from the railyards. However, one fell onto Fg Off Murray's aircraft from above and hit the port inner engine, causing major damage. The propeller worked forward on its shaft, fouled the port outer and wrapped itself around the engine cowling. Some eight inches were knocked off the port outer prop tips but the Captain was able to maintain control and land safely at base.

Training at base filled the days until the next call on 19 May, when there were two. Twelve crews were detailed for Gardening that night off the Brest Peninsula and an additional twelve for an attack on railyards at Boulogne. There were no significant unserviceabilities and all of the crews completed their missions, with the bombing in Boulogne reported and assessed as accurate. Further Gardening was undertaken on the nights of 22/23 and 23/24 May. Twelve crews were detailed for the first night when, apart from Flt Lt Evans' aircraft losing an engine immediately after takeoff and having to land back, all crews laid mines successfully off St Nazaire. On the following night, the 'garden' was in the German Bight – ten crews were briefed, one failed to take off, and the others carried on successfully. For the first time, three targets were allocated for the next night, 24/25 May. Nine crews joined a raid on rail installations at Aachen requiring over 400 aircraft; five joined a group attacking coastal batteries, and six went Gardening. The Aachen raid was generally successful, but WO Blackford's aircraft was hit

by flak and crashed in Holland - all seven crewmen baled out and became POWs. The battery allocated was at Colline Beaumont, south of Boulogne and, whilst haze prevented a full assessment, bomb bursts in the target area were seen. The six crews allocated to the Gardening group all reported mines dropped off Lorient, as briefed. Eight crews did more successful Gardening off St Nazaire on the night of 26/27 and on the following night, two targets were allocated. A group of ten crews was briefed for what was a very accurate attack on the Army camp and gun site at Bourg Leopold deep in Belgium, whilst three crews did further Gardening off Lorient. Training filled the rest of the month as, after standby on 30 and 31 May, Ops for both nights were cancelled, thus closing a busy month.

Moving into June, it may be worth recalling that D-Day was planned to be on 5 June 1944; that the Invasion was postponed on account of weather; and that the forecast for 6 June provided the brief window of opportunity on which the order to proceed was given. As just one part of the preparation for the day, a Halifax attack on the radio/radar station at Ferme D'Urville was mounted on the night of 1/2 June, for which the Squadron detailed twenty three crews, almost a quarter of the force assembled. Many bomb bursts were seen through extensive cloud in the marked area but, unfortunately, not much damage was done. (The site was marked very accurately two nights later and put out of action.) On the next night, twelve crews joined a raid on the rail installations at Trappes. This went reasonably well, but the raid incurred a high loss rate that included two Squadron crews shot down by fighters. (Sgt Albert Fickling, the Bomb Aimer on FS Bowmer's crew, later recalled:

> "As soon as we crossed the French coast there was very heavy night fighter activity, and we ran out of ammo (some 12,000 rounds) twenty minutes from the target. The only remaining defence was evasive action.....We got home by aerobatics alone. I sat with Pete Ward (Navigator) on the way back – we saw lots of aircraft shot out of the sky."

Sgt Hunter, Sgt Kumar's WOP, was the sole survivor from his crew and he managed to evade capture thanks, initially, to help given by a Mme Moreau and other villagers of Raizeux. (The aircraft crashed in a meadow close to the home of an uncle of M Bernard Leroy, a boy at the time. He believes that the official burial details for one or two members of the crew may be in error and, at the time of writing, his local investigations continue.) Fg Off Murray's aircraft was also shot down – he and his WOP, WO Williams, both died, but the others baled out safely to become POWs, apart from Sgt Hallett, the Mid Upper, who managed to escape. Five crews also dropped mines off the Hook of Holland that same night. On the following night, ten crews were detailed for more minelaying, some in the same area and others further south, off the Brest peninsula.

There is a particularly significant story behind the fate that befell two members of Fg Off Murray's crew on the night of 2/3 June – navigator, Fg Off Stan Booker, and flight engineer, Stan Ossleton. After baling out, they were initially picked up by a Resistance group but were later betrayed to the Gestapo by a collaborator, Jacques Desoubrie. No longer in uniform, they were not taken into the customary POW processing system but went instead to the large prison at Fresnes, to the south of Paris, where the Gestapo was holding captured Resistance personnel and other 'undesirables.' They were held in close confinement, in poor conditions. As the Allied advance got ever closer to Paris, on 15 August 1944 the Gestapo emptied the prison to move all its inmates into Germany. Prisoners were able to associate with one another during this process and it became clear that a total of 168 Allied airmen had been held in Fresnes, in contravention of the Geneva Conventions. These were survivors from both bomber and fighter operations over France, with over half being American, whilst the remainder were from the RAF, RCAF, RAAF and RNZAF. Most are believed to have been betrayed by the same individual, tried and condemned to death in 1949.

As bad as conditions had been in Fresnes, they were about to be much worse. After being bussed across Paris, the prisoners were herded into rail boxcars, upwards of ninety at a time, for transportation to the concentration camp at Buchenwald, deep in Germany, near Weimar. Badly delayed by bomb damage to the French and German rail networks, this journey took five days. On arrival, they were stripped, deloused and shaved and, for two weeks or so, lived in the open air. The conditions were almost unimaginable – one man spoke of being "surrounded by death and by the smell of death" – although survivors believe that they were not as bad as at Auschwitz or Belsen. The airmen grouped together to become as far as possible a formed military unit, under command of Lancaster pilot, Sqn Ldr Phil Lamason, RNZAF, a process that served them well and caused great annoyance to their captors. It emerged that there were Russian and other underground groups at work in the camp and it became possible to establish contact with the Luftwaffe on the outside. It is still unclear how these contacts worked but, on 20 October, the airmen were moved out to Stalag Luft 3, where they finally achieved proper POW status. Two airmen had died of illness during their captivity, and it later emerged that all had been scheduled for execution only days after their release. Post-war, the men met a degree of official unwillingness to believe or publicise their stories. In 1999, Stan Booker was joint author with Arthur Kinnis of "168 jump into hell," a detailed account of their captivity and surrounding circumstances, published in Canada.

Jacques Desoubrie, the man who betrayed most of the 168 who went to Buchenwald, would no doubt have received his 10,000 Francs.

Bekanntmachung

Jede maennliche Person, die notge landete oder durch Fallschirmabsprung gerettete feindliche Flugzeugbesatzungen direkt oder indirekt unterstützt, ihnen zur Flucht verhilft, sie verbirgt oder ihnen sonstwie behilflich ist, wird sofort standrechtlich erschossen.

Frauen die derartige Unterstützungen leisten, werden in Konzentrationslager nach Deutschland abgeführt.

Personen, die notgelandete Flugzeugbesatzungen oder Fallschirmabspringer sicherstellen oder durch ihr Verhalten zur Sicherstellung beitragen, erhalten eine Belohnung bis zu **10.000** fr. In besonderen Faellen wird die Belohnung noch erhoeht.

AVIS

Toute personne du sexe masculin, qui aiderait directement ou indirectement les equipages d'avions ennemis descendus en parachutes ou ayant fait un atterrissage force, qui favoriserait leur fuite, les cacherait, ou leur viendrait en aide, de quelque façon que ce soit, ***sera fusillé*** sur le champ.

Les femmes qui se rendraient coupables du même delit, seront envoyées dans des camps de concentration situes en Allemagne.

Les personnes qui s'empareront d'equipages, contrains à atterrir ou de parachutistes ou qui auront contribue par leur attitude a leur capture, recevront une prime pouvant aller jusqu'à **10.000** francs. Dans certains cas particuliers cette récompense sera encore augmentée.

Paris, den 22 September 1941

Der Militärbefehlshaber in Frankreich

VON STUELPNAGEL,

General der Infanterie.

German Poster warning French civilians what to expect if found helping Allied Airmen: males to be shot, women sent to concentration camps: 10.000Fr reward for informers.

On the night of 5/6 June, over one thousand aircraft were launched against coastal batteries in Normandy. The Squadron sent twenty-three as part of the attack on the guns at Mont Fleury. There was considerable cloud cover, but twenty-two crews attacked on markers, backed by H2S and Gee fixing – the other aircraft was in a cloud layer and the crew could see no markers at all. The immediate priority after the landings on the morning of 6 June was to disrupt rail and road routes that might bring German reinforcements to Normandy. Accordingly, over 1000 bombers were deployed on the night of 6/7 June against a range of such targets and the Squadron detailed twenty-two crews for an attack on troop concentrations at St Lô, an important town on the line of advance for US forces that would suffer badly in July. The attacks were made at relatively low heights from 2000 to 3000 feet, but still using markers over thin cloud. One crew could see no markers as its ETA approached and went on to jettison most of its bomb load.

On subsequent nights, the Squadron was sending crews on Gardening sorties whilst others continued the assault on communications targets. On 7/8 June, six out of seven crews laid mines off the mouth of the Loire and St Nazaire, with an electrical fault preventing a release in the seventh aircraft. At the same time, twelve crews joined an attack on rail targets at Juvisy, south of Paris. One had to abandon the mission on losing an engine, whilst the others were able to release at around 5000 feet on the Master Bomber's instructions in what was assessed as a more accurate raid than that at St Lô. Two crews claimed German fighters as destroyed after brief encounters on the return leg. (There was no Op on the next night, but Fg Off Leitch crashed whilst attempting to land at base in poor weather, after a night cross-country exercise. The crew were unscathed but the aircraft was written off.) There was more mixed targeting on the night 9/10 June. Ten crews were sent Gardening at two locations coded Beeches and Artichokes off the Brest peninsula. All dropped as briefed using H2S fixing to establish datum points and those flying closest to Lorient reported moderate but accurate flak. The interdiction targets that night were airfields that might be used for reinforcements and the Squadron briefed thirteen crews for an attack on the field at Laval, some sixty

miles behind the Invasion beaches. Nine crews reported completing attacks using the Master Bomber's instructions over thick cloud cover, including Flt Lt Burcher's crew who had pushed on after losing an engine over the Thames Estuary. Two crews did not bomb for, whilst they could hear the Master Bomber, they could see no markers. Two crews did not return, the only Halifax crews lost that night. Plt Off Henderson and Fg Off van Stockum's crews were shot down near to Laval, with no survivors. (Willem Jacob van Stockum was Dutch, with a significant academic background that included early work in the field of general relativity.) The last night with mixed targets in this period was on 12/13 June. Nineteen crews were detailed for a raid on rail installations at Amiens. All returned to report bombs dropped on markers, as instructed by the Master Bomber. FS Bowmer's gunners claimed two German fighters as downed from separate encounters. Fg Off Rosen's crew also claimed a fighter as destroyed, one of a pair – the other was seen to attack and shoot down another Halifax. (There were no Squadron losses from that raid.) Five crews also went Gardening that night off the Brest peninsula – four dropped as briefed and the fifth had to abandon the mission with a failed H2S set.

Some two hours or so after those aircraft headed back over the Channel towards Melbourne, another airborne machine followed. Launched from a site in the Pas-de-Calais, this was the first V1 flying bomb to land in London, in Grove Road, E4. Six people were killed as houses were demolished and a further 6000 would follow, with some 18000 badly injured, before all of the launch sites were destroyed or captured in the Allied advance. These sites would be targeted in the days ahead, but there were still raids to be flown in support of the Invasion.

On the night of 14/15 June, over 300 aircraft, mainly Halifaxes, were sent on attacks against rail targets, and the Squadron detailed twenty crews for the raid on the rail depot at Douai. One crew had to abandon the mission after losing an engine, but all of the others reported bombing on markers as directed by the Master Bomber whose Lancaster was later shot down. Within minutes of bomb release, Sgt Bond was descending through cloud and collided with a FW 190. His aircraft lost about five feet of the port wing, sheered off by the impact, and the fighter was seen to go down in flames and hit the ground. He was later able to make a safe landing at RAF Woodbridge, in Suffolk, one of three airfields built with longer and wider runways to help returning aircraft with control problems. On the next night, eighteen crews were detailed as part of a raid on German ammunition storage at Fouillard, near Rennes, inland from the beach areas. All returned, reporting attacks completed over cloud cover, as instructed by the Master Bomber.

Responding to the V1 threat that had so recently been made real, a new campaign against their launch sites was initiated, with four chosen as targets on the night of 16/17 June, and eleven Squadron crews were detailed for a raid on the site at Domleger, referred to in the unit record as 'Enemy constructional works.' Unfortunately, Plt Off Leitch's aircraft had a fire in the starboard outer engine some thirty minutes after takeoff. This proved uncontrollable and the aircraft crashed at Rawcliffe, with only FS Coleman, the Tail Gunner, managing to get away before the bombs on board exploded. The other crews reached the target area, covered by cloud, and those who saw them bombed on markers whilst the others used the glow of fires as an aiming point. (It may be simply a matter of a new compiler having taken over the recording task, but entries around this time make clear that an active training programme was being run in parallel with whatever Ops were tasked. That was probably the case before, but it was not recorded in the same detail.) For some days afterwards, the weather over Northern France stopped most Bomber command activity but, on 19 June, thirteen crews took off with Domleger as the target once again, only to be recalled by HQ 4 Group after being airborne for around thirty-five minutes. Crews were placed on Standby each day but it was not till the night of 22/23 June that another Op was considered feasible. Twenty crews were briefed that afternoon as part of a force to attack the marshalling yards at Laon, in Picardy. The weather there was clear and crews reported successful attacks and active defences, with two also claiming night fighters destroyed.

The next Ops were flown in daylight, no doubt with the need for accuracy taking precedence over other considerations. On 24 June, twenty-two crews were detailed as part of an attack on a V1 storage depot at Noyelle-en-Chaussée. (The Clerk who typed the F541s entered 'Doodle

Bug Depot' only to have that crossed out and the less demotic 'Flying Bomb' written above!) A hydraulics failure prevented one crew from raising flaps and undercarriage after takeoff, so they had no choice but to jettison and return to base. The others reached the target, readily visible in clear conditions, and bombed as instructed in a successful raid. Another daylight raid followed on the next morning, when the target was a V1 depot at Montorgueil, with twenty-one crews detailed for what was another worthwhile raid. Regrettably, Fg Off Rosen's crew returned to report that they had seen two of their bombs hit another Halifax beneath them. That aircraft blew up, with pieces hitting a second, causing both to crash with three parachutes seen. (W R Chorley's record for that day has two aircraft from 77 and 102 Squadrons colliding, with no survivors.) After this incident, it was decided that 4 Group aircraft should fly in an open formation with an experienced pilot leading.

On the night of 27/28 June there was a return to mixed targeting when eight crews were sent on a successful Gardening sortie in the approaches to the port of Brest, whilst fifteen were detailed for a raid on another V1 site at Mont Gandon. One crew had to abandon the Gardening mission when an engine fire could not be extinguished. Height was lost and the crew prepared for ditching before the fire went out and a safe recovery made. All of the crews on the Bombing mission returned to report bombing on markers in good visibility, with no opposition. The final Op in June was flown on the night of 28/29, when nineteen crews were scheduled as part of an attack on marshalling yards at Blainville. Seventeen returned to report a successful raid in good conditions, with most using markers and Master Bomber instructions. Flt Lt Hewitt's crew debriefed on three separate encounters with enemy fighters, from which they claimed a Me 210 as destroyed, and three other crews also made 'Destroyed' claims after a number of encounters. However, two crews were less fortunate and did not make it home. Plt Off Livesey's aircraft was shot down, with no survivors amongst the crew. Fg Off Taylor's aircraft crashed in France, but with only the Tail Gunner, Sgt Cuffey, killed. Ralph Taylor made contact with the Resistance, only to be betrayed by the notorious collaborator, Jacques Desoubrie. Like Stan Booker and Stan Ossleton who baled out earlier in June, he was improperly taken to Fresnes prison and became another of the 168 airmen later transported to Buchenwald. Two colleagues became POWs and the three others evaded capture.

The campaign against V1 launch and supply sites continued into July and provided the targets for Squadron crews for the first half of the month. Raids were flown against the site at St Martin L'Hortier in Normandy. The first was flown on the afternoon of 1 July, when seventeen crews were tasked. Cloud covered the site, several crews saw no markers, and not all heard the Master Bomber, so it is likely that the target was only partially damaged – thus leading to the return raids on subsequent days. Fg Off Rosen's aircraft was hit by predicted flak and crashed, with only Sgt McKinnon, the Tail Gunner, surviving. He evaded successfully and was reported as safely back in the UK in October. Writing in 1994, Fg Off J Hullah recalled seeing Fg Off Rosen "raise his hand as in farewell as he lost speed and height, trailing an ominous cloud of greasy smoke."

The first return raid was on the afternoon of 4 July, when twenty crews joined an attack on the site. Weather conditions were better, markers were seen, some crews could bomb visually and good concentrations of bursts were observed. The third raid was in the early hours of 6 July, with nineteen crews engaged and good concentrations reported once again. That evening, whilst it was still light, twenty crews were detailed for a raid on a site at Croixdalle, also in Normandy. The bombing was reported as well concentrated and six aircraft came home with some degree of flak damage. Attacks on other sites followed. Six crews had gone Gardening off the Frisians in the early hours of 12 July and, later that evening, twenty others took off for a V1 depot at Thiverny, north of Paris. One had to return early but the others continued to the target area, largely covered by cloud. The reports are variable as regards markers seen and the audibility of the Master Bomber's instructions, and several suggest that the bombing was rather scattered. The next Op was on the night of 15/16 July, when the target for sixteen crews was a V1 supply site at Nucourt, to the northwest of Paris. Technical problems sent two home early, but the remainder completed a successful attack. Twenty-two crews briefed for a return to the site at Mont Gandon on the night of 17/18 July and all returned to report what

appeared to have been another successful raid.

There was an attack on rail yards at Vaires on the afternoon of 18 July in which nineteen crews participated but, earlier that morning, six crews were detailed as part of the 900+ Bomber Command element of a joint RAF/USAAF operation in support of Operation GOODWOOD. The Allied advance through Normandy had been slowed-down and General Montgomery needed to break out of the area facing Caen held by British forces and move more deeply into France. GOODWOOD was devised as a mass armoured assault that would be preceded by an enormous aerial bombardment, probably the first use of heavy bombers in an essentially tactical role. The aircraft went over in a continuous 45-minute stream in good visibility, using the Master Bomber's guidance, and significantly reduced the ability of German Divisions to participate in the day's action.

After a break of some three months' duration, Squadron crews returned to Germany on the night of 20/21 July, when twenty-four crews were detailed to join a raid on the synthetic oil plant at Bottrop. Engine problems forced two crews to turn back but twenty went on to bomb through cloud and fairly heavy flak. Two crews did not come home. Fg Off Hadley's crew was brought down over Germany with only Sgt Grant, the Mid Upper, surviving as a POW. Fg Off Bond's aircraft was hit by flak, setting the starboard wing fuel tanks on fire, and he had to make a forced landing at Wijk en Aalburg in Holland. Sgt Tough, the Tail Gunner, was killed by a flak splinter but his colleagues survived as POWs. Germany remained the focus for the next two Ops. On 23/24 July, twenty crews were detailed as part of a force of 629 aircraft targeting the port area at Kiel. All of the Squadron crews reached the target and all bombed successfully using the PFF markers in what was a particularly damaging raid for the city and dock areas. For the following night, twenty-one crews were briefed for another raid with over 600 aircraft on the city of Stuttgart. Two crews had to abandon the mission and return with technical problems, but the others reached the target and bombed on the markers as instructed for what proved to be another very effective raid. Two further large raids on the city followed, but 10 Squadron found itself tasked once again with V1-related targets. On 25/26 July, nineteen crews took part in an attack on a site at Ferfay in the Pas-de-Calais, not too far from the Squadron's first WW1 airfield at Chocques. A fault meant that one crew could not release its bombs, but all of the others reported having done so on the markers, or as directed by the Master Bomber. Sixteen crews were listed for the force sent to attack the V1 storage site at Forêt de Nieppe on 28/29 July. One had to return when their radio failed, but the others were able to continue and to bomb the target in good visibility. A successful follow-up attack was flown just a few hours later by a force including eight Squadron crews.

This busy period continued into August with twenty-four crews detailed for an attack on another V1 site at Prouville, again in the Pas-de-Calais. 777 aircraft had been listed for attacks on launch sites that afternoon but the weather conditions were such that only 79 were able, or permitted, to do so. A Squadron crew had to abandon the mission with engine trouble and, a little over two hours after takeoff, the Squadron contingent was ordered back to base by the Master Bomber. Around noon on 3 August, twenty-five crews set off for a V1 storage site in the Bois de Cassan, one of three sites targeted that afternoon by over a thousand aircraft. The weather was much improved and a concentrated bomb pattern was achieved, with later crews effectively aiming into the smoke raised by earlier bombs. A short break for most crews followed on 4 August when just seven were called for a Gardening mission in the approaches to Brest. Six were able to complete the task whilst the other crew had to abandon it as its H2S radar, essential for establishing an accurate datum for the dropping run, became unserviceable. There was then a return to maximum effort on 5 August, when twenty crews were detailed for another raid on the V1 storage site in Forêt de Nieppe. There was one early return after an engine failed, but the other crews found the target in good conditions and bombed as instructed. (The unit record also notes that, with effect from that day, all ground personnel on Squadron strength, other than the Adjutant, were posted to No 44 Base. A first separation of sorts had occurred in November 1943 when ground crew had become part of No 9010 Servicing Echelon and, all in all, this looks very much like an early version of the 'centralised servicing' that would be familiar in post-war years. It is unlikely, however, that this admin change had much effect on the daily rhythm of those working on the dispersals around Melbourne.)

On the night of 7/8 August, Bomber Command tasked over 1000 aircraft to attack a number of areas in front of British positions in Normandy. The need for accuracy was vital and the waves of bombers were strictly controlled. In the end, only 660 bombed and, after being airborne for two hours en-route to a position coded TOTALISE 3 at May-sur-Orne, south of Caen, the twenty Squadron crews were instructed by the Master Bomber to return to base. After a quick rest, five of those crews, plus ten others, were detailed for an attack on a V1 site at St Philibert Ferme, again in the Pas-de-Calais, later that evening. FS Thorne's aircraft suffered a control failure on takeoff, swung off the runway and was severely damaged, albeit that the crew emerged uninjured. The other crews found the target in fine weather conditions and all reported having bombed successfully, with most going on comment on the amount of flak encountered towards the coast on the way home. Seven crews reported hits, with one having to shut down an engine as a result. Another daylight raid was planned for the next day, 9 August, when the target for 160 aircraft was a fuel dump in the Forêt de Mormal, near the Belgian border. The Squadron briefed twenty-one crews for this mission, one of which abandoned the task after losing an engine. However, the others continued to the target where visibility was good and the bombing was seen to be well concentrated.

Twenty-one crews were detailed for the next Op, a return to night interdiction on 10/11 August, where the targets were a rail junction and marshalling yards in Dijon. One crew was late as a result of a navigational error and had to abandon the mission, but all of the others reached the target where considerable damage was done. In the evening of 11 August, ten crews left to attack a V1 Site at Le Nieppe, where several crews reported seeing the launch ramp as they attacked. Shortly afterwards, Fg Off Peacock's aircraft was damaged by flak and his Bomb Aimer was injured. The port inner engine was badly shot up and the crew had prepared to ditch when the propeller sheared off, and a three-engine recovery became possible.

On the night of 12/13 August, eleven crews made a return to Germany as part of an experimental raid on Brunswick. There were no Pathfinders or markers and each crew was instructed to use its H2S radar to establish a release point. Ten crews returned with debriefs not greatly different from those seen over the summer to date, but the official assessment was that the raid had not been a success and it emerged that towns up to twenty miles away had been misidentified and attacked in error. Fg Off Saynor's aircraft is thought to have been fired on by another Halifax soon after crossing the German coast, an incident that caused a fire in the port outer engine and a fuel tank. The crew had to bale out before the aircraft crashed and all became POWs. A cruel irony is that they had completed over 30 missions and were likely to have been screened from further Ops on return to Melbourne.

Shortly before noon on 14 August, twenty-three crews took off from Melbourne on another close-support operation in Normandy, this time in support of Canadian troops advancing on Falaise. Seven targets amongst the opposing German positions had been selected and the Squadron contingent was scheduled to attack one coded as TRACTABLE. Despite close control by a Master Bomber over each target, a confusion over the use of yellow markers meant that some aircraft bombed a Canadian Regiment near one of them. All Squadron crews bombed that day, guided by a mix of red and yellow markers and, with their releases made in good visibility from 6-7500 feet, there is no reason to believe that these crews, out of a force of 805, were involved in this incident.

With the Allied advance into France starting to gain momentum, and although it was still subject to control by General Eisenhower's SHAEF HQ, Bomber Command was now able to contemplate a return to the assault on Germany. In preparation for that, a series of counter-air attacks on seven German night fighter airfields in Holland and Belgium was launched during 15 August. Twenty Squadron crews were allocated for the raid on the airfield at Tirlemont/Gassoncourt (Tienen/Goetsenhoven) to the east of Brussels, and all bombed using visual identification, target markers and a Master Bomber's instructions. Oddly, there was no flak or opposition met over the target, possibly because the airfield was a reserve base. The next Op was an attack on the port at Brest by a force of 79 Halifaxes in the afternoon of 17 August, for which twenty-three crews had been briefed. The target was largely covered by cloud

and there is no indication in the debriefs of it having been marked and, in consequence, seven crews were unable to see enough to bomb, despite some making more than one run, and another found the bombsight had failed. The fifteen others made their releases using a mix of glimpses through cloud, Gee and H2S fixing.

The night of 18/19 August saw more mixed targeting – four crews completed successful Gardening sorties to lay mines off the mouth of the Gironde, whilst eighteen joined an attack on the synthetic oil plant at Sterkrade in Germany. One crew had to abandon the mission after compass and Nav equipment failures, whilst the others went on to contribute to what was assessed as a successful raid. An Ops break followed until 25 August, when crews were detailed for two distinct missions. Seven crews joined a daylight raid on a blockhouse at Watten, near St Omer, part of the support infrastructure for the V2 weapon. It had already been put out of action by a Tallboy dropped by 617 Squadron in July, but a little construction work was continued as a diversion and this raid appears to have been planned on a 'just in case' basis. Six crews returned reporting accurate bombing but one did not – Fg Off Walton's crew was brought down near the target with only two crew members surviving to evade capture. A little later that evening, seventeen other crews took off to attack gun positions at Ronsceval, near Brest. The target was covered by heavy cloud and only two crews managed to bomb after descending to relatively low levels, whilst the others were turned back by the Master Bomber. On the next night, 26/27 August, eleven crews were detailed for a Gardening sortie off Kiel whilst a Lancaster force attacked the town and port itself. They were all able to lay mines as briefed in good conditions, using H2S fixing, but with most having to divert on return due to fog at Melbourne.

The next Op is notable in that it was the first significant daylight Bomber Command raid over Germany since August 1941. The target was another synthetic oil plant in Homberg, and seven crews were detailed for the force that also had a significant Spitfire escort. One crew was forced to abandon the mission when the aircraft fuselage filled with fumes, but the others were able to complete the mission in good conditions, using markers. The last Op of the month was on the night of 29/30 August when twelve Squadron crews went Gardening further east than hitherto in the Baltic, off Rostock. All laid their mines as briefed with no fighter opposition – the main attack that night was further east, at Stettin.

Three months after the D-Day landings there was still a good deal of business to be done in the general area of the Pas-de-Calais, on the British Army's left flank, enough to keep 10 Squadron and others engaged there on several days when Air Chief Marshal Harris would no doubt have preferred to see all of his units exclusively focussed on

The Seaton Ross Mill, close to Melbourne, was a welcome sight after a long night on Ops.

the assault on the German heartland. However, by mid-September, progress on the ground was sufficient to permit operational control of Bomber Command to revert to the Air Ministry, always provided that its employment chimed with Allied planning overall. Against that background, the last V1 site capable of hitting England was not captured until October and on the morning of 1 September, twenty-two crews took off to attack a launch site at Lumbres, well inland towards St Omer. Conditions around 0830 appear to have been near-perfect and the target could be clearly seen, although it was also marked, with a Master Bomber in charge. One crew had already returned early, another found that the master bomb switch had failed, so twenty crews completed the attack which was assessed to have been particularly successful. WO Thorne, who had turned back after losing the port inner engine, lost his port outer about thirty minutes later. He then headed for RAF Manston for an emergency landing but crash-landed three miles west of the field. The port outer then caught fire, but that was extinguished by the crew who escaped without injury.

On 3 September, another counter-air day was organised, this time with six airfields in Holland as targets for nearly 700 aircraft, and the Squadron contributed nineteen crews to the attack on Soesterberg, near Utrecht. All of the crews bombed accurately in middling weather, some making several runs until a secure release could be assured. There was moderate flak but no air opposition and two crews diverted to Snetterton Heath on return, having had to shut engines down after the target. There was then a week's Ops break until 10 September when two sorties were flown as part of a campaign over some days against German positions around Le Havre. Seventeen crews were sent on the first, morning, mission and all bombed from medium level on the markers, save one that experienced a complete hangup. Later, in mid-afternoon, another six crews went across, with all bombing as instructed. On the following night, there was another Gardening expedition in the Baltic involving ten crews. Two had to abandon the mission with technical failures but the others were able to continue and complete the mission in good conditions. On the way in, Plt Off Peacock's aircraft was attacked by an unidentified fighter but, after two bursts from his Tail Gunner, it was not seen again.

The continuing campaign against Germany's synthetic oil plants, essential to the ability to wage war, provided targets on 12 and 13 September. Twelve crews were detailed to participate in an attack on the Scholven facility in Gelsenkirchen, where the weather was clear but a smoke screen had been created. As two had to return early after engine failures, ten crews completed the mission using markers and the Master Bomber's instructions. There was a good deal of flak and five crews reported varying levels of damage, but no injuries. An hour or so after these crews landed, ten others set off for Munster, with the main station as their aiming point. Nine bombed as briefed, whilst the tenth had a hangup but was later able to bomb a rail junction in Stadtlohn. Two aircraft suffered severe flak damage – one reported then being escorted by two Halifaxes to the Dutch coast and beyond, whilst the other had two Lancasters and six Spitfires to see it safely clear of the target area. (The first of those was Plt Off Winter's aircraft with Sgt Smith as Bomb Aimer. In 1983, Arthur Smith published a short memoir of his time on 10 Squadron – 'Halifax Crew, The Story of a Wartime Bomber Crew' – and this contains a graphic description of the flak damage sustained and of their landing on the emergency strip at Woodbridge to find a Lancaster, flown by the Air Bomber as his pilot had been killed, landing in the opposite direction! The second aircraft was flown by Fg Off John Bowmer's crew on the final Op of their 10 Squadron tour. John would rejoin the Squadron in 1967 as one of its first VC10 captains.)

The 13 September raid was on the Nordstern oil plant in Gelsenkirchen, for which the Squadron briefed seventeen crews. Despite making two runs, Fg Off Brown's crew was unable to see a marker so they attacked a marshalling yard that could be seen, whilst the others were able to bomb as instructed. Flak over the target was heavy and only four aircraft escaped with no damage at all. Shrapnel wounded FS Davies' Flight Engineer, caused fifty holes in the fuselage and a fuel fire that, though extinguished, meant the loss of a starboard engine. He landed on the emergency strip at Woodbridge 'on three engines and one good tyre.'

For the night of 15/16 September, sixteen crews took off to participate with over 450 others in yet another raid on Kiel. All were able to bomb on the markers in good

weather over the target, but FS Frey's crew had an eventful flight. Flak over the target hit a port engine and made the hydraulics unserviceable, and they were also hit by a number of 4lb incendiaries, nine of which remained in the aircraft with one found largely burnt-out in a fuel tank. A starboard engine began to give trouble, the bomb doors fell open as a result of no hydraulics, and the port undercarriage leg fell down. Nonetheless, "Aircraft made a normal landing at Middleton St George." In the morning of Sunday 17 September, a substantial force of over 750 aircraft was tasked against German positions around Boulogne. The Squadron contribution was eighteen crews, all of which were able to bomb as marked or instructed. (Further round the coast, a Lancaster force attacked German flak positions at Flushing in support of the first wave of aircraft on Operation MARKET GARDEN that would be heading for Nijmegen and Arnhem that afternoon.) An Ops break followed until the night of 23/24 September, when the target for some 550 aircraft was Neuss, on the western bank of the Rhine, opposite Dusseldorf. Twenty-two Squadron crews took off but one had to abandon the mission when the Bomb Aimer, Sgt Smith, became ill in-flight. (In his memoir, Arthur Smith admitted frankly that he had 'frozen with fear' that night, and that his return to flying was due to a penetrating interview with a Medical Officer and the readiness of his crew to have him back.) Visibility over the target was poor, no debriefs mention marking of any kind, and it appears that bombs were released on DR-based ETAs. Nonetheless, the raid is assessed as having been quite successful. Plt Off Kite's aircraft was one of just two Halifaxes lost that night, when it came down at Zwier Wimburg in Holland. He and his WOP, FS Saunders, were both killed, whilst their colleagues survived as POWs.

The remaining Ops in September were all in support of the continuing ground battle around Calais. Twenty crews took off in the morning of 25 September but, after being airborne for some ninety minutes, the Master Bomber instructed them to abandon the mission on account of low cloud over the target. Things went much better on the following morning when eighteen crews were able to carry out accurate attacks in good visibility on German coastal gun positions at Cap Gris Nez. On the next morning again, 27 September, when fifteen out of sixteen crews carried out further successful attacks on positions around Calais. The other crew twice overshot the target due to smoke and cloud conditions and was finally instructed to abandon the mission.

On the evening of 4 October, nine crews were detailed for another Gardening mission in the Kattegat area. The allocated area was largely cloud covered but drops were made using H2S fixing, and with one crew making three runs before their mines released on time. A further five crews flew a similar mission off Heligoland two nights later. Earlier that afternoon, fourteen crews had taken part in a return, marked raid on the Scholven oil plant, though one had an early return. On 7 October, eighteen crews took part in a raid on the town of Kleve, close to the Dutch border, to help protect the right flank of forces near Nijmegen that were threatened after the "Bridge too far' failure of MARKET GARDEN. On the evening of 9 October, twenty crews joined 415 others for another raid on Bochum. There was one early return, but the others completed the mission using markers and Flt Lt Evans' aircraft had a sharp encounter with a fighter in which he lost an engine.

There was also another change of command on 9 October when Wg Cdr U Y Shannon, a New Zealander, took over from Wg Cdr Radford. It is worth recording that, some six months after the arrival of the Halifax Mk III, conversion to type for new pilots and crews was still being done on the Squadron and the new CO completed his final cross-country on 23 October.

14 October was notable for two 1000-Bomber attacks made on Duisburg on the same day. These were planned in response to a new directive (Operation HURRICANE) given to Bomber Command and the USAAF's Eighth Air Force, requiring them to demonstrate the "generally overwhelming superiority of Allied Air Forces in this theatre." Twenty-three crews were detailed for the morning raid, but there was one early return to land at RAF Grimsby when Fg Off Sifton's aircraft lost both starboard engines. Crews bombed on target indicators and could see that their results were well concentrated. Two aircraft were damaged by flak – WO Davies' aircraft in particular was badly damaged, his Flight Engineer was wounded, and he landed at RAF Manston on

three engines. Seventeen crews took part in the night raid, including several from the earlier attack. One was forced to turn back after a hydraulics failure but the others reached the target and many remarked on the number of fires still burning from the earlier raid as they bombed on the markers.

For the evening of 15 October, eleven crews were listed for another raid on Wilhelmshaven. There was one enforced return, whilst the others pressed home their attacks. Plt Off Bishop's aircraft was hit by flak both approaching and leaving the target, damaging his starboard undercarriage. He prudently waited until all of the others had landed before making an approach for, on selecting 'Undercarriage Down,' the right wheel fell off and a wheels up landing ensued, writing off the aircraft but with no injuries to the crew. FS Owen's crew was lost without trace that night. About an hour after these crews had departed, another six went Gardening at the northern end of the Kattegat. Five returned to report mines laid as briefed, with Flt Lt Body adding that a Halifax had been seen shot down – possibly that of Sqn Ldr Hart whose crew was lost, apart from his Canadian Navigator, Flt Lt Parks, who got back to England on 10 November.

The next Ops break followed, with training carried out whenever the weather at base allowed, and the next Op was called for the night of 23 October when over 1000 bombers were sent to Essen. Nineteen crews were detailed for that raid, with one early return after an engine failure. The other eighteen reported bombing on markers over heavy cloud cover. Another raid of almost 800 aircraft returned to Essen in the afternoon of 25 October, for which the Squadron briefed twenty-two crews. One was unable to retract the undercarriage on takeoff whilst two others each lost an engine during departure so all three had to proceed out to sea to jettison their bombs before returning. Fg Off Gibbs lost his port inner over Aachen and was also losing power from his port outer. Losing height, he jettisoned part of his bomb load over Stommeln and another 2000 lbs over Haan, and was then able to continue to Essen to drop what was left. These two raids effectively finished the Krupps works as a significant industrial force for the rest of the war. Then, three days later, on the afternoon of 28 October, it was back to Cologne for twenty one crews. Engine failures caused two early returns but, in addition, Fg Off Bleakley had to ditch after two engines failed and he was unable to keep a straight heading. All of the crew were picked up by HMS Middleton after four hours in a dinghy and landed at Portsmouth. The other eighteen crews all completed attacks as part of an extremely damaging raid. Nonetheless, two further raids followed. Twenty-one crews were detailed for the first on the night of 30 October, when a mix of engine and Gee failures caused six crews to return early – Fg Off Scott had an starboard outer engine fire, followed by the propeller shearing off, so he wisely elected to land at Manston as quickly as possible. The others all bombed on markers, except for Flt Lt Janes' crew who could not bomb as the Nav had not armed the Master Bomb switch! The next return to the embattled and partly deserted city was on the next night, 31 October. Fourteen crews were detailed – one returned early and the others attacked on the markers against moderate flak.

The assault on Germany did not let up and sixteen crews were detailed for yet another raid on Dusseldorf on 2 November, the last major raid there of the war. Engine and instrument failures brought two back early, with the others bombing on markers in excellent visibility. Bochum was another target revisited for the last time on 4 November, with nineteen crews detailed together with 730 others. There was one early return, but seventeen completed attacks on the city. Hydraulic lines on Fg Off Daffey's aircraft fractured on the way home so flaps, bomb doors and tail wheel deployed so, with the extra drag seriously affecting airspeed, he diverted to RAF Woodbridge. Also, soon after leaving the target, Fg Off Sifton's aircraft was attacked by an unidentified twin-engine fighter. Both Gunners fired at it and it was seen to break away and disintegrate only to be followed by a second fighter that was driven off. At noon on 6 November, sixteen crews took off to join over 700 others in a daylight attack on Gelsenkirchen. When they reached the target city, they found fairly extensive cloud cover plus smoke from earlier phases of the raid and bombing was later assessed to have been scattered around the area. Fg Off Brown's crew saw so little that they elected to follow other aircraft and bombed rail yards at Zollverrein, to the southwest of Gelsenkirchen. Flak was relatively heavy and accurate - a number of crews reported minor damage and

Fg Off Sifton's Bomb aimer received a finger wound from a splinter that disabled the bombing panel. The next Ops break followed, mainly filled by fighter affiliation and bombing practice, until the next op was called for the afternoon of 16 November. Bomber Command had been asked to bomb three towns close to German positions that were about to be attacked by American troops, and a 10 Squadron contingent of twenty-one crews was to be part of the 413-strong force sent to Julich. There was one early return, but all other crews bombed successfully, either visually or using markers. Fg Off Taylor's aircraft was hit by flak over the target, after which a starboard engine began to prove troublesome and resisted all attempts to feather the propeller. The aircraft gradually lost height until, just after crossing the English coast, the engine caught fire ad the prop sheared, resulting in a landing at Manston. Other crews had different engine problems – one landed at Woodbridge, whilst the others were able to get back to Melbourne.

Another daylight raid on Munster followed on 18 November, when fifteen crews were detailed to join a force of 479. An engine problem sent one back early, whilst the remainder pressed on and attacked on markers over cloud. The evening of 21 November saw a return to the oil plant at Sterkrade. One crew abandoned the mission after having difficulties with two engines, and another had to return when the Mid Upper Gunner became ill. The others bombed on markers but it would appear that, whilst damage was inflicted in the area, the oil plant survived once again. Essen was attacked once again in the early hours of 28 November by a force of 316 aircraft, to which the Squadron contributed twenty-three crews. All of them reached the target to bomb through cloud cover using markers and H2S. (The bomb loads appear to have been exclusively of high explosive weapons, with no incendiaries in the mix – in distinct contrast to the type of loads earlier in the year. An indication that there was thought to be little left at such a frequent target that would burn?) However, the raid was not without incident. Plt Off Rebick's aircraft was attacked three times by a Me 110 and suffered damage. The Tail Gunner returned fire and, on the third attack, saw the fighter's starboard engine catch fire as it broke away. Plt Off Tudberry lost a port engine en-route to the target but carried on, only to find that the aircraft lost some 9000 feet in height as the bomb doors opened and continued to lose height on the return. The amount of vibration experienced finally dictated that a forced landing was inevitable and he came down outside Guines, near Calais, well behind Allied lines, in a field that had just been cleared of mines. The crew were returned to the UK on 1 December, and Plt Off Tudberry was awarded an immediate DFC in recognition of his efforts that night.

Another familiar target, Duisburg, was listed for the evening of 30 November, when the Squadron provided twenty-two crews towards a force of 576. There was heavy cloud cover and bombing was on the markers, with Gee or H2S backup when possible. Flak was reported as weak to moderate and there was one inconclusive FW 190 encounter. The campaign continued into December, with a first Op on the night of 2/3 December, when the target for 504 aircraft was Hagen. Twenty-two Squadron crews were involved and all bombed on markers over total cloud cover without serious incident. On the night of 5/6 December, rail installations around Soest, beyond the Ruhr, were the target. Twenty-one crews were involved and all reached the target. Weather was clear and crews could see that bombing was well-concentrated in the desired areas. On the following night, rail installations around Osnabruck became the target, with sixteen Squadron crews involved. Two crews reported fighter encounters – Fg Off Scott's crew mention the first 10 Squadron brush with what may have been a jet-powered Me 262, and it is known that some were being tried out at night by this late point in the war. It did not fire upon the Halifax and, whilst he took evasive action, Scott's Tail Gunner fired at it and it was seen to fall out of control, with a heavy explosion occurring. Fg Off Dark's crew returned claiming a Me 110 as destroyed on the evidence of two crew members. Fg Off Welch's crew was lost without trace at some point in the night.

Another substantial raid on Essen took place on the night of 12 December with ten Squadron crews involved. Cloud covered the city and all crews bombed on the markers for what was considered a very damaging raid – indeed, in a postwar interrogation, Albert Speer, Hitler's Minister for Armaments, made a point of how effective it had been. Whilst most crews had a short break from Ops over

Germany, four went back to the Kattegat Garden on the evening of 14 December, with no opposition, and a further four returned to a similar area on the next night. Then, in the early hours of 18 December, twenty-one crews set off to join some 500 others to attack Duisburg once again. One crew had to return early and nineteen returned to report bombing using markers, Gee and H2S. Flt Lt Body's crew did not come home, having had a midair collision with a Halifax from the Canadian 432 Squadron en-route to the target from which the only survivor was the Canadian pilot, Fg Off Krakovsky. (The aircraft came down near the Franco/Belgian border and the remains of the crewmen are buried in various French and Belgian cemeteries. Fg Off D J Mole, WOP with the Squadron crew, and navigator Fg Off J H Waldron are buried in the cemetery at Taillette in the French Ardennes where, in October 2011, a memorial plaque was inaugurated. Air Cdre J Maas, Air Attaché in Paris and a former 10 Squadron Flight Commander, and Gp Capt Gunby, a former Squadron CO, were present and the Squadron Association standard was paraded.)

Whilst it may be easy some seventy years later to think that, by this stage in the war, RAF and USAAF bombers were being used in a way that was strategically inappropriate, it should be recalled that the war was by no means over. The major German counter-offensive in the Ardennes (The Battle of the Bulge) had just begun and was gaining ground, and the Nazi ability to wage war, though damaged, was by no means defeated and there would be many more losses amongst the bomber forces before victory was assured.

As the German offensive developed, the next Op was an attack by 106 aircraft on rail yards and facilities at Bingen, located on the Rhine and on a route into the Ardennes. Twenty crews were detailed for an early evening raid on 22 December and things went awry quickly after takeoff for some. Fg Off Yates suffered a hydraulics failure overhead base, went out to sea to jettison his bombs and, on landing back, overshot the runway and wrote off the aircraft. The crew were uninjured, but the blockage created was no doubt responsible for at least one other returning crew having to divert elsewhere. Two other crews had to abandon the mission whilst a third, having taken off amongst the others, was twenty minutes late in setting course and eventually accepted that they could not be over the target within the stipulated time. Nonetheless, fifteen crews did reach Bingen where their bombing seemed accurate, and it is assessed that there were no more rail movements from there into the Ardennes. On the afternoon of 24 December, raids were launched against German airfields and the Squadron provided fifteen crews for the daylight attack on the field at Mülheim, southwest of Essen. There was just one technical return that day, and the others all bombed the airfield accurately and had to divert elsewhere in Yorkshire on return due to fog at Melbourne.

By 26 December, weather over the Ardennes had improved sufficiently for bombers to be sent to attack German positions near St Vith, in Belgium. With aircraft still scattered around diversion airfields, the Squadron could only provide three crews to that effort. Two bombed accurately but ended up at RAF Dyce in Aberdeenshire on return – the fog at Melbourne was persisting. The third had probably bombed too but, having survived the crash landing at Melbourne on 22 December, Fg Off Yates' crew went missing, probably crashing into the sea off Margate shortly after two W/T contacts made at 1633 and 1636 hours, and the body of his Tail Gunner, Plt Off Addyman, was recovered on 1 January 1945.

On 28 December, there were five aircraft still available at Melbourne to despatch for a raid in the early hours on the rail works at Opladen and all crews returned, having bombed on markers. The recovery of the scattered aircraft appears to have been possible that day for twenty crews were detailed the next day for an attack by two bomber waves on rail yards at Koblenz, supplying the Ardennes battle. All reached the target that afternoon and bombed on markers and visual checks. On departure, one crew discovered that nothing had released so they had to jettison once back over the sea. Seventeen crews were than detailed for their last Op of the year as part of a 470-strong force attacking the Kalk-Nord marshalling yards at Cologne on the night of 30 December. All reached the target, completely cloud covered, and bombed using markers plus Gee and H2S backups for what was later known to have been a highly effective raid. Notably, however, Fg Off Daffey's aircraft lost its 2000 lb High Capacity bomb

when it fell through the bomb doors into open country just after takeoff. Tom Treadwell, his Bomb Aimer, recalled the circumstances in a Squadron Association Newsletter in 1996. The customary Board of Inquiry had established that there was a fault in the bomb gear mechanism, a circumstance aggravated by Fg Off Daffey putting into practice his theory that building up extra airspeed before rotation would see him safely into the fields beyond the runway should an engine cut out, and the resultant bouncing put too much strain on the system. The crushing of the bomb fuse inside its thin casing would have reduced the power of the explosion that had no doubt caused the crew "a feeling of being shot upwards at a much greater speed than usual." The reaction of those living near the airfield is nowhere recorded!

There were no Ops called for the Squadron on 31 December, so the year ended with five practice bombing details, a Fighter Affiliation exercise, and possibly a pilot reflecting on whether it might be best to accept the Halifax handling techniques he had been taught in training, particularly when carrying volatile bomb loads! And with Allied forces now firmly back in continental Europe, it must have been possible to take a rosier view than a year previously, but there were still hard rows to hoe as the first days of January would make abundantly clear.

1945

Eighteen crews too off during the afternoon of 1 January to join others in an attack on a coking plant in Dortmund. There were two engine-related early returns and, unfortunately, Fg Off Winter's aircraft overshot on landing at Melbourne and crashed beyond the airfield and his tail Gunner and WOP died in the accident. The other crews reached the target where one discovered that no bombs had released despite their photo flash and camera operating as usual. The others dropped on the markers but it emerged that the damage done was around the target plant which remained in operation. On the next day, another eighteen crews were detailed for what was a more successful attack on the IG Farben chemical factory in Ludwigshaven by 389 aircraft that forced it to cease production. One crew had to return after an engine failure but the others attacked on markers over light cloud or haze. The next Op was called on 5 January, when twenty-one crews were detailed for a large raid on Hannover. There was one early return and another bomb hangup over the target, but eighteen crews reported releases on markers – Flt Lt Wood's crew felt a severe bump during their bombing run and, on landing, it was found that a bomb door had separated and damaged the fuselage. Fg Off Sifton's aircraft was shot down by a fighter and crashed north of Detmold with no survivors.

(The death of FS Coates, an eighth crew member on Sifton's crew, draws attention to the fact that, from around mid-December 1944, some aircraft were carrying a Mid-Under Gunner. It seems likely that, to permit the fitting of the cupola required, the very few aircraft concerned were not fitted with H2S as the radar's antenna housing would almost certainly have interrupted the gunner's field of view. Various references suggest that this additional cupola was fitted to a very small number of Halifax aircraft, and the likelihood that this was essentially a squadron trial may be strengthened by noting that whilst Sqn Ldr Turner flew MZ948 (ZA-E) on the Duisburg raid of 18/19 December, Fg Off Tudberry's crew flew the same aircraft on 22 and 29 December with a standard seven-man crew. And besides the loss of Fg Off Sifton's aircraft, another with the fit was lost when Fg Off Yates crashed at Melbourne on 22 December.)

On the evening of 6 January, a mixed force of 482 aircraft was sent to attack the major rail junction at Hanau, east of Frankfurt. Seventeen crews were detailed, with one early return, and fifteen reported bombing on markers through continuous cloud cover. Flt Lt Harrow's crew was lost that night, with no survivors, after a collision with a 415 Squadron Halifax over Oberscheid. Weather at base prevented further flying until 12 January, when five crews were detailed for Gardening at the mouth of the Langelandsbaelt, north of Kiel, where all went well. On 14 January, crews were detailed for three separate missions. The fourteen crews detailed to join a daylight raid on rail yards at Saarbrucken in clear conditions all reported having bombed accurately, having used visual identifications, markers, and the Master Bomber's instructions. (The continuing hazards of bombing in a stream were evident when Fg Off Taylor had to take evasive action after "Bombs Gone" to avoid being hit by others from above.) That night, a second group of three crews joined a raid on Luftwaffe fuel storage at Dulmen, whilst a further seven successfully laid mines east of Aarhus, in sealanes being used to ship German troops from Norway. A good day's business – twenty four launches with no returns or crew losses.

Unfortunately, the same could not be said about the next Op to Magdeburg, on the night of 16/17 January. Twenty-two crews were detailed, with two early returns, and eighteen reporting accurate bombing on markers, backed by visual checks. Two crews did not return that night. Fg Off Marshall's aircraft was brought down over Germany – he, his Flight Engineer, Sgt Griffiths, and his Mid Upper, Sgt Thornley, were all killed with their four colleagues taken as POWs. Fg Off Whitbread's crew was lost without trace and all are commemorated on the Runnymede Memorial.

There was then a lengthy break, in part caused by bad weather, until 28 January when ten crews were called for a raid that night on an Aero Engines factory at Zuffenhausen, in Stuttgart, and as part of a second wave attacking the city that night. An engine failure brought one back early, and the others reported bombing as directed. The post-attack assessment was that bombing was scattered in what had been the last large raid on that city.

On 29 January, there was another change of command when Wg Cdr A C Dowden arrived to replace Wg Cdr Shannon, promoted and posted out to be Station Commander at RAF Full Sutton. Unusually, the new CO was not a pilot but, having first qualified as an Observer in 1941, he was by then flying as an Air Bomber.

Another busy month started straightaway on the evening of 1 February when twenty-three crews were detailed for an attack designed to hinder German troop movements around Mainz, and there would be other missions in support of the Allied advance on the ground in the days that followed. On this occasion, there were two early returns after both crews had to shut down the starboard outer engine. For the others there was complete cloud cover over the target, with all bombing on markers against little opposition. On the next night, the allocated target for over 300 aircraft was another synthetic oil factory, this time at Wanne-Eickel in the northern Ruhr area. Nineteen crews were detailed, with one early return caused by a combination of technical problems. Seventeen reported bombing on markers, once again over complete cloud cover, and it appears that most bombs fell around a local coal mine. Fg Off Gibbs' aircraft crashed near Roermond with the loss of the crew, except for Fg Off Cook, the Bomb Aimer, who survived as a POW.

Six crews went Gardening in the approaches to Bremerhaven on the evening of 3 February. Four laid their mines as briefed, whilst release and H2S failures caused abortive sorties for the others. The following night saw a return to Germany, with nineteen crews detailed for a raid on a 'military road junction' at Bonn. Fg Off Davies' aircraft had a hydraulics failure soon after takeoff, jettisoned his bombs over the sea, and landed at the emergency strip at Carnaby. His colleagues continued to find another cloud-covered target that was attacked using markers, backed by Gee fixing where possible. Crew reports on the concentration of bomb fall varied and the official verdict was that most fell to the south of the required target. The next Op was again in support of a planned British attack across the German border, and the stated aim was to interrupt German troop movements around Goch. 464 aircraft were tasked that night but, after the first waves, the Master Bomber instructed crews to abandon the mission as the volume of smoke had made control of the attack impossible. Four Squadron crews did bomb and their quoted release times suggest that they did not hear the instruction in time, although their colleagues had done so. Fg Off Dade pushed on into Germany and bombed around Wesel, on the Rhine. A second group of eight crews went minelaying that night in German sea lanes being used to ship troops back from Norway. All were successful, although temporary hangups made two take a second run.

In the early hours of 9 February, another unsuccessful attack was made to stop all oil production at Wanne-Eickel. Fifteen crews were detailed for the raid and one had to return after losing an engine. Another crew also lost an engine and bombed Mönchengladbach as an alternate target to cut short the mission once it was clear that they could not maintain height. The next Op on the night of 13/14 February was on another oil plant at Bohlen, for which twenty-three crews were allocated. One returned early after an engine failure and, for the others, yet again the target was obscured by cloud and the markers were

scattered. The Master Bomber was doing his best to rescue the situation but it seems unlikely that much damage was done that night. Several crews reported severe icing on the return and, as an aside, advances on the ground meant that it was now possible for one crew to refuel in Belgium on its return after lingering over the target, trying to select an aiming point. Bohlen is some seventy miles from the centre of Dresden, the target that night for two waves of Lancasters in an attack that remains a matter of controversy to the present day. Earlier in the year, an outline plan – Operation THUNDERCLAP - had been drawn up for coordinated attacks at a suitable time on Chemnitz, Leipzig and Dresden to assist the Soviet advance from the East and that time had come. On the next night, 14/15 February, over 700 aircraft were sent to Chemnitz, with the Squadron detailing fifteen crews for a target that, apart from Berlin, was further east than any allocated hitherto. All reached the area to bomb on the markers. Whilst this raid was in progress, four other crews went on a Gardening mission to lay mines in the approaches to the port of Lubeck. Three did so successfully, but Fg Off Grayshan's aircraft was attacked by a fighter and crashed in Denmark. He and his Navigator, FS Berry, were killed; his Bomb Aimer and WOP became POWs; and, aided by the Danish Resistance, the three others managed to reach Sweden.

On 17 February, twenty-two crews were detailed for a daylight raid on rail communications and marshalling yards at Wesel, the Rhine town bombed by Fg Off Dade earlier in the month. Unlike the raid there by a Lancaster force on the previous day in clear conditions, this one met complete cloud cover and, on arrival, the crews heard the Master Bomber give instructions to abandon the mission. Two crews went on to bomb alternate targets – Flt Lt Bridge chose Gladbeck and Flt Lt Dark (?) hit Homberg. Both bombed over cloud with no results seen. Fg Off Dade's aircraft had lost an engine cowling soon after departure. This allowed the slipstream to enter the wing and, eventually, the wing skin burst so, on reaching Wesel, he effectively jettisoned his bomb load to begin the return leg and landed at the RAF Eye, a USAAF base in Suffolk. The other aircraft all landed at Mepal or other airfields due to fog at Melbourne, with two exceptions. Sqn Ldr Janes and Flt Lt Cook both landed at base with the help of FIDO. (The 'Fog Investigation Dispersal Operation' system was installed at a number of airfields from early 1943, and used the ignition of large quantities of fuel in pipes along the runway sides to generate enough heat to lift fog sufficiently so that aircraft could land in comparative safety. Aircraft had been landing – and possibly taking off - at Melbourne using FIDO before February 1945 but, for some reason, this was the first occasion on which specific mentions appeared in the unit records.)

The next Op was another attack on German oil production, this time at a refinery in Reisholz, a district of Dusseldorf, on the night of 20/21 February. Sixteen crews were detailed to join and all returned reporting well-concentrated attacks on markers over cloud in what was later assessed as a very accurate and damaging attack. (From time to time, the records show staff officers from HQ 4 Group or Bomber Command flying on sorties over Germany and, on that night, a Maj Southgate, presumably from Ack-Ack Command and described as '4 Group AA Specialist' had flown with Flt Lt Atkins' crew. Opposition that night was reported as light, so he may not have seen as much as he may have hoped for!) Sixteen crews were also briefed for the following night's Op, an attack by 349 aircraft on the town of Worms. Fourteen returned to say that the weather over the well-marked target was clear and that accurate attacks had been made, something supported by subsequent analysis. Unlike on the previous night, flak was reported to have been moderate to severe and Fg Off Bastard's aircraft, in particular, suffered considerable damage but recovered safely to base. Two crews were lost over Germany that night. Flt Lt Hurrell and Sgt Whyles, his Tail Gunner, were both killed, with their colleagues becoming POWs. FS Parsons' crew survived as POWs, save for Sgt Jones, his WOP, who evaded capture and made it back to the UK in April.

A return to Essen by a force of over 340 aircraft was scheduled for the afternoon of 23 February, when seventeen Squadron crews were detailed. Once again, the target was obscured by cloud and the marking involved use of green smoke puffs. Crews used H2S and Gee to assist aiming, but one crew could see no markers or other aircraft bombing so they aborted the mission. The other sixteen were able to bomb, and considerable damage was assessed to have been done to the Krupps factory. On the next afternoon, another 340-strong raid was flown against the oil plant at Kamen. Nineteen crews were briefed, with one having to abandon the mission due to a hydraulic failure shortly after takeoff. Cloud covered the target once again, with crews bombing as instructed by the Master Bomber. The final Op for the month was as part of a large 458-strong force sent on a daytime raid on Mainz. Nineteen crews took part and all dropped through cloud as instructed by the Master Bomber for what was seen as the worst raid of the war for the city.

(A taste of what was to come for the Squadron can be seen in the posting of eighteen aircrew personnel to No 187 Squadron as of 21 February. These moves were part of the formation of a Transport Command unit, initially with the Halifax III, but soon to get Dakota III and IV aircraft for operations in the Far East.)

By early March, the advance into Germany had progressed such that Cologne was virtually a frontline city, and a very large raid by 858 aircraft was planned for the morning of 2 March, aimed at disrupting enemy troop movements. Twenty-two crews were detailed, with one abandoning the mission soon after takeoff when an engine had to be shut down. For once the weather was clear, and crews bombed visually under direction from the Master Bomber. The raid was very destructive and the city was occupied by the US Army four days later.

The evening Op on 3 March is notable for what occurred as crews returned to base after another raid on Kamen, with twenty-one of twenty-two completing the mission in an attack that halted oil production there. German night fighter commanders, aware that their force was much less effective than it had been a year or so previously, had decided to demonstrate that it could still carry out intruder operations over England, as it had done in 1941. They elected to launch operation GISELA that night, sending some 200 Ju 88 fighters to surprise bombers returning to their bases. Once it was clear that German aircraft were over England, the Squadron's homebound crews were told to divert to Leeming, Skipton and High Ercall. Unfortunately, one of the twenty aircraft caught by the intruders was Flt Lt Laffoley's, shot down near Knaresborough. He and four of his crew were killed and the others, including a Second Pilot, Plt Off Palmer, were all badly injured. The Luftwaffe mounted a small-scale repetition on the following night without success, and GISELA was the last occasion on which any significant fighter threat was posed to returning aircraft over England.

'Cook's Tours' were flown over Germany immediately after the war to show ground crews the results of their past labours, but Cologne Cathedral's towers are left still standing.

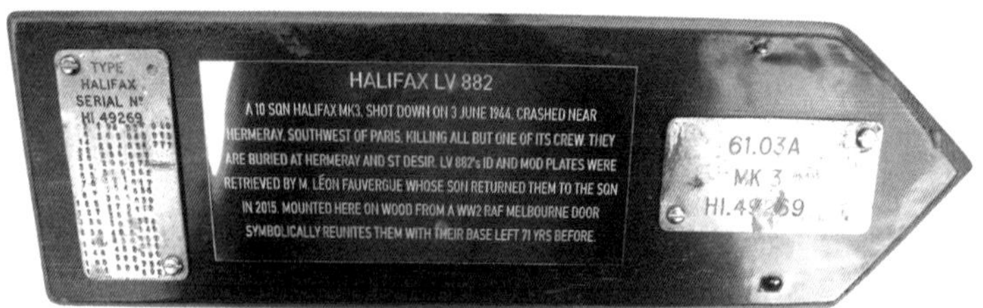

Above: A local French villager, Léon Fauvergue recovered LV882's ID & modification plates from the wreckage at considerable risk. In 2015 his son Pierre sent them to the Association. They have been mounted on wood from a derelict wartime RAF Melbourne building with a plaque describing the event and were returned to the Squadron on Remembrance Day 2015.

Left: Sgt Vijendra Kumar, shown here with his mother Chandra, was the captain of Mk3 Halifax LV 882, which crashed near the villages of Hermeray and Raizeux, France on 3 June 1944. All but Sgt Alec Hunter, the WOP who escaped, were killed. Sgt Kumar and A/B Fg Off C.D. Taylor are buried at Hermeray. Nav, Fg Off E. Heyworth, F/E Sgt J. Archer, M/U Gnr Sgt T. Blacklock & T/G Sgt G. O'Leary are buried at St Desir, near Caen. The crew had only flown one previous Squadron mission which was on the night before they were lost. They had only joined the Squadron in May 1944.

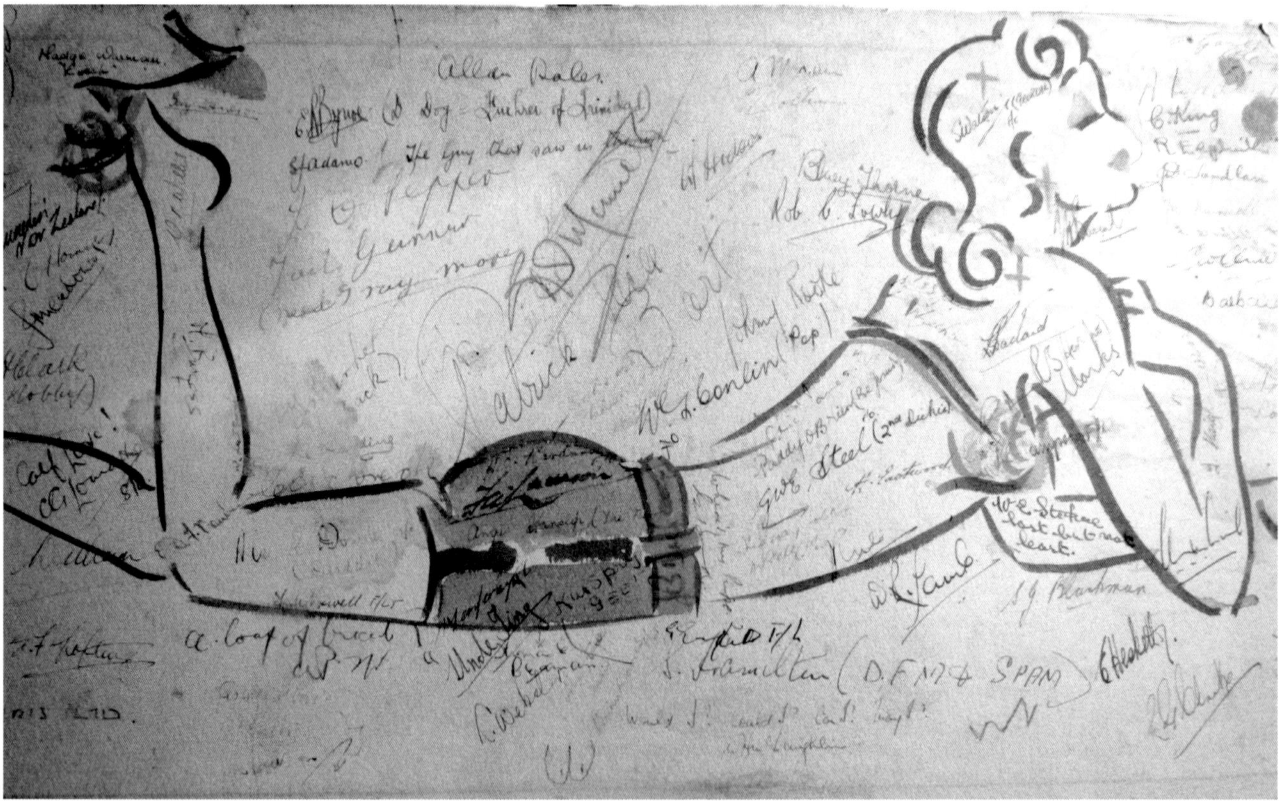

During Canadian Flt Lt Leo Murphy's crew 'end of ops tour' party in February 1945 he was presented with this picture probably drawn by the same artist who did similar pictures above a bar in Melbourne, and signed by his crew and others serving at that time. Murphy had half of his moustache shaved off during the celebrations. After his later death, it was returned to the Squadron by his son Paul and was then framed and entitled 'The Painted Lady' in 2015.

Right: Inside a Halifax. From left to right: legs of Sid Stephens FE, Stan Somerscales Pilot, Jim Lewis Nav.

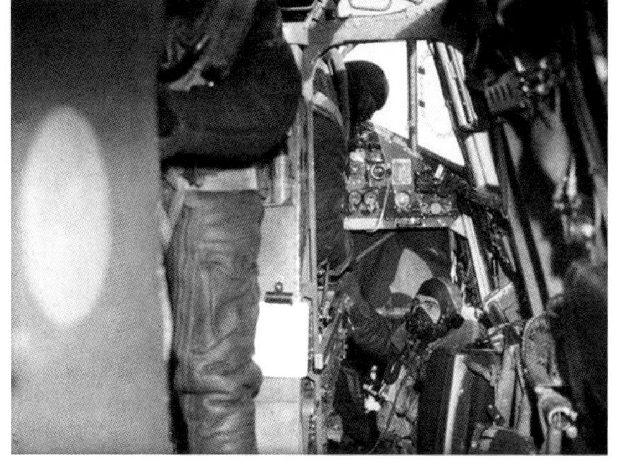

Left: The pull-down seat of the flight engineer above the navigator and bomb aimer. In this view, Sid Stephens is on the 'dicky' seat, while Jack Hounam, A/B, confers with Jim Lewis, Nav.

Mark ll Halifax similar to LW 344 ZA-J that Jack Walker flew when shot down. Note both starboard engines shutdown.

The Mk 3 Halifax had more rectangular tail fins than those of the Mk 2 and Bristol Hercules Engines replaced the older Merlins.

Sgt Martin Gilbert, Halifax rear gunner on Sgt J. Saynor's crew, May 1944. A cold and lonely place of work

Six crews were sent on a Gardening mission on that next night. All laid their mines at the western entrance to the Kiel Canal and, on the return, were issued with a diversion order on account of intruders having been spotted, but this was later cancelled and all returned to base without incident. Then, on the night of 5/6 March, a very large force of 760 bombers was mustered to attack Chemnitz once again, still in support of Operation THUNDERCLAP and the Russian advance from the east, and twenty-two crews were detailed for this raid. Nineteen are known to have attacked the target, over total cloud cover, and as directed by the Master Bomber and the raid, in general, is understood to have been successful. Plt Off Currie had to bomb an earlier alternate target and return when an engine failed and he found difficulty in maintaining height. One crew reported seeing 'jet fighters' in the target area, another sustained some upper turret damage in a brush with a Ju 88, and two did not return. The aircraft of Flt Lt Moss and Flt Lt Stephen were brought down over Germany, with only Flt Lt Moss and two of his crew surviving to be the last 10 Squadron personnel to become POWs in the last weeks of the war. (The Squadron ORB for April records both Flt Lt Moss and his WOP, FS Fowler, as safely back in the UK by the end of that month, with the award of an Immediate DFC for Fg Off Moss. With an American tank driver and two British Army officers captured at Dunkirk, Desmonde Moss had managed to escape and make contact with advancing US forces whilst being moved between camps.) Weather diversions were ordered for all returning crews that night, with most sent to Manston, so no Ops were possible later that day.

Having recovered to base, crews were back in action on the next two nights. Nineteen were detailed for a raid by over 280 aircraft on another oil refinery at Hemmingstedt where, despite clear conditions, the marking must have been inaccurate, as post-attack analysis showed the bomb fall some 2-3 miles away. On the night of 8/9 March, twenty crews were part of another raid on Hamburg by over 300 aircraft, with the Blohm und Voss U Boat yards as the main aiming point. The city was cloud covered, all crews bombed on markers under direction from the Master Bomber but it appears not to have been a highly successful raid.

Twenty-two crews were detailed to participate in the largest RAF raid of the war to date when 1079 aircraft launched a daylight raid on Essen, the last there by the RAF during the war, on 11 March. The intention was to aid troops on the ground, the raid was accurate, and American troops entered the city shortly afterwards. One crew had to abandon the mission early on after losing an engine before setting course. A new record for force size was set the next day, when 1108 aircraft were sent to Dortmund to attack marshalling yards. All of the Squadron's nineteen crews reached the city, covered by cloud, in a devastating attack, with one crew claiming that smoke was still visible 100 miles away on the return leg. The assault on the Ruhr area continued on the next afternoon when another twenty-two crews joined 330 others in an attack on Wuppertal. Once again, all were able to complete the mission, bombing as directed through cloud, in a raid that caused considerable damage for no losses. Fifteen crews were detailed for the next evening, 14 March. The target was Homburg, to support General Patch's US 7th Army by causing damage to block troop and supply movements through the town. The weather was clear, and all of the crews crossed the target area and returned safely after an effective raid.

What was almost certainly the busiest month of the war for the Squadron continued without letup. On 15 March, two groups of crews were detailed. The first involved seven crews in a moderate-sized daylight raid on the Stinnes benzol plant in the Ruhr, described in the Squadron ORB as the "smallest target of Bomber Command." Perhaps so, but the conditions were good, and the attack was considered to have been accurate. Later that evening, another group of nine crews joined a successful raid on Hagen to disrupt rail facilities being used by German troops facing the British advance. The attack went well and, whilst considerable fighter activity was seen, no Squadron aircraft were attacked. A slight break in the tempo of operations followed, with six crews sent Gardening in Danish sealanes on the night of 16 March, another attempt to hinder the shipping of German troops from Norway.

The break was short and nineteen crews were detailed for an attack by some 320 aircraft on rail facilities at Witten in the Ruhr in the early hours of 19 March. All of the Squadron crews reached the target, where conditions were good and a successful raid resulted. On the next morning, fourteen crews were briefed to join an attack on railway yards at Recklinghausen. Although there was only cloud half-cover, PFF marking was not accurate and a scattered raid resulted. Two of the Squadron crews reported having attacked alternate targets as they had been unable to identify the specified target, in one case after three orbits of the target area. On 22 March, eighteen crews were detailed to participate in a comparatively small-scale attack on Dülmen. This went well in good conditions, and, whilst controlled by the Master Bomber, crews were able to check aiming both visually and by internal navaids.

After bombing attacks overnight and an artillery barrage, coordinated crossings of the northern Rhine by British and US units of General Montgomery's 21st Army Group began in the early hours of 24 March. Bomber Command continued to support the operation, attacking several targets during the day, and 10 Squadron detailed twenty-two crews for a daylight raid at Sterkrade, where the aiming point was not now the synthetic oil plant but the railway yards. Conditions were favourable once again, and all bombed as directed using markers and visual identifications, contributing to what was assessed as a highly damaging raid. After landing, an 'unofficial' stand-down was agreed and a party was held that night for two crews at the end of their Ops tours, so it was with considerable surprise that many of those involved found that they were being awakened at around 0330 hours to brief for another early daylight attack on marshalling yards in Osnabruck. Twenty crews took off around 0700 hours but an engine failure broke the run of excellent serviceability of recent days and forced one crew to jettison and return to base. All nineteen others completed the mission in clear conditions, with another successful outcome and, whilst some crews reported no opposition at all, others came back with minor flak damage. One of these had a damaged Airspeed Indicator and elected to land on the emergency runway at Carnaby, leaving the other eighteen to arrive back at Melbourne in style, executing a large formation beat-up of the runway before landing.

Training continued from Melbourne and no other Ops were called for the Squadron until 4 April, when there was a return to night flying in a medium-sized raid on the Rhenania oil plant at Harburg, a district of Hamburg, for which twenty-two crews were detailed. One crew had to abandon the mission immediately after departure when its airspeed indicator failed, but the others reached the target. Conditions were clear, the target was readily identifiable and it was severely damaged. There was then a short break until the night of 8/9 April when 440 aircraft were sent for a further attack on the Blohm und Voss shipyards in Hamburg, for which the Squadron detailed twenty-one crews. Twenty returned to report bombing on markers over a continuous low cloud cover and this was the last significant raid on Hamburg of the war. Two crews with hydraulic system difficulties made precautionary landings on the emergency strip at Carnaby, and FS Hicks' undercarriage collapsed on touchdown at Melbourne, writing-off the aircraft but with no injuries to the crew. One crew did not return as Plt Off Currie's aircraft had come down in Germany, in an area held by Allied forces. Two of the crew survived but with injuries and they were moved to hospital. The five airmen who died - Plt Off Currie, Sgt Parkin, Sgt Squire, Sgt Switzer and Sgt Fortin - were the last members of the Squadron to be killed in action in WW2.

Crews were called for two missions on the night of 9 April. Seven went Gardening, leaving a chain of mines along the Little Belt, north of Kiel. At the same time, eleven crews provided a half of those sent on a diversionary raid on the airfield at Stade whilst almost 600 Lancasters attacked Kiel. All bombed on markers over haze with negligible opposition. However, about thirty minutes later, FS Beaumont experienced a problem with his starboard inner engine. The propeller would not feather, vibration increased alarmingly, the engine was seen to be on fire and, as the aircraft became uncontrollable, he ordered his crew to bale out. All seven men landed in an area of France held by the British Army and were returned to Melbourne, although the Navigator was not found until sometime later.

On 11 April, a predominantly Halifax force was sent to carry out an afternoon attack on the railway yards at Bayreuth in northern Bavaria for which the Squadron detailed twenty-seven crews, the largest number sent on any single mission. Twenty-five reached their designated target, having parted company with two others that found themselves bound

for Nuremberg. One realised too late that they had formed up with Pocklington-based aircraft after takeoff, whilst the other had engine trouble, lost formation whilst descending to cure the problem and then re-joined a mixed Halifax/ Lancaster formation flying above it. Both bombed with the formations that they had joined, whilst those that reached Bayreuth found it visible in clear conditions and bombed as directed for what was deemed a successful attack. After that, the night of 13 April saw what would be the Squadron's last Gardening mission when six crews laid mines in Aabenraa fiord in an attempt to bottle-in elements of the German Navy. Flt Lt Davies lost an engine en-route to the briefed area and, although losing height, he continued to complete the mission on three engines.

(For anyone wondering how despatching twenty-seven aircraft squared with using twenty-six letters for individual aircraft identification, there were two aircraft that flew as ZA-W that afternoon, whilst a further three used 'Bar A/B/D' – ie the letter underscored by a painted bar.)

On 18 April the Squadron was warned to prepare to move to RAF Full Sutton, at a reduced establishment of two Flights, with effect from 25 April. The instruction to move was countermanded by HQ 4 Group on 23 April, although the reduction in establishment was confirmed and was implemented on 25 April. The end was in sight.

Nonetheless, there were still two more Ops to be flown. Another maximum effort was called for on 18 April, when twenty-six crews were detailed for a daylight attack by 969 aircraft on the naval guns and other installations on Heligoland. One crew had to turn back on finding that a port engine feathered itself and would not maintain an operable condition. The others pressed on to bomb both visually and using markers and, overall, the force created what was later described as a "crater-pitted moonscape." (Notwithstanding that description, a force of 9 and 617 Squadron Lancasters returned the next day, armed with *Tallboy* bombs.)

The Squadron's final Op was on 25 April, another attack on coastal guns, this time on the Frisian island of Wangerooge, at the head of the approaches to Bremerhaven and Wilhelmshaven. 482 aircraft were involved, and there was no larger raid in the remaining days of the war. Twenty crews were fielded that afternoon, there were no early returns, and all returned safely to Melbourne despite reporting moderate to heavy flak from adjacent islands. The weather was good and aiming was accomplished visually and on markers as directed.

Although that was the occasion of the last bombs dropped by the Squadron in anger, there were still two weeks to go until the partial surrender signed on Lüneburg Heath on 4 May, followed by the complete unconditional surrender on 7 May – and training continued throughout that period. VE Day was celebrated on 8 May 1945 and so, after almost six years of unremitting effort that had started with a Nickel sortie on 7 September 1939, it was all over, but at a cost. 835 of the 55,573 deaths on Bomber Command operations had been from 10 Squadron. There had been recognition, of course – 9 DSOs, 333 DFCs, 173 DFMs and 2 CGMs plus Bars – totals that ought not to be surprising for a unit that had flown on more raids than any other 4 Group squadron since the first week of the war. Those of us who came later can only hope that, in similar circumstances, we would have done as much.

Epilogue

In late 2011, co-author Dick King fulfilled an ambition when he took a voyage as a passenger on a container ship across the Atlantic and through the Panama Canal from Hamburg to Lima. There was another passenger, Werner Baltrusch, and he and Dick became good friends. Herr Baltrusch grew up in Hamburg, where he still lives after an international career in banking. At the time of the Hamburg raids and the Firestorm of July 1943, he was nine years old and living with an aunt outside the city, but his mother was still working there. His father was in the Wehrmacht on the Eastern Front, where he was killed. Dick later asked if he felt able to say anything about his memories or feelings concerning that time, and this is an extract from his reply:

> "This is a combination of memories of the kid and the old man, and shall be a more or less forgiving and conciliatory trial to understand the British and Germans in WW2.
>
> It was not a very pleasant time for the British and German populations and what both parties did towards the population

of GB and Germany had nothing to do with heroism. But the aim was to bring to an end the terror of WW2 by weakening the morale of the population on both sides. But that was a mistake, the staying power became stronger.

For me as a kid it was not a pleasant time of course, but never traumatic. The kid heard by radio that Goebbels, as Minister of Propaganda, before a big audience, all selected members of the Nazi Party, asked "Do you want the Total War?" and they shouted "Yes!!" Germany started the Total War by bombing London, Coventry and other cities. German

The Bomber Command Memorial in Green Park, London, unveiled by HM The Queen on 28 June 2012.

pilots got some experience during the Civil War in Spain when German bombers destroyed Guernica, responsible the Legion Condor. The answer of bombing British cities was that the RAF started bomb attacks on German cities. In July 1943 began the bombing of Hamburg, the most terrible one was the night from 27th to 28th I guess, nearly totally destroyed by the RAF. My mother was alone in Hamburg helping the victims of former nights – God gracious! The house where we lived was destroyed by a heavy bomb, all inhabitants killed…….The aim of my mother was to be together with her sons and therefore we got a special train to Bavaria and lived there like refugees…..For me, the most painful was: no longer a home, no longer a childhood in peace.

When Hamburg became occupied by British troops we were nearly the first coming back to Hamburg, in an ocean of destroyed houses. We lived for a longer time in the flat of my grandparents before a nearly normal time started………Hamburg was the centre of the British Zone. The Headquarters of the Military Government started very quickly with introducing a democratic system, using the advice of former German managers – one of them was the former Manager of our Bank. He helped to create a new financial system and Trade. By the help of the Marshall Plan Germany became a new future …..

At the end, the old man thinks that the victims and losers of such a terrible WW2 were on both sides, that we should never forget!!"

On instruction from Whitehall, and much to the displeasure of their CinC, bomber crews dropped leaflets and associated material into Europe well into the war. Some surviving examples from the Squadron archives are shown here.

B 21

For Belgium in August 1942, the back and front covers of 'Air Mail,' a 32-page mini-magazine in French and Flemish.

Left: Adapting the gladiatorial salutation to Caesar at the start of a day's entertainment in the Colosseum, this 1942 leaflet to the Italian population shows Mussolini as Hitler's puppet, supplying Italian troops for the German war effort.

Right: A leaflet from the summer of 1944 advising French civilians to move away from major rail junctions and installations, as these would be subject to heavy bombardment.

AVIS

Le Commandement Suprême Interallié a fait radiodiffuser par son porte-parole l'avis qui suit :

LES attaques dirigées contre les centres ferroviaires en France et en Belgique, qui se succèdent depuis quelque temps, et au sujet desquelles nous vous avons donné des avertissements répétés, seront intensifiées au cours des semaines qui vont suivre.

Nous sommes contraints par des nécessités d'ordre militaire de bombarder, de nuit et de jour, les centres d'une importance primordiale aux communications de l'ennemi, tant en territoires occupés qu'en Allemagne même.

Les moyens dont dispose l'ennemi pour concentrer des troupes et du matériel doivent être disloqués. Pour atteindre ce but il faut accroître et maintenir le poids de ces attaques.

C'est à contre-cœur que nous avons pris cette décision militaire grave. Nous savons que l'intensification de nos attaques est susceptible d'entrainer des risques pour la population civile à proximité de ces objectifs, risques encore plus grands que ceux qu'elle a encourus jusqu'ici.

Nos pilotes se rendent parfaitement compte qu'il y va de la vie et des foyers de nos amis. Ils prendront donc le plus de précautions possible ; mais, inévitablement, l'étendue de ces attaques doit aggraver les souffrances que vous, nos amis loyaux, endurez avec tant de courage dans cette guerre.

Nous connaissons les difficultés extrêmes qui, à l'heure actuelle, entravent une évacuation de la population civile. Malgré tout, nous vous demandons avec insistance de faire l'impossible pour vous éloigner immédiatement du voisinage de toutes les installations ferroviaires importantes.

Avec la même insistance nous faisons appel à ceux d'entre vous qui se trouvent dans des régions abritées pour accueillir chez eux ceux qui sont obligés de se déplacer.

Tous les points vitaux des chemins de fer en France et en Belgique vont être soumis à de lourdes attaques aériennes au cours des semaines qui suivront. Eloignez-vous du voisinage de ces objectifs.

F.46

Schlag auf Schlag

Ostfront: Vernichtung der 6. Armee bei Stalingrad

Feldmarschall Paulus und sein Stabschef General Schmidt im Verhör. Paulus kapitulierte mit seinem Stabe 15 Minuten, nachdem sein Hauptquartier unter Beschuss kam.

Die Russen machten bei Stalingrad 91.000, an der ganzen Ostfront in 3 Monaten 349.000 Gefangene.

Ein Toter und ein Überlebender. Generalleutn. von Daniel (377. Inf.-Div.) geht in die Gefangenschaft. 24 deutsche Generäle ergaben sich bei Stalingrad.

240.000 Mann der 6. Armee gingen bei Stalingrad zugrunde. In drei Monaten fielen im Osten über 700.000 Mann.

In drei Monaten verloren die Deutschen 7000 Panzer und 17000 Geschütze.

Generalleutnant Sanne (4. leichte Division) mit den Überlebenden seiner Truppe als russischer Gefangener.

Two leaflets from 1943.

Top: 'Blow upon blow', the first plays up the destruction of the German Sixth Army at Stalingrad. It follows up inside with photos of damage to the Krupps factories in Essen in March 1943, and speculates that events in North Africa may turn out as badly as in Russia.

Bettom: One side of a leaflet revealing some of the destruction wrought upon Hamburg in July 1943 – its reverse tied this in to the German defeats at Stalingrad, Tunis and in Sicily that year.

RAF Melbourne has now become Melrose Farm and the old main runway is often used by a motor sports club. In an old taxiway light mounting this plaque outside their HQ encourages present-day enthusiasts to reflect on past times.

CHAPTER 9

THE DAKOTA YEARS

As the war in Europe ended, the war with Japan in the Far East had still to be won. In late 1944, to support Admiral Mountbatten's theatre invasion plans, the Chiefs of Staff had approved a large-scale policy of reinforcement air trooping once Germany was defeated. This, and the sheer distances involved, meant that an increase in Air Transport capability was a necessity. So, once Bomber Command's immediate work in Europe was done, its aircraft and, more importantly, its aircrews became resources that could be deployed elsewhere. Thus on 7 May 1945, all 12 stations and 14 squadrons of No 4 Group were transferred to Transport Command, pending the delivery of Douglas Dakota (C 47) aircraft under the Lease-Lend scheme. (Transport Command had been created on 25 March 1943, on the amalgamation of the former Ferry Command with a number of UK and overseas-based transport Groups.)

The Operations Record book for 7 May reads as follows:

> "No operations. With effect from 0100hours on this date No 10 Squadron ceased to be incorporated in Bomber Command on transfer to Transport Command.
>
> Four air to sea firing details were completed during the day, Weather at base: Fair. North Westerly wind.

And after listing Postings In and Out:

> 2359. Operations Record Book closed on 7th May 1945."

10 Squadron's war in Europe was over and it had now to take on a new role.

Despite the administrative change, the new aircraft were not immediately available in numbers and the unit retained its Halifaxes for some weeks. A good number of cross-country exercises were programmed, with a proportion of these flown over the Ruhr area with unit groundcrew on board to gain some idea of the end result of their efforts over the past years. (Two of these Continental or Ruhr Cross-Countries – referred to by crews as 'Cook's Tours' - were laid on each day for some weeks, and the last flight was programmed for Thursday 12 July 1945.) Also, for a short time, sorties were flown to dispose of now-surplus bombs over designated areas in the North Sea. (Interestingly, a number of new pilots also underwent a short Halifax conversion, for it would appear that the possibility of using the Halifax in the transport role had not been ruled out.)

However, the Dakota aircraft began to arrive – some were Mk IIIs that had seen considerable previous service but, by early August, most were new Mk IVs. By early June the substantial conversion programme required by this major change in role was in hand, with the records showing it to have been a 7-days per week business. Starting with basic type familiarisation, and moving on to a series of

The Douglas DC 3 - Dakota

low-level cross-countries by day and then at night, and including sorties for the Navigators to become familiar with their Rebecca (for homing and distance measurement) and Loran fixing aids, some 2500 hours were flown in June and July from Melbourne. The Gunners from the Halifax days were not now required, but the Air Bombers were retained for some time into the Dakota era and became Map Readers and then Second Pilots.

There are signs that crews were 'decompressing,' to use a contemporary term, now that they no longer had to face flak and night fighters, somewhat to the frustration of a training staff required to get them ready for overseas deployment as quickly as possible. Ex-Service readers may therefore find familiar reminders of the Squadron having become effectively a training unit in some extracts from the daily schedules:

> "Break Period: Aircrew MUST return to lecture rooms at 1105 hours. Tea is supplied specially at Intelligence and there is no reason for unpunctuality. Failure to observe this will result in the break period being cancelled."
>
> "Any further cases of absenteeism from lectures will receive strong disciplinary action."
>
> "The attention of Pilots is drawn to the fact that they MUST take-off at the time stipulated and an improvement in this respect is expected by the Squadron Commander immediately."

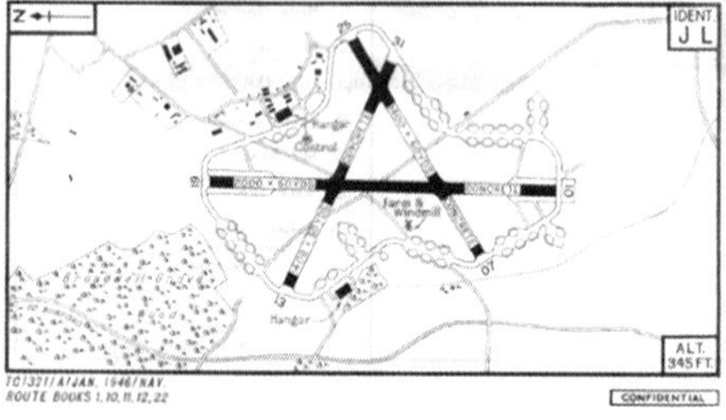

Broadwell 1945 – Now disused it is but 1 mile to the west of Brize Norton.

These daily programmes also included regular indications that an overseas move was imminent, with all personnel being progressively scheduled for medicals and inoculations. All of this culminated in a 14-day pre-embarkation block leave period taken from 21 July, after which the Squadron was to move south to Broadwell, in Oxfordshire, little more than a mile west of the current base at Brize Norton, for applied tactical support training. However, as crews returned from leave, a significant change in plan was in train. On 6 August 1945, the USA dropped

10 Squadron July 1945 prior to leaving Melbourne for Broadwell and India

Prop Repairs at Broadwell.

ZA, the 10 Sqn code letters, were omitted later in India, where the aircraft's single identifier letter was underlined.

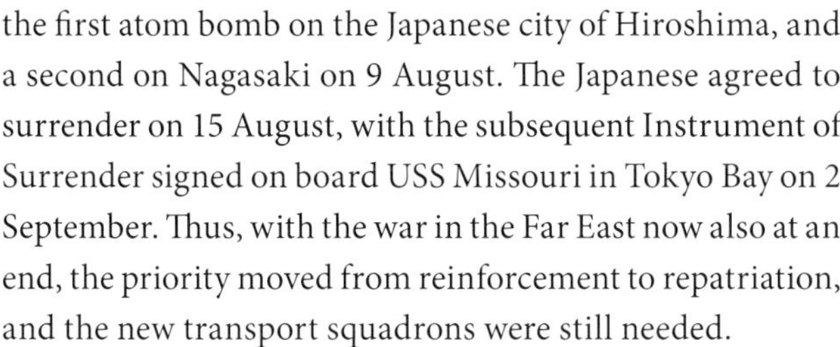

the first atom bomb on the Japanese city of Hiroshima, and a second on Nagasaki on 9 August. The Japanese agreed to surrender on 15 August, with the subsequent Instrument of Surrender signed on board USS Missouri in Tokyo Bay on 2 September. Thus, with the war in the Far East now also at an end, the priority moved from reinforcement to repatriation, and the new transport squadrons were still needed.

As a result, the move to Broadwell went ahead in early August and training continued from there throughout the month. The surviving logbooks of Flt Sgt Stanley Hobbs (Navigator), Sgt Richard Buckingham and Sgt Charles Brignell (Wireless Operator) confirm that trainee crews flew a Glider Towing sortie, a handful of paratroop and supply dropping sorties, and a Radio Range navigation trip before preparing for departure. So it was with only some 45 to 50 hours on type that they flew to RAF St Mawgan near the tip of Cornwall, the starting point for the deployment to India, towards the end of August. Departures would have been staggered to avoid overloading staging posts and, as the official record has a gap from this point, a useful memoir written by Ken Stewart, a squadron navigator at the time, helps fill out the picture:

"Training continued and from 19th August onwards, we moved to St. Mawgan, Cornwall, prior to departure for India. Flights to Poona in India commenced on 1st September 1945, with aircraft leaving at regular intervals with 2 crews per aircraft, ground crews and spare parts.

The route from St. Mawgan was across France via Bordeaux to Elmas, Sardinia, then to Benina and El Adem in Libya, and along the coast to Cairo and Lydda, with the intention of continuing via the Persian Gulf. However, the plan was changed because the Gulf route and associated staging posts were overcrowded with other traffic. Hence, from Lydda, back to Cairo and on to Wadi Halfa, thence to Aden, Masirah Island (N.E. tip of the Hadramaut coast), thence to Drigh Road, Karachi. After a brief stay at Karachi, we moved to Bilaspur, Central Province, then to Poona by early October."

Fg Off Fen Charlesworth, with Ken in his crew, deployed with Fg Off Finding's crew. They left St Mawgan on 3 September 1945 and finally reached Bilaspur on 12 September. The routes flown by others varied a little, depending on the degree of congestion at staging posts, and all were in place by mid-month. (One crew believed that it had been fired upon by Yemenis on its way along the coast of the Arabian peninsula and had to check for small arms damage at Masirah.) The last aircraft carried the Engineering Officer, Flt Lt Joe Soper, plus 11 SNCOs, and had orders to assist any crews who had encountered problems along the way. (A flight to Naples was required to get spares for an aircraft at Malta, and to collect an ENSA party.)

With Wg Cdr Dowden still in command and flying as Copilot on various crews, work in-theatre began at Bilaspur. B Flight Commander Leslie Holes provided extracts from letters home to his wife to a 1996 Association Newsletter, and these give a glimpse of life over the first months in India. He had flown the last peripatetic aircraft to arrive and, as he started to settle in, he remarked:

> "What I like about Ten Squadron is its spirit. Everywhere we've been en-route, life has been rather snooty, but Shiny Ten has a very chummy air, and it was good to get back to them."

He remembers plans for potential basing in Burma and that he carried out a recce of facilities at Mingaladon, near Rangoon – "living in tents, water brought by bowser, not enough to eat." However, on 28 September word came that the final move would be to Poona, to fly repatriation trooping sorties, and that the Squadron should be in place by 5 October. The move took 46 flights using all 24 aircraft. There was also a MT party that had an eventful journey over difficult terrain, including a 135 mile detour when a hoped-for fordable stream turned out to be a river some 8 feet deep. Leslie wrote home to his wife:

> "It's a lovely climate here, cool and clear. The monsoon is just going, just odd storms like an English summer. The country is flat, 2000 ft high, with peculiar hills, three or four of 'em sticking up, and with a lovely blue range of hills, quite high, to the west, making a perfect setting for the greenery of the town. The camp, what I've seen of it, is horribly dispersed."

From Poona the Squadron flew trooping flights, carrying time-expired troops and some ex-prisoners of the Japanese who were brought from Singapore to Madras by flying boat. 10 Squadron with other squadrons then flew them from Arkonam, Madras via Poona, to Karachi, for return to the U.K. by Liberator bombers. As a typical example, during October, November and December 1945 and January 1946, one of the many squadron Dakota crews made 23 trips from Madras to Karachi, carrying a full load of 24 troops each time.

From late October, Leslie's letters home began to mention changes to come in unit organisation. He wrote prophetically, if not enthusiastically:

> "It's a very bad show. We certainly seem to be going for the complete Transport Command organisation in a big way, and I can see Shiny 10 flying transport kites for many a moon."

One change meant that the Halifax Air Bombers who had been retained as Second Pilots on the Dakota were to be replaced by pilots and, within a fortnight the first of them arrived. Leslie's full complement was in place by the end of November:

> "over half of them are glider bods who haven't flown since Service school, and who were put onto gliding when they returned to England."

The practical realities of running a passenger service began to become clear:

> "19 Nov: This trooping is very delicate, and it doesn't take much to upset it. Now we have some more duff weather in the Middle East and everything is held up. Mauripur is solid with kites waiting to go, the trooping camp is full up. A couple of weeks ago I had too much work for my blokes, and now I can't find enough. We don't have all that number of kites either, as they are all coming due for big inspections. The chaps seem pretty keen, although they are not as passenger-conscious as I should like.
>
> 23 Nov: Having been scrubbing trips for about a week, now we rush. Yesterday, we had four extra on as, when

KN 235 – Q awaits a crew and baggage loaders

weather did clear, Mauripur was cleaned right out."

And the need for a measure of control in a unit with a passenger-flying task possibly seemed more necessary when he wrote on 1 December 1945:

> "Today has been pretty bad, as a couple of my bods had a mid-air collision, entirely the fault of one of them, and while they were doing local flying. He got his rudder sawed completely off, by the port propeller of the other aircraft when pulling up in front of it. A foot or more and it would have been his elevators. He came in and landed beautifully, but he is living on borrowed time from now on…..The WingCo is raving, or was. He has quietened down a little."

1946

Leslie noted a reduction in the rate of trooping flights during December, and a Government announcement to this effect came in January 1946. He wrote on 15 December:

> "Trooping is very nearly at a standstill, and the chaps in our trooping camp are getting a little fed up too. Nothing extraordinary, but it's understandable that chaps who once were rubbing their hands at the possibility of being home for Christmas, but seeing the odds against mounting every day, get a little fed up."

On 11 January, he noted that *"trooping is running at about half strength. Three aircraft from a squadron with 25 kites and twice that number of crews"* and matters came to a head in late January. Wartime personnel became concerned that participation in the trooping task would lead to their losing jobs back in the UK as others got home ahead of them, and a strike took place over a few days at various RAF Far East Command locations, including Poona. Nonetheless, 10 Squadron operations were maintained over that time by regulars, NCOs and officers.

In the interim, more detail had become available about what would be involved in the planned Transport Command categorisation scheme. Additional flying training began and Leslie began to carry out pre-Cat checks both in the air and on the ground:

> 6 Feb: Poor old Jock Kerr has been my guinea pig on a test categorisation. It took an hour and a half yesterday morning, an hour last night, and another ninety minutes this morning, but it's finished and done with now, until the real one comes along in about a month. [Events would force a delay in that plan.]
>
> 8 Feb: After organising things at the flight this morning, I did an hour and a half of local training. An instrument take-off, lots of instrument flying, flying on one engine,

simulation of fire in the air and finally, what we call a QGH, a letdown procedure used in bad weather, a bad weather circuit and landing.

The morning's flying appears to have set him up for what was to follow, for which it will help to know that Leslie was also by then the President of the Mess Committee, and clearly taking his responsibilities seriously. He added:

"Later, when I went into the Mess, I discovered a lot of flap over a pi-dog, one which had been bothering us for a long time. In the end we got him, and I nearly blew his head off with a .38. We dumped the body in my truck and disposed of it on the airfield."

Further developments were in the wind:

"16 Feb: It would appear that we might revert to our original close-support role at any time, and we will probably gradually give up trooping. Starting yesterday, we've had a bit of a flap on…. Nothing has happened yet, but we have had aircraft and crews standing by to go somewhere in a hurry, and we don't know where.

22 Feb: I notice that Poona has been mentioned in the newspapers concerning the Bombay do [trouble in the Royal Indian Navy[…. We've all the aircraft standing by. These have been used to a certain extent to bring up air reinforcements and, if anything does happen in Bombay, these ships will have had their chips in no uncertain manner. These troubles are the reason for our being out here, as I've said all along, and I'm glad we are being used as now it's proving to the blokes, both ground and air, that we are of use. The Wingco and I agree that this is an excellent thing.

26 Feb: I've been busy …. sorting out how we are going to do this close-support training. The morale of the Squadron is quite high now we are really working for our livings. The Bombay riots have died down."

Japanese PoWs were used as loaders at Myitkyina and Meiktila, during the Spring 1946 Famine Relief flights to locals who had denied the Japanese their crops during the war.

Rice dropping from Myitkyina, north-east Burma. It was in this area that 10 Squadron lost 3 Dakotas on the same day on 29 March 1946.

Leslie's release for repatriation came in early March, but not before he had taken other unit personnel to Santa Cruz, Bombay, for the same reason:

"1 Mar: The Wingco told me beat the place up, so I did, three times, and apparently it was well received. 220 mph is quite good going for a Dak."

And then:

> 12 Mar: To answer your question, trooping was just a job, one which had to be done, and so we did it, but at the back of the high-ups' minds was and is the fact that to cover any trouble, airlift is very useful, whipping men and material around the country."

The validity of Leslie's latter point would emerge again and again as violent unrest surrounded the approaching Indian independence and the separation of Pakistan as an independent nation. But there was a more immediate need for 10 Squadron's capabilities. On his way home, Leslie wrote from Bombay:

> "23 Mar: I left the Squadron on Monday afternoon, certainly at the right time as they are all nicely ensconced in tents at a place called Meiktila, south of Mandalay, in Burma. There was quite a mad panic over the weekend. The whole squadron has gone, including erks and supplies, and what has made it a little more difficult is that the Adj is new. The signal came through on Friday, and the advance party left on Saturday. There is a dreadful food shortage in N. Burma, by the Chinese border, and an extra squadron was needed to drop rice. So Ten went, good old Ten. This has made most of the Squadron quite happy."

And that is exactly what happened. In mid-March, the unit was detached to Meiktila, in Central Burma, for Operation HUNGER II. (Repairs and maintenance would be carried out at Chukulia, and drops were also carried out from an airfield at Myitkyina.) Flt Lt Doug Newham LVO DFC was Squadron Navigation Leader at the time, and has vivid memories of the period:

> "The operational requirement arose in the first instance from the fact that the invading Japanese largely "lived off the land", and sequestered food from the local village. In order to minimise this the District Commissioner (Brit) of the local Civilian Burmese authorities (I subsequently met him) was infiltrated behind Japanese lines to persuade the local villages to destroy their stocks of rice, thereby denying the invaders of local food and forcing them to supply from elsewhere. Anything to make life difficult for the enemy.

Japanese prisoners loading salt.

4 × 1 ton trucks – Burmese style.

Out in the mid-day sun.

The Kachin peoples were already waging a bitter guerilla war against the Japs and kindly obliged. Once the Japs had been driven south, and finally beaten, the authorities had a moral and humanitarian obligation to resupply the Kachins who were facing a serious famine and already existing only on their seed-rice.

My records quote a requirement for 6,000 tons of rice and 180 tons of salt. The Army and civil authorities distributed about 50% via rail and road, but it was only by air-drops that the remote mountain villages could be reached in time and before the onset of the monsoon. Two main supply bases were established – Meiktila in the central plain and Myitkyina in the far north-east - both supplied primarily by rail. Shiny Ten was moved from Poona to Meiktila to undertake the supply dropping using "free-drop" technique – parachute supply would have taken far too long and been insufficiently accurate in the terrain concerned. Liberators and Halifax aircraft engaged in supplying China via the "The Hump Route" were also used, but I believe that this was primarily to build-up the supplies at the two main depots rather than being employed on the tactical drops.

We started arriving at Meiktila on 17 March and established a tented camp on the earlier Japanese landing-strip that had been bombed and shelled when the Allies advanced south towards Rangoon. A replacement landing strip had later been constructed by the British Army Airfield Construction units when under fire as they moved south – bulldozing by day, but retreating a few hundred yards at night. We were camped amongst the bomb-craters which became small lakes full of frogs when the rains came! The flying strip was just a few hundred yards away and ran parallel. (Major Willie Kerr was i/c this unit. Post-war he became a close friend of my younger son and I discovered the near coincidence of our time in Meiktila.) I flew from Meiktila to Rangoon on 20 March to collect a supply of topographical maps of the dropping zones. As far as I remember one of the Flight Commanders later did a recce of some of the sites, primarily to verify that they were each displaying an identification, and to satisfy himself on the approach-paths. Certainly we were all briefed on the need to approach at a safe altitude, circle down to execute the repeated dropping-run, and then climb up to a safe height before leaving the area.

The rice was pre-packed in 75lb sacks. Three sacks inside each other, the outer might tear and the inner possibly burst, but very rarely would you see a burst of white rice showering the ground. After the first pre-monsoon rains the sacks then weighed well over 100lbs and were warm as the rice started germinating. (One severe thunderstorm flooded the shell-holes around our tents and caused probably about 60% of them to collapse as the tent-pegs pulled out when rock-solid ground suddenly turned to liquid mud.) When on a dropping operation the crew routinely included a Sepoy or "squaddy" from the Royal Indian Army Service Corps to help in man-hauling the sacks of rice from forward in the fuselage to pile them neatly and head-high at the very threshold of the open doorway. We flew the Dakotas with the main door removed, although reinstalled when back at base at night to prevent rain ingress In all cases we departed Meiktila with a full load of rice. In some cases this was to assist in building up the necessary stock at the forward base of Myitkyina and, if adverse weather prevented a successful drop, we would return there to unload. If the drop went OK we would return to Myitkyina to collect a second or even third load for repeated operations, not necessary to the same dropping zone/village. The terrain to the east and north of Myinkyina was particularly inhospitable rising to well over 10,000ft in a jumble or razor sharp peaks and ridges with precipitous gorges and valleys in-between. Eight villages were selected by the authorities for use as dropping zone, with the local head-men responsible for further distribution. These villages were for the most part located along the crests of the mountain ridges."

The terrain into which crews were flying was indeed mountainous and challenging, and these conditions took a toll. The Squadron lost three aircraft and crews on the same day, 29 March 1946. Pilot Flt Lt Len Broadhurst was at Myitkyina that day and in 1996 wrote for an Association Newsletter:

"During the early afternoon, having checked out the the aircraft that should have landed at Myitkyina for a second load, Roy (Phillips, WOP) and I became increasingly concerned for the three aircraft that were by late afternoon well and truly overdue. Due to the heat we were unable to get in touch by radio with either our Squadron base at Meiktila or Group HQ at Rangoon. Eventually we made contact and reported that the three aircraft and their crews were "Missing.' I made contact with a retired Army officer living in Myitkyina and he enlisted the help of some local people to search for our missing aircraft.... An investigating officer arrived from Group HQ and took command over what had by now become a major search operation."

However, Doug Newham was in one of the crews airborne that day:

"On 29 March four aircraft flew first to Myitkyina where we had a final briefing on the selected village – Htawgaw. To my knowledge this was our first visit to this Dropping Zone (DZ) and I was in the first aircraft with Fg Off Palmer as the pilot. The approach necessitated climbing to about 12,000ft over multiple sharp mountain ridges. Having identified the village we circled down in the confining valley and made our repeated runs at less then 30ft above tree-tops, running along the ridge rather than across it (for to do otherwise would have caused height judgement to be considerably less accurate). Only the pilot and co-pilot would be in the cockpit, the rest of the crew would be aft, hauling, stacking and dispatching the sacks when the signal light above the door said "Go". As soon as the stack of sacks had been toppled, pushed and kicked out of the door, one of us would put his head out to look behind and down, to ascertain the accuracy of the delivery. (My favourite position was to lie on the floor with my shoulders against a spare sack on the starboard side and heave the sacks out with my feet, whilst the radio operator and RIASCO squaddy would be toppling the pile sideways. I'd then roll onto my stomach and put my head out of the door at floor level.) Meanwhile, and no easy task, the pilots would increase power and wriggle their way round in that narrow valley ready for the next run. Having completed our drops we then climbed in fairly tight circles up to safety height, and I distinctly remember that aircraft numbers two and three were in orbit at safe height, waiting until we had finished before taking their turn.

We returned to Myitkyina to collect a further load for a repeat sortie, but were unexpectedly reallocated to deliver a load of freight to an airstrip at Putao (Fort Hertz) in the extreme north of Burma, in a narrow valley bordered by very high snow capped mountains. (We unloaded our freight directly into panniers attached to the sides of elephants which were gently edged right up to the door threshold.) It was whilst airborne on our back to Myitkyina (with four gun-running Chinese who were in chains, plus the same District Commissioner who had earlier been behind Jap lines) that we received radio requests querying if we had any clues as to three missing 10 Squadron aircraft. Missing? We knew nothing! Number Four never returned from his first sortie and Numbers Two and Three did not return from their second. After landing at Myitkyina we immediately flew back to the original dropping zone at Htawgaw to search for the aircraft, but with no signs of crash, smoke or fire, until darkness forced us to abandon our search and return to Meiktila.

Search action started next day and the Jungle Rescue units were also alerted Detailed search areas were allocated to a number of squadron aircraft, each diligently searching a small designated area before moving to the next. It was during one of these searches that one crew, at relatively low level in a narrow valley, turned a corner to find an almost blind-end. I learned from the crew that it was only with difficulty, full power and a large amount of luck that they avoided crashing into the mountain wall. But in doing so one of the crew, standing in the doorway, saw the remains of aircraft Two and Three, with a surviving figure trying to wave to them! Jungle Rescue finally got into the site. The survivor was one of the RIASCO Sepoys who was either thrown out of the door or jumped at the very last second and landed in bamboo. He was severely damaged, but alive. There were no other survivors."

The Indian Army Sepoy was spotted on the morning of 3 April by a searching 10 Squadron crew, and supplies were dropped to ground parties. (Extracts from the logbook of Sgt John McCubbin, Wireless Operator with Flt Lt Moss' crew, have proved helpful in providing detail over this period, and it was this crew that found the survivor.) Further searching and patrolling was carried out – bits of an aircraft were seen on 4 April, and more supplies were dropped on a number of following days. There were search sorties each day but nothing more was found, and Sgt McCubbin's logbook

Time carried forward :— 455 : 45 | 159 : 10

Date	Hour	Aircraft Type and No.	Pilot	Duty	Remarks (including results of bombing, gunnery, exercises, etc.)	Flying Times Day	Flying Times Night
8·4·46	0800	KN 450 DAK 'O'	F/LT. MOSS	W/OP.	JUNGLE SEARCH AROUND BROADWAY STRIP FOR OTHER A/C NOT FOUND - NO LUCK.	4 : 15	
8·4·46	1600	KN 450 DAK. O	F/LT. MOSS	W/OP.	SUPPLY DROPPING TO GROUND PARTY.	1 : 10	
9·4·46	0600	KN 450 DAK 'O'	F/LT. MOSS.	W/OP.	FURTHER SEARCH FOR A/C NOT FOUND — NO LUCK	4 : 30	
10·4·46	0600	KN 450 DAK. 'O'	F/LT. MOSS.	W/OP.	FURTHER JUNGLE SEARCH FOR A/C NOT FOUND - NO LUCK	5 : 00	
11·4·46	0620	KN 450 DAK. 'O'	F/LT. MOSS.	W/OP.	FURTHER JUNGLE SEARCH FOR A/C NOT FOUND - NO LUCK	4 : 10	
11·4·46	1130	KN 450 DAK 'O'	F/LT. MOSS.	W/OP.	AIR TEST.	:20	
11·4·46	1600	KN 450 DAK 'O'	F/LT. MOSS.	W/OP.	SUPPLY DROPPING TO F/LT LUNAN ON SEARCH	2 : 00	
12·4·46	0630	KN 450 DAK. 'O'	F/LT. MOSS.	W/OP.	DROPPING TOOLS TO MAKE AIR-STRIP FOR L5.	1 : 40	
12·4·46	0900	KN 450 DAK 'O'	F/LT. MOSS.	W/OP.	PATROL OVER GROUND PARTY — [INTERCEPTED MESSAGE]	1 : 45	
12·4·46	1300	KN 450 DAK 'O'	F/LT. MOSS.	W/OP.	PATROL OVER GROUND PARTY — TO RECALL THEM	1 : 30	
13·4·46	1030	KN 450 DAK. 'O'	F/LT. MOSS	W/OP.	SUPPLY DROPPING TO W/O. PEW AT NEW STRIP.	3 : 05	
					TOTAL TIME	485 : 10	159 : 10

John McCubbin's logbook showing the search for the aircraft lost on 29 March 1946.

records a final sortie on the afternoon of 12 April to patrol over the ground party and recall them. With records for the period lost, it has not proved possible to match all names to aircraft. The best-known listing of those missing that day is as follows:

> KN 643: Fg Off H A Beer; WO D Main
>
> KN 644: Fg Off R Stimson RNZAF; Fg Off W K Cooper; Fg Off G Stedman; WO L Welch
>
> KP 270: Flt Lt F B Shaw; Fg Off G F Humes
>
> Others lost include: WO E C Cook; Flt Lt W Gorley; Plt Off J Paby; WO D C Redfern; FS J E L Rhodes; WO F Robinson; WO F B Robinson. There were also four Indian Army Sepoys named: Bakhtawar, Dhaneshwar, Kanshi Ram, and Atar Singh.

The weather and Dacoit bandits in the area were amongst the hazards likely to be encountered whilst flying in Burma as Fg Off Al Alcroft recalled in a contribution to the Association Newsletter in 1996. (Leslie Holes had written that he was heartily glad that the Squadron had not been based at Mingaladon, as once contemplated, after hearing that 8 men had been knifed by Dacoits whilst asleep.) His crew had completed their last rice drop of the day and were en-route back to Meiktila:

> "We were clearing the hills but the cloud lowered and lowered, and the thunder and lightning festered all around us. WOP Stan Mitchell finally wound in the aerial, finding communications impossible. Knowing that we were near the Irrawaddy and that, if we didn't act quickly, we would be caught in cloud in the hills with no navigational aids, we decided to do a dive through a hole through which I recognised an airstrip we passed on each trip – Schwebo, a deserted ex-Jap fighter strip.
>
> Conditions were now so bad that the only thing to do was to effect a precautionary landing.... It involved flying low over the terrain to ascertain the nature of the surface prior to landing. We did two or three runs before 'lobbing down' and securing the Dak in the Japanese revetments – and not without some difficulty. A Jap Zero was a lot smaller than a Dak. What to do now? Each of us had a Smith & Wesson .38 pistol. It was decided that two should stay on guard, and the other two do a recce for signs of life, hoping there might be some British troops around. By this time, the storm had nearly subsided but the day was too far gone for us to unshackle the aircraft and return to base."
>
> A truck carrying a number of villagers with an English-speaking headman now appeared. After having a good look at the aircraft, he mentioned that a detachment of British officered Kachin troops was in the area, as were Dacoit bandits, pillaging and killing. A jeep with Kachin soldiers now arrived and the Corporal in charge deployed his armed men to secure the aircraft – and matters began to inprove:
>
> "He invited us into the jeep and quickly drove us through the jungle to an Army Mess, built in the trees and entered by a ladder. We were greeted in a most enthusiastic and friendly way by the 'brown jobs' and, after I had managed to ring Meiktila on a field telephone, were right royally entertained. Imagine – dining off pheasant and drinking the best of brandy in a tree house in the Burmese jungle! .... We took our leave the following day, and the only thing our hosts wanted from us, if possible, was a soccer ball. The troops were mad on the game and had no real ball to play with. This was soon remedied – another crew and the PE section combined to fix a drop of three soccer balls on Schwebo airstrip."

When the task in Burma was complete – 2855 flying hours in 40 days - the Squadron returned to Poona in late April, some 20 days earlier than expected thanks to the intensive flying rate achieved. But it would not be long before another more permanent move was ordered. On 24 May, instructions came that the Squadron was to move to Mauripur (Karachi) and an Advance Party left two days later. There had been a RAF station there since 1942, and it had been extended to cope with the dramatic post-war movement back to the UK. The task remained one involving general transport flying but, with Indian independence not far ahead, the ramifications of 'partition,' involving the creation of a separate Muslim state of Pakistan, would increasingly affect matters. On the admin front, it should be noted that wartime Squadron personnel were completing their service during this period and being repatriated, usually by troopship, whilst replacements were coming out from the UK. Also, with the main trooping task at an end, a decision had been taken to decentralise command of overseas-based Transport Command units. As a result, in March 1946, control of No 229 Group (including 10 Squadron) transferred to HQ Base Air Forces SEA, administered by Air Headquarters (India). Finally, coincident with the move to Mauripur at the end of May, Wg Cdr A J Chorlton took over command of the Squadron.

The official record resumes from 1 October 1946, giving every sign that the Squadron was well established at Mauripur by that date:

> "The 'LJ' Service was flown by Flt Lt Bumford. One aircraft was used for day training; a full lecture programme continued with unabated boredom."

And on 9 October:

> "The 'water run' was the only trip; training continued with its wonted weariness."

(The 'water run' was to the base at Jiwani, close to what is now the Pakistan/Iran border. It had a complex water supply system pumping water to the airfield, and it appears that this system had to be supplied from outside during the very dry months of the year.)

We therefore know that a number of regular routes were being flown, and that the delayed first visit of the Transport Command Categorisation Team - 'The Trappers' - was imminent. Sure enough, they arrived on 22 October to the remark: "*To B or not to B, that is the question!*" And when the Team left on 11 November, three Pilots, a Navigator, and a Wireless Operator had attained that category, with others assessed at C or D – all of which will ring loud bells with 10 Squadron aircrew of the later VC10 era.

Having carried out a famine relief operation earlier in the year, 10 Squadron crews now found themselves involved in another. Since August there had been serious rioting in the predominantly Muslim areas of East Bengal, accompanied by burnings and killings, together with allegations of forced conversions. The situation deteriorated markedly in October and to provide transportation in an area increasingly brought to a halt by the communal strife, RAF aircraft were called for. A first aircraft left for Dum Dum, on the other side of India, near Calcutta, on 10 October, prepared to drop food. Two more were sent on 24 October, another two on the next day, and a further two on 27 October. Two more were put on Standby from 29 October, and one of these deployed on 1 November and another on 8 November. The crews and aircraft recovered to Mauripur between 17 and 22 November.

(Wg Cdr A J Ogilvie DSO DFC AFC took over command of the Squadron on 1 November 1946.)

On 18 November, the unit received confirmation that it was now to deploy to RAF Chaklala, a parachute training facility near Rawalpindi in the Punjab. Crews would be away on a 4-week detachment, carrying out paratroop and supply dropping and general Army cooperation. (Aircraft and crews of No 62 Squadron based at Mingaladon in Burma would also participate.) An advance party left on 21 November, with a further 9 aircraft and all available crews following over the next two days. It would appear that the dispersal allocated was 'very satisfactory' but there was some difficulty in finding office accommodation. As crews and groundcrew settled in, an unusual task was received on 24 November when a crew had to fly an Iron Lung to Dum Dum for use by a RAF sergeant in a Calcutta hospital. It returned to Chaklala on 26 November where training had begun in preparation for a Company-scale drop at Quetta on 1 December.

Eight aircraft were used for the Company drop. It went

Loading a field gun into a Dak

Aircrew Tents at Meiktila after heavy rain

Landing during the Monsoon

SNCO's Transit Billets at Chaklala

10 Sqn and 31 Sqn Dakotas at Mauripur 1947

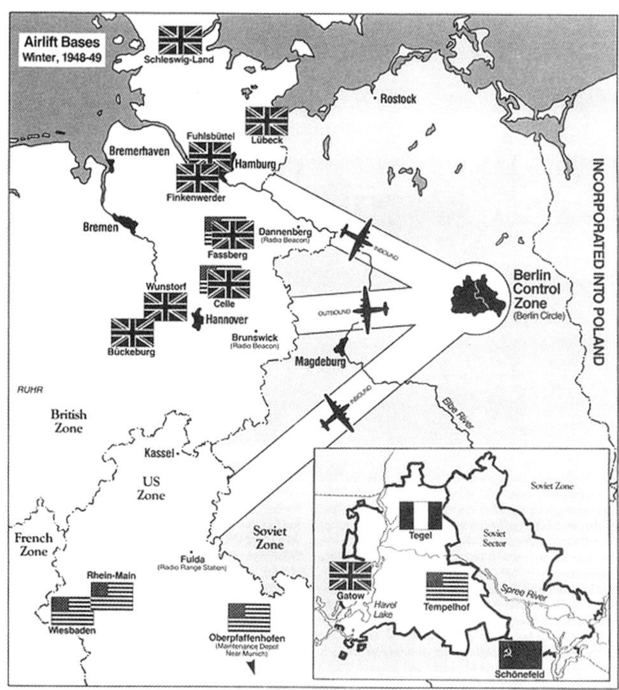

The 3 Berlin Corridors established in 1948 linking Berlin to the West, continued to exist until the 1990s. 'Operation Knicker' was changed to the more well-known 'Operation Plainfare' in July 1948.

Ground crew at Lübeck

Carrying mainly food, milk and coal into Berlin the Daks' outbound loads frequently comprised those wishing to leave the city, with hungry mouths to feed.

Gatow in the British sector and Templehof in the US sector were the main airfields used in Berlin during 'Operation Plainfare' but Tegel in the French sector was used from November 1948 onwards. Gatow handled 20 aircraft per hour in fine weather.

Refueling at Lübeck . Boxing Day 1948 saw 10 Sqn move for a three week period to Buckeburg, near Minden, as a temporary base during the Airlift.

Turnround in Berlin. In the early days of the Airlift coal for the city was carried in white RAF kitbags owing to of a shortage of coal sacks.

well, with all troops landing on the DZ, and the first Time Over Target (TOT) just 40 seconds late, but things went less well administratively at Samungli, near Quetta, where the aircraft landed to refuel and pick up the paratroopers. A significant lesson learned was that troops are not at their best after four hours in a Dakota with its doors removed – and hesitant exits, creating a need for second runs over the DZ, were the result. Much useful training and further Company-size drops took place in the subsequent weeks, and an exercise with an operational setting was held, Exercise HILL PARTY, from 12-14 December. (The recording officer at this period, Fg Off J C Nicholls, who demonstrated a distinctly personal style in his monthly musings, took delight in describing in some detail a para drop carried out on the afternoon of 19 December by Wg Cdr Ogilvie, 13 other squadron aircrew, and their Parachute Jumping Instructor (PJI). "Unorthodox exits," and "astonishingly painless landings" are reported, save in the case of the PJI who appears to have hit the DZ face down.) The standard of formation flying and dropping appears to have improved over the period of the detachment, although the record does not hide the few occasions when map reading errors or the like took place. The detachment ended on 23 December, when Wg Cdr Ogilvie led a 9-ship formation back to Mauripur, with a rear party following later in the tenth aircraft, for what appears to have been a convivial Christmas. An post-exercise washup and some flying occupied 30 December, after which most adjourned for an equally convivial New Year.

1947

Flt Lt Johnson and his crew plus, of course, sundry groundcrew, were unable to party with their colleagues as they had a trip on 1 January, to Palam and Peshawar. (The task took until 9 January and is intriguingly described as a 'Commitment of a Secret Nature from Peshawar," but with no additional detail recorded.) Nor did Flt Lt Wight, the Nav Leader, have a sparkling start to the year. Whilst expecting a release from hospital that day after an appendix removal, it emerged that he now had amoebic dysentery – and he was joined by the Squadron CO on 3 January. Clearly something in the water!

On 7 and 8 January, 6 crews and 34 groundcrew left for another detachment at Chaklala, again with 62 Squadron participation, but his time for only two weeks. The exercise culminated in a 2-Company line-astern drop of some 300 personnel onto a small DZ (500 x 100 yards), followed by a POL drops later in the day and on the next. Two crews returned on 24 January, with the others remaining at Chaklala to watch a demonstration loading and drop of a 75mm gun by Sqn Ldr Flemons, and all returned by 26 January. Over the month, the Squadron took over scheduled flights from 31 Squadron. Three new crews were posted-in, bringing crew strength to 17, and February saw more aircraft at Chaklala and Poona, mixed with scheduled flying and local training and the continuing instruction of RIAF aircrews.

The political situation in India at this time was particularly difficult and, as the year progressed, would increasingly affect unit tasking. The desire for independence had a long history in colonial India and, although support for the UK in both World Wars had been very great indeed – 10 Squadron had worked with the Indian Army from its first months in France in 1915 - Gandhi's Quit India movement had also been a strong influence at home. After the war, the move towards independence became unstoppable and, with varying degrees of enthusiasm, was accepted in London. In February 1947, Prime Minister Clement Attlee announced that India would have self-government by mid-1948 at the latest. We have already seen that serious inter-communal strife between Hindus and Muslims had occurred in 1946, and more and worse would come in 1947. It became clear that a partitioning of India into essentially Hindu and Muslim states would be inevitable, and a plan for this was announced by the Viceroy, Lord Mountbatten, on 3 June 1947. Both Bengal and Punjab would be divided as a result, with the new boundaries published just after partition to create West and East Pakistan (now Bangladesh), separated by the width of India.

It was against this developing situation that the Squadron took on a significant task from January – the training of crews of No 12 Squadron, Royal Indian Air Force (RIAF). The unit had formed in December 1945 at Kohat; moved to Risalpur and then Bhopal with the Spitfire Mk VIII; and spent much of 1946 flying the Airspeed Oxford and adding a Navigation School, in anticipation of becoming a transport

squadron with the Dakota. The first aircraft arrived in September, a crew from 10 Squadron and another from 77 Squadron were posted-in for instructional duty, and the unit came under functional control of 229 Group. In late December the Dakota elements of the unit were detached to Mauripur to join 10 Squadron and complete its conversion as the RIAF's first transport squadron.

In early March, serious rioting in Multan resulted in a number of tasks to transport troops into the area. 9 aircraft left in the early hours of 10 March; on their return, a further call came, this time for 10 aircraft; and another 8 on 14 March. With these reinforcement tasks complete, flying for the remainder of the month reverted to the customary mix of schedules, local and airdrop training, with the training of 12 Squadron RIAF personnel still in hand. The aircraft strength at the end of March was noted as 12, plus 8 in storage. (Incidentally, the first use of the new 1946 terminology for non-commissioned aircrew occurs in March 1947 – eg Pilot II Smith and later, Pilot 2 Smith. The scheme was very unpopular and ended in 1950 with personnel reverting to their previous SNCO ranks.)

As the year progressed, with the training of 12 Squadron RIAF completed in May, the Squadron was tasked from June with the training of personnel of No 6 Squadron RIAF, converting from Spitfire Mks VIII and XIV. Post-partition, this squadron was destined for the Pakistani Air Force and, in mid-July, unit personnel had to choose to serve with the Indian or Pakistani services and a substantial majority elected to serve in India. The 6 Squadron record for August is headed 'Royal Pakistan Air Force' for the first time and shows just three pilots, a navigator, and two WOPs in training with 10 Squadron, with the Indian contingent moving out to Agra early in the month. No equipment or aircraft were held on charge, there were only seventeen airmen and two SNCOs remaining, so it was a barely viable unit that emerged after partition – in stark contrast with 12 Squadron and its fully-manned ten Dakotas. Matters were not helped by the level of 10 Squadron tasking in the immediate aftermath that severely curtailed training, with only two sorties recorded.

In the same period, another RAF squadron began to form at Mauripur, No 62, gradually drawing personnel from both 10 and 31 Squadrons whilst, administratively, on 16 June 10 Squadron passed from No 1 (I) group to control by RAF Base, Karachi. Operationally, the Jiwani schedule reverted to 31 Squadron in mid-June, and the records suggest that 10 Squadron was then used increasingly for personnel or trooping tasks. For example, on 9 July, 10 aircraft brought airmen from Santa Cruz to Mauripur, presumably for repatriation. Conversely, 4 aircraft then flew Nurses, families and airmen to Santa Cruz, and a number of flights were tasked for the movement of airmen of 5 Squadron, recently converted to Tempest fighters, to Peshawar. Also, with partition approaching in mid-August, crews were tasked with delivering 'Viceregal' and other secret documents around the country throughout the last week of July. Partition occurred around midnight on the night of 14/15 August 1947. Nehru's celebrated rhetoric that "At the midnight hour, when the world sleeps, India will awake to life and freedom" was slightly adrift. In fact, Pakistan was declared as a separate nation at 2357 hours on 14 August, thus permitting the new Indian state to be declared independent 5 minutes later at 0002 hours on 15 August.

With the colonial era now at an end, the return of British forces to the UK could not be long delayed and, for the moment, recognising the changed circumstances, the Squadron was now attached to a HQ entitled Armed Forces Pakistan. (Records of postings-in reveal that it had also formed a dedicated crew for occasional flights carrying the new Governor General of Pakistan, Muhammad Ali Jinnah, leader of the Muslim League in the years before the grant of independence.) However, before any return, there was much work to be done. As the borders of the two new separated nations became known, inter-communal strife increased and an immense movement of populations began.

John Webb was an early National Service AC Fitter who joined the Squadron in early 1947. Writing for an Association Newsletter in 1997, he recalled instances showing how desperate the situation was. During the training of the Indian squadron, its airmen would not leave their billet for fear of being shot by local guards and were finally evacuated. He recalled supervising their departure and arranging for trucks

to back up to their hut doors whilst he and colleagues kept the Pathan guards at bay. And on another occasion, he was with a crew about to depart on a Sikh repatriation flight. The engines were started when a large Sikh man appeared with baggage, frantically pleading to be allowed on board. The captain shouted that he could come as he stood, but without bags, and John helped him onboard. He was in tears and relieved to be alive, and his reason for this became clear as the aircraft taxied to the runway for a rapid takeoff – a group of mounted Pathans appeared, firing at the aircraft, clearly frustrated that this individual had escaped them.

The unit records for September are missing, but those for October give the impression that Service transportation was provided essentially for officials, their families, and troops. An Operation codenamed BEAVER II lasted from 8 – 16 October and involved 12 aircraft at Chaklala, moving Sikh troops. 23 sorties from Miranshah are recorded on 10 October, and 28 from Bannu on 12 October. The first explicit mention of the carriage of refugees occurs for 16 October, when the last 4 aircraft on this operation returned to base via Palam, where these were emplaned. More such moves would follow, and it is likely that normal load restrictions were frequently put aside in the pressing circumstances. For instance, on 20 October, 2 crews are recorded as having lifted "69 refugees and 2400 lbs of kit to Palam" and, even allowing that many of those may have been small children, numbers of that sort plus freight go well beyond the aircraft's normal capacity. On 24 October, 8 aircraft are recorded as having left for Palam, each carrying "a full load of Indian refugees" and it would be interesting to know exactly how many were carried that day. On the following day, they took Pakistani refugees to Lahore, from where they then moved British personnel and families to Mauripur.

Airlift of refugees continued into November, where it is also clear that the movement of British personnel in advance of repatriation was gaining in pace. On 19 November, confirmation of long-running rumours was received – No 10 Squadron would be returning to the UK, to disband by 31 December 1947, and the recording officer wrote:

> "The melancholic atmosphere of a squadron about to be disbanded after twenty years continuous service was dispelled by the order to fly the aircraft to the UK and the prospect of Christmas at home."

Operations would continue, as did training with *"crews perfecting their flying and landing aids for the British winter weather."* On 29 and 30 November, the first 3 crews detailed to ferry aircraft to the UK carried out air tests and cleared customs, prior to departures on Monday 1 December. The end was in sight.

In customary Service style, a Movement order for the recovery to the UK was written and, regarding the aircraft, was *"not to be considered complete until the aircraft have been delivered to No 12 MU, Kirkbride."* (Kirkbride, 10 miles west of Carlisle, had been a Ferry unit and held over 1000 aircraft of all types in storage after the war.) The plan was that:

> 3 aircraft were to leave on 1 December.
> 5 were to leave on 15 December for delivery to Air Command Middle East at Aquir in Palestine, and known to 10 Squadron in 1942. The crews concerned would then take 5 other aircraft to the UK. (John Webb flew back on one of these flights, and recalled that the Engineering Officer at Aquir refused to take the aircraft as it was in a worse condition than the one intended for return to the UK!)
> 6 aircraft were to depart on 17 December, and 1 on 18 December.
> The final 6 aircraft were to depart on 19 December.

The planned route involved 5 days flying, with overnight stops at Shaibah, Aquir, Luqa, and Lyneham. The first 3 pilots and navigators had to return to Mauripur to ferry the last aircraft, and some quick arithmetic shows that leaving Mauripur on 19 December was unlikely to guarantee a Christmas at home. Navigator Fg Off Bob Binks, flying with pilot, Fg Off Keith Sergeant, later recalled:

> "The Squadron had 3 more Dakotas than crews. Keith and I volunteered to do two trips flying two virtually empty Dakotas – what a pity we didn't have the money to invest in Indian Carpets, we could have made a fortune.
>
> On our last trip Keith in particular was very keen to get back to his lovely wife Bertie. They had not been married long before he went to India. We were told on the second trip that

> if we did not get into Lyneham by 23 December we would not be able to return in time for Christmas. We decided to compress two days flying into one day and flew from Habbaniya Iraq to Luqa Malta. We arrived in dirty weather. I could see, sat in the second pilot's seat, that Keith was very tired. I tried to follow his checks. On 21 December we were diverted to Bordeaux because of the Mistral. The poor old Dak hardly had enough petrol to get us to Bordeaux as our groundspeed had been low flying into wind. After refuelling we pressed on to Lyneham." (Association Newsletter No 35.)

At the start of December, the RAF's withdrawal from India and Pakistan to Karachi was virtually complete, with the exception of RAF Palam. A limited number of tasks was flown in early December by crews and aircraft departing later, including a final lift of Indian refugees using 5 aircraft and a flight for the Governor General. And perhaps inevitably, Sod's Law came into play. At Palam, during the refugee task, an aircraft needed a port engine change; the replacement at Palam proved unserviceable *"necessitating a further sortie, the exercise of great patience and self control by all concerned."* It recovered to Mauripur on 11 December, and the aircraft that had flown the unplanned sortie with a second replacement engine got back on 12 December, thus completing the Squadron's operational task. The pilots and navigators of the first 3 crews also returned that day, and all attention turned towards air tests and preparation for the main unit recovery phase.

In most cases, three groundcrew flew home on each aircraft. The majority of personnel would have returned by troopship, though some were destined to stay on or to be posted to duties in the Mediterranean or Middle East. A Disbandment Party was planned in some detail for the evening of 9 December, at which the CO was *"authorised to make a short, guarded speech, touching on the miracles wrought by the company present and underlining the Squadron's accident free record."*

Wg Cdr Ogilvie clearly wrote the last words of the 540 personally. He did not fly out and was the last unit member to head for the boat on 20 December, the date he recorded as that on which physical disbandment of the Squadron was complete. He also wrote:

> "With the departure of these aircraft ends the operational record of No 10 Squadron during its second lease of life. All hope that the number plate of one of the first RFC and RAF Squadrons will not be allowed to hibernate in Air Ministry Historical Records section"

He was not to be disappointed....

The Berlin Airlift

> "For they intended evil against thee: they imagined a mischevious device, which they are not able to perform." (Psalm 21, Verse 11 – inscribed on the nose of the final Dakota into Berlin, 23 September 1949.)

The uneasy wartime alliance with the Soviet Union had given way to the peacetime reality of a divided Europe and, as early as 5 March 1946, Winston Churchill was able to declare that "an iron curtain has descended across the Continent." Specifically in Germany, the Red Army's advance in the last months of the war had given the Soviets effective control of much of the country, now separated from the remainder occupied by the UK, France and the USA. Berlin, the former capital, lay well inside East Germany and was itself divided into four zones run by the wartime allies. Access to the city from the west was possible by road, rail and river/canal routes and via three 20-mile wide air corridors. Significantly, only the latter were the subject of formal agreement with the Soviets.

By early 1948 relations had deteriorated badly and, after a period of partial restriction of ground access and a midair collision between a Soviet fighter and a BEA Viking airliner near Gatow, on 24 June the Soviets announced that all land and water communication with Berlin would be stopped. On the next day, all supplies of food to the western sectors of the city were ended. A clear political challenge to the Western powers had been made, with every expectation that they would acquiesce in the Soviet position, aimed at gaining practical control over the entire city.

Confounding the Soviets, what was decided instead was the aerial resupply of Berlin with food and materiel to sustain the population in the western zones. What we now know as the Berlin Airlift began on 28 June, with the

most-remembered codename for the RAF element of the operation, PLAINFARE, adopted in mid-July.

In the first weeks, the Dakotas and Yorks of RAF Transport Command played a major part in the operation until the much larger US effort could be brought fully to bear. RAF aircraft used the northern corridor inbound to RAF Gatow, returning to base via the central corridor and, from late-August, the Dakotas were based at RAF Lübeck. The operation there was well underway by 4 October when No 238 Squadron, whose home base was RAF Oakington near Cambridge, was renumbered as 10 Squadron. (238 Squadron had itself been renumbered from 525 Squadron on 1 December 1946.) Overnight 238's CO, Sqn Ldr T F C Churcher had a new command, but business went on as before.

Apart from essential currency training and occasional Special Flights to a variety of destinations, including Oakington, that business meant sorties day after day to Gatow predominantly, perhaps 18 per day, with some sorties to the US terminal at Tempelhof or the French airfield at Tegel. Two significant exceptions to that pattern occurred from 28 December 1948 to 14/15 January 1949 and again from 27 April 1949 to 16 May 1949. For both periods the Squadron moved to RAF Bückeburg, an airfield close to the RAF and Army HQs in Germany at that time. (The Joint HQ at Rheindahlen, significantly further west at Moenchengladbach, near the Dutch border, was not built until the mid-1950s.) Whilst there, crews flew what were designated the P3 and P19 Services - flights that appear to have been laid on for official passenger traffic between the HQs, Berlin, and London, as the number of passengers carried in and reported in the records increased considerably on each occasion. The P3 Service ran to RAF Northolt, and the P19 to RAF Gatow.

The Russian blockade ended on 12 May 1949, as suddenly as it had begun, but the Airlift continued. However, with the blockade lifted, some easing of the pressure on unit personnel could be permitted, and 10 Squadron returned to its UK Base at Oakington on 2/3 June *"for two months in accordance with the Plainfare policy."* The pace of life there slowed down markedly with a mix of Special and Ferry flights being flown, together with continuation and categorisation. However, PLAINFARE was still going and, from 27 July, the Squadron returned to Lübeck from where 40 sorties to Gatow were flown in the last days of the month. The Airlift was once again the unit's main preoccupation until it was ended in late September. The last RAF Dakota sortie was flown on 23 September 1949 by a 10 Squadron crew, Pilot I John Brown, Nav II Eric Pearce, and Sig I James Batson. They had flown together throughout the operation and had amassed 222, 220 and 192 sorties respectively. The Squadron returned finally to Oakington between 26 and 28 September to go on leave, and as the Operational Record fairly stated:

> "Thus ended a very successful chapter in the history of the Squadron."

When flying resumed in mid-October, the Squadron faced a period of reorganisation and adjustment after months with a near-single focus on PLAINFARE. No operational tasks were received and an extensive local training programme was put in hand. Scheduled services were resumed from 10 November, and a good deal of route training had to be planned as many crews had no experience whatever of the routes involved. However, with that done, more tasks could be flown in December, but it is noteworthy that only 5 crews were required and that the number of postings-out since October had been considerably in excess of new arrivals.

A little more flying was achieved in January 1950 and on 10 January, Sqn Ldr I D N Lawson DFC took over as CO from Sqn Ldr Churcher. However, his period in-command was to prove very short indeed. With the Berlin crisis over, the RAF needed to resume its planning for a post-war bomber force and the resources devoted to Transport Command were to be reduced as a result. In late January, a decision was announced that would reduce medium-range transport units to just two as quickly as possible, with 10 Squadron one of 4 Dakota squadrons to be disbanded in February. As a result, no flying was undertaken that month and after a *"convivial evening"* at The Red Cow in Cambridge on the evening of 8 February, the unit was formally disbanded on 20 February 1950.

The re-equipment of Bomber Command was now the RAF's priority – and it would not be too long before a squadron that had flown in the bomber role from 1928 to 1945 would do so again.

FROM BROOKLANDS TO BRIZE

PART 3

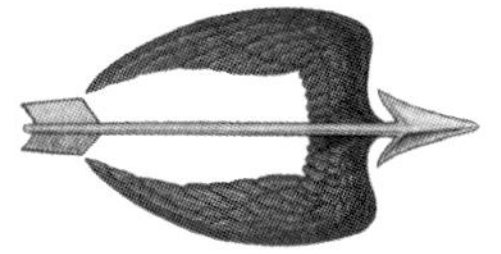

THE JET AGE ~ 1953 →

English Electric Canberra
Handley Page Victor
Vickers VC 10
Airbus Voyager

CHAPTER 10

BACK IN BOMBER COMMAND

As the Russian blockade of Berlin was lifted in May 1949, the prototype English Electric Canberra bomber was rolled-out and made its first flight some days later. That blockade had made very clear the competition for strategic control of central Europe that would dominate Cold War defence planning for the next 40 years. However, with the immediate threat to Berlin lifted and Transport Command rapidly reduced in size, the RAF could now give priority to the re-equipment and expansion of Bomber Command – and 10 Squadron would feature in two manifestations of this.

THE CANBERRA, 1953 – 1957

After the war, Bomber Command's main equipment was the Avro Lincoln, a derivative of the Lancaster that came into service as the war ended. From 1950, and the outbreak of the Korean war, these were supplemented by eighty longer-range Boeing B-29 aircraft, acquired from the USA under Loan arrangements, and named the Washington in RAF service. These were essentially seen as a necessary stop-gap whilst the twin-jet English Electric Canberra medium bomber was in development and later into quantity production. A Canberra prototype first flew on 13 May 1949 and the first definitive version produced for squadron service, the B2, flew in April 1950. With a maximum speed of some 470 knots and a ceiling well above 40,000 feet, it took the Bomber force into a new realm, effectively beyond the reach of contemporary fighter aircraft.

Canberra deliveries to squadrons began with 101 Squadron at RAF Binbrook in May 1951. A Wing of 5 squadrons built up at Binbrook from then until late 1952

10 Sqn Canberras and others lined up at Honington, in September 1956 for a visit by the Commander of the Russian Air Force.

and established the new aircraft as a high level bomber. (Trials as a low-level night intruder had also been carried out.) A second Wing was created at RAF Scampton, starting with a revived No 10 Squadron in January 1953. A group of pilots and navigators was posted-in from the Conversion Unit at Bassingbourn just before Christmas 1952. They were allocated space in a hangar and, anticipating what was to come, they painted a large X on the office doors. The OC-designate, Sqn Ldr D R Howard DFC, who had gained experience on type as a Flt Cdr with 101 Sqn, arrived in 14 January and, on the morning of Thursday 15 January, was able to announce to all and sundry that No 10 Squadron had officially re-formed.

The unit had 8 established crews – each comprising a pilot, a Nav/Plotter and a Nav/Radar – two with experience from Binbrook and the others new to type from the OCU. With time the crew establishment would grow to 12, with 10 aircraft, but at the start there were no aircraft at all. Fog at Scampton prevented any deliveries from Binbrook until 17 January, when Sqn Ldr Howard was able to deliver a first aircraft. The next week was little better, and the 2 remaining aircraft were not in place until 22 January. However, time was not wasted as work was begun, with a fair degree of self-help, to alter the available accommodation for the new unit. The weather was still poor but the CO was finally able to carry out a first check ride with Sgt Hardy on 29 January, and with Plt Off McCormack the following day. The official record wound up the month:

> "January 31 1953: The Station Commander's Parade arranged for this morning was cancelled because of bad weather. Ground crew attended educational classes and aircrews carried on with the preparation of the Flight Planning section. Work finished at 1200 hrs for the weekend."

Things picked up to a certain extent in February. Two more aircraft were delivered from English Electric; CO's checks continued; and continuation flying and cross-countries began. Snow, ice and fog interrupted progress in the mid-month but some instrument flying and checks were possible using a Meteor aircraft on temporary loan. The weather continued to limit flying in the first half of March, but improved to permit the Squadron to participate in its first exercise, 'Jungle King,' from 16 – 22 March, and for which the aircraft had tip tanks fitted. 24 sorties against ATAF targets in Germany were flown from Scampton over 4 nights, with bad weather requiring crews to divert as far as Kinloss and St Eval on two of them. (A footnote to the exercise is that an un-named WRAF reservist officer was attached to the unit as an Ops Officer, and was 'adopted' by the Squadron thereafter.)

Regrettably, the Squadron suffered its first fatal accident with the new aircraft on 27 March, when Fg Off Reeve, Plt Off Woods and Plt Off Owen were killed during a continuation training flight. Two successful QGH and BABS approaches had been completed and the aircraft climbed to the west for another. Some 11 minutes later it crashed in a near vertical attitude near Dilhorne in Staffordshire. The crew had tried to abandon the aircraft, and no clear cause for the incident could be found.

On 16 April, Sqn Ldr Howard and a single navigator, Sgt Davey, left for Khartoum for the first of a number of overseas trial flights of an aircrew pressure suit for use at high altitudes. Trials also began of a synthetic engine oil. Its corrosive nature meant that the engine cowlings and a strip of the wings on each side of the engines were painted with a special silver paint, and silver nacelles were subsequently retained on 10 Squadron aircraft. Otherwise, the Squadron was settling into a routine of visual and blind bombing training flights and exercises, cross-countries at home and over Europe, interspersed with display flying rehearsals. And recalling that June 1953 was Coronation Year, it is no surprise to see that related activities began to occupy a good deal of unit time. Several four-ship rehearsals were flown for the Queen's Birthday Flypast over London on 11 June but weather on the day prevented aircraft taking-off. Later in the month, rehearsals began for squadron participation in the Flypast of some 640 aircraft planned for the Coronation Review of the RAF to be hosted at RAF Odiham in Hampshire on 15 July 1953 – almost certainly the largest ceremonial flypast ever staged in the UK. Forty-eight Canberras took part, with six provided by 10 Squadron. The CO's aircraft (WH 672) was specially modified for the purpose for, in addition to the changes already made for the pressure suit trials, it was carrying BBC Outside Broadcast commentator, James Pestridge.

Above: Crew transport and kit.

Left: With Canberra WH 674, 6 December 1954, at Idris, about to leave for Gibraltar: Tony Clayphan, Obs; Terry Hoare, Nav; Dick Hayward, Pilot.

As the year wore on, the Scampton Wing took shape as Nos 27, 18 and 21 Squadrons were formed and in September, 10 Squadron participated fully in Battle of Britain Week events. Four aircraft took part in the first all-jet flypast over London early in the week and, on the Saturday, the CO led a four-ship around a number of the airfields with At Home Days and, on return to Scampton, a dive-bombing set piece was carried out using 8 lb plastic bombs. (A prior rehearsal with 25 lb bombs had gone awry as these hit the ground and bounced up to 1000 yards across the airfield, ending up on an outside road and amongst dispersed aircraft.) Bombing practise appears to have been paying-off for, later that month, the Squadron emerged as First among Canberra squadrons and Fifth overall in the Lawrence Minot Bombing Competition, with an average error of 109 yards.

In early October, a highlight was participation in Exercise Mariner involving six day and eight night attacks on shipping in the Atlantic. Despite the lack of navaids to the west of Ireland these were carried out successfully, with two in particular causing much disruption amongst US Navy ships caught refuelling at sea. Later in the month and, indeed, for a good deal of the winter that followed, bad weather severely restricted the training that was possible and set back those crews needing a combat classification. The Meteor aircraft on charge was handed over to the Station Flight, which also had an Oxford, a Tiger Moth and a Chipmunk, where it could still be used as required by squadron pilots. For November the records show a number of exercises to test a new Standard Letdown at Scampton and to help calibrate the Scottish Sector of the Ground Controlled Interception environment. In December, Sqn Ldr Howard escaped the weather that was hampering training at home. The latest in the series of pressure suit tests involved the addition of an air-ventilated suit for use in tropical climates, and this provided an opportunity for a pre-Christmas flight to Ceylon (Sri Lanka). This was followed by a longer one in January to Singapore, and another still in February. In March and April, there was a return to formation practise, this time to prepare for a RAF flypast on 15 May 1954 as HM The Queen and The Duke of Edinburgh returned to the UK from a six-month long visit to Commonwealth countries around the globe. Something else a little out of the ordinary occurred in April when Associated British Picture Corporation crews arrived at Scampton to start filming "The Dam Busters." The Canberras were hidden, the Station became a film set, and 10 Squadron crews found roles as extras.

Four-ship formation displays appear fairly often in the records of the time. There were two in June, for example, with the display augmented by having a single aircraft break out to fly a solo section, a feature that was retained in later years. Crews were also rehearsing in June for a firepower

demonstration for NATO Service Chiefs, using live bombs to be dropped on the Larkhill ranges on Salisbury Plain on 2 July. Later in the month, crews flew fifty one sorties over 10 days on Exercise Dividend after a slow start when poor weather caused a number of cancellations. Display work came to the fore once again for Battle of Britain events in September. *Flight* magazine reported that 56 stations hosted 'At Home' Days that year – a reminder of how much more firmly the Services were integrated into national life in those days – and noted that *"Formations of many different kinds made tours, so that each station would be able to provide a view of the maximum variety of aircraft types."* Six crews were allocated to join others in a flypast over London on 15 September, four crews participated in the 'At Homes,' whilst another formation took part in flypasts at nine airfields, finishing up with a demonstration over Scampton. Training had to suffer a little as a result of these commitments but four squadron crews were still able to win a Station Bombing Competition during the month and the number of 'Combat' rated crews rose to eight in all, and to nine (out of twelve) in October. In November, Sgt Brice's crew was the first from the Scampton squadrons to be rated 'Select.'

The first mention of a Lone Ranger flight also appears in the record for October 1954. Although this was to Wunstorf in Germany, these Bomber Command training flights for a single crew would normally be to somewhere further afield and somewhat warmer, providing a welcome change for UK-based crews otherwise accustomed to a diet of cross countries and visits to the Bombing Ranges. Two would be flown in November – another to Wunstorf, whilst the CO had a look at Nairobi. A reshuffle of crews was needed during the month after a policy decision emerged requiring that no commissioned navigator should be crewed with a NCO pilot. (From late 1950, Air Council policy had been that all pilots and navigators should be officers and recruitment had been targeted accordingly, and this was one consequence of a transition that continued well into the 1960s.) December was busy – three Lone Rangers, retention of the Station Bombing Trophy, and a range of training exercises. Flt Lt Collins had a minor accident on landing with a retracted nosewheel. He landed the aircraft perfectly in poor visibility, with the aircraft sustaining minimal damage, and the incident did not prevent his being promoted to Sqn Ldr on 1 January 1955.

1955

Snow proved a serious hindrance at the start of the year, with no flying possible between 11 and 23 January. Fortunately for No 27 Squadron, with a parade and Standard Presentation scheduled just beforehand, 10 Squadron was able to put on a four-ship formation display for visitors as part of the day's events. A Lone Ranger left for Idris in Libya and Gibraltar on 8 January, went to Marham on return as the airfield there was not closed, with the aircraft finally recovered to Honington on 24 January. Snow also covered the airfield for some three weeks in February, albeit that extensive effort put into runway clearance allowed some limited flying, including the delivery of two additional aircraft, bringing the establishment up to 10. (A pay increase awarded only to NCO aircrew resulted in the Junior Officers building a snowman that looked like a squadron Master Pilot outside the squadron offices. It carried a sign saying "Sorry as I am for thee young Jack, I am doing very nicely, thank you.") Two Lone Rangers managed to get away to Gibraltar and to return on time, and four crews participated in a Station Limited Aids Cross Country on 8 February. This was a lengthy navigational challenge, mainly over the sea - from Base to Rockall, to a point midway off the coast of Norway at 65N 10E, to the Faeroes, and Home.

Dave Bridger's aircraft arrives on a Lone Ranger to Gibraltar in 1954.

One crew completed it successfully; another shortened it due to lack of fuel; a third completed it but only by refuelling at Leuchars; and the fourth had to divert to Kinloss as the Nav Plotter was suffering from hypoxia.

The weather improved in March and the Squadron was able to catch up on its training, flying more hours than in any other month since its formation with the Canberra. A full spectrum of training was possible - 259 practice bombs were dropped, a variety of Cross Countries and exercises and four Lone Rangers were flown, one as far as Aden. The Bomber Command Bombing Competition began on 28 March and the four 10 Squadron crews entered went on to win the Armament Officers' Trophy for the best combined Gee-H and Visual bomb scores. April saw new arrivals – two more aircraft bringing strength tempoaraily to 12, against an establishment of 10; and, on 18 April, a new Navigator CO, Sqn Ldr G. Sproates.

The new CO's first major task was to oversee a unit move to RAF Honington. By 1955 Bomber Command was in the early stages of creating its strategic 'V-Force,' and all four Canberra squadrons at Scampton would be moved out to permit work to begin on preparing the station to become the home for the first Vulcan squadrons. Consequently, very little flying was carried out as efforts were concentrated on the mass move that took place on 23 May – nine aircraft flew out that morning and a road party moved most airmen and equipment. (The move also involved a change from 1 Group to 3 Group or, in contemporary terms that may well mean nothing to those who today know only the Premier League, from 1st Division North to 3rd Division South.) An advance party had gone to Honington a week beforehand to make arrangements to ensure a rapid settling-in and, as a result, the unit was quickly operational, with five aircraft participating in an exercise three days after arrival. With the move behind it, the Squadron settled to the accustomed round of training, exercises, cross countries and Lone Rangers. The training appears to have been effective for, on 8 September, Fg Off B J Warwick carried out a deadstick landing after both engines flamed-out at 40,000 feet and refused to relight. In November, the Squadron formation team, led by Flt Lt Ernie Lack, was designated as the official Bomber Command Demonstration Team for 1956.

1956

As the year turned, a crew had a Lone Ranger that turned out to be less fun than expected as, full of eastern promise, they left winter behind and set off for Amman. On landing at Nicosia, the aircraft was directed to a barbed-wire protected dispersal, and they were told to refuel quickly and get airborne - the EOKA terrorist campaign in Cyprus was active. Then, on landing in Amman, they found that they were directed to another protected dispersal and confined to the Mess for the duration of their stay – there was rioting and unrest in the city. Meanwhile, some colleagues were having a somewhat better time in Kenya, participating in an Air Show. Fg Off Dick Hayward, with colleagues Terry Hoare and Tony Wilson-Pepper set off on 3 January for another Lone Ranger to Nairobi, via Libya and Aden, to participate in Kenya's first Air Show at Kitale, some 200 miles northwest of Nairobi, and home to the Trans-Nzoia Flying Club. After a rehearsal on Friday 6 January, they displayed on the Saturday and Sunday – as did three Venoms from 8 Sqn in Aden – and *Flight* magazine reported:

> "The aircraft drew a tremendous roar of approbation from the crowd as it flew over in a series of fast, low runs, followed by a run with bomb bay open, another with wheels down and a third with flaps down. On the final fast vertical climb-away the power levers were suddenly cut, with a resultant 'deathly hush' after the deafening roar."

The monthly record shows that "The crew spent an entertaining evening as guests of the Flying Club on Saturday," thanks to the availability of a RAF Pembroke that flew all concerned there and back. They recovered to Honington on 11 January.

Early in the year the Squadron was notified that it was eligible for the presentation of a Standard. (Squadrons qualify for a standard after twenty-five years service, and 10 Squadron was one of a number so recognised in the London Gazette as far back as 11 January 1952.) It appears that the CO collected it from the makers and brought it to Honington, and a Presentation ball was arranged in Bury St Edmund's in April, with unit silver and Mayoral regalia on show, whilst fruitless negotiation of a date for the formal presentation took place as nothing could be arranged around Squadron

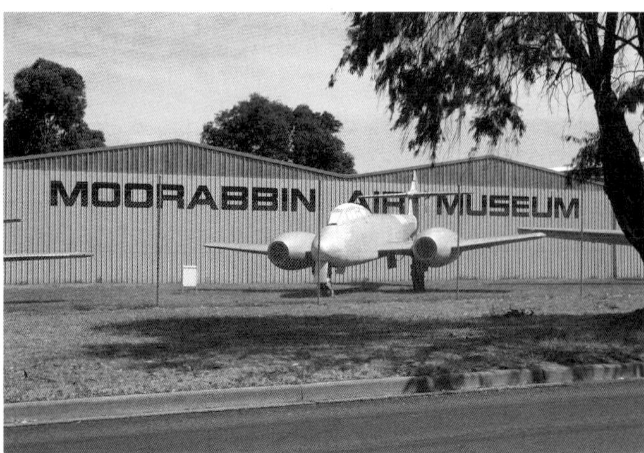

10 Squadron's Meteor WH 118 was used by crews for practice assymetric-engine training during the Canberra's early days. It ended its long career with the RAAF and is now a static exhibit at Moorabbin Airport , Melbourne; home of the Australian National Aviation Museum. In September 1953 Mstr Plt V. Brown won an air-race in this aircraft during RAF Scampton's Battle of Britain Open Day events.

10 Squadron was a 'Dambuster ' Squadron too.......when, in 1954, Scampton became the set for the film and the Squadron's crews were used as extras. The film's star, Richard Todd, who played Wg Cdr Guy Gibson, is shown here with 'real' aircrew.

Gibraltar from the air in the 1950s. The large concrete water-catchment area can be seen below the cloud on the east of the Rock.

Fg Offs: Dave Bridger, Basil Williamson & Dicky Halford at Scampton

At RAF Eastleigh, Nairobi with WH666, for the Kitale Air Show, early January 1956. Terry Hoare, Nav; Dick Hayward, Pilot; Tony Wilson-Pepper, Obs.

Fg Off Alan Cottle (kneeling) by his Canberra with navigator, possibly Fg Off P.C. Cullen - circa 1956

Ernie Lack (centre) and his crew.

........and today the nose of Ernie Lack's aircraft WH 646 is in the Coventry Airport Museum, with ECM equipment nose changes from its time with 360 Sqn.

Flt Lt Ernest N.H. Lack was the leader of the 10 Squadron Demonstration Team, whilst based at Honington in 1956, and also took part in 'Suez' raids on Egypt.

The Suez Crisis of September 1956 when the Squadron deployed to Nicosia, Cyprus for the attacks on Egypt, codenamed 'Operation Musketeer'.

Commander of the Russian Air Force, Chief Marshal Zhigarev inspects the Guard of Honour on his visit to Honington early in September 1956, shortly before the Suez Crisis.

Crew Transport Turkish style whilst on detachment to Eskisehir Turkey in mid-1956.

commitments. The Standard was returned to the makers and was eventually presented by HRH Princess Margaret in 1958, when the Squadron had begun to fly Victors at RAF Cottesmore.

In early April, Canberra B2s were grounded for some two months to permit modification of tailplane actuators. (Instances had occurred on which these had runaway, forcing the aircraft fully nose up or down.) 10 Squadron aircraft were amongst the first to be modified and pilots were given access to some T4 training versions to keep current, and it was with a flight of these that it carried out its display commitments over the period. The first arose during the official visit to the UK made by First Secretary Nikita Khrushchev and Premier Nikolai Bulganin of the Soviet Union. (They arrived in Portsmouth aboard the Sverdlov class cruiser Ordzhonikidze, investigated by Cdr 'Buster' Crabb who 'went missing.') The visit lasted from 18 to 27 April, and included a day at RAF Marham on 23 April, when 10 Squadron provided crews for four of the sixteen Canberra T4s in the flypast. Modifications were complete before the next significant commitment, a NATO Southern Region exercise, codenamed THUNDERHEAD. For 10 Squadron, this involved a detachment to Eskisehir, Turkey, from 23 – 30 June 1956. The purpose of the exercise was to test the NATO Control and Reporting system by flying high level routes, with interceptions flown by Turkish and Greek F84 and F86 fighters. A ground party left by air using a Transport Command Hastings and an Italian Air Force C-119 on 21 June. Eight aircraft left on 23 June, routing via Luqa. Loose formations were used en-route, with aircraft closing to standard formations for descents, approaches and landings. All forty pre-planned sorties were flown, initially at 20,000 feet and later at 30,000 feet. Eight additional sorties were arranged for 45,000 feet and .75 Mach, but it became clear during the first five that the Turkish radar could not plot these, and a later three were flown at 30,000 feet and 250 knots. (25 June – 12 sorties; 26 June – 12 sorties; 27 June -11 sorties; 28 June – 13 June.) The visit went well, with a good deal of reciprocal hospitality, and flights to Istanbul by Turkish Dakotas for both air and ground crew.

News of a likely near-term disbandment emerged in mid-1956 as it became known that RAF Honington was earmarked as a base for new squadrons equipping with the Vickers Valiant, the first of the 4-jet strategic bombers that would comprise the 'V Force.' In the meantime, there was participation in another Royal event at RAF Marham on 23 July, just before world events would intrude on any rundown plan.

On 26 July 1956, Egyptian President Nasser nationalised

the Anglo-French Suez Canal Company, planning to use its revenues for construction of the Aswan High Dam once external Western aid had been denied. The confused situation that followed in Whitehall as to what should be done emerges clearly in the record of a RAF Historical Society Seminar held in 1987 with contributions from many of those involved at differing levels. However, as is now known, by October, the UK and France had established a secret agreement with Israel whereby Israel would attack Egypt, providing a pretext for an Anglo-French invasion of Suez. These events would influence 10 Squadron for the remainder of its Canberra flying months.

From 2 to 11 August, other than a handful of Air Tests, all flying was devoted to Operation ACCUMULATE, a weapons outloading that involved most Bomber Command Canberras at the time. For 10 Squadron, it involved four aircraft on most days flying six 1,000 lb bombs to RAF Luqa, unloading, and returning to base – some 31 sorties in all. There was no clear plan so soon after the Canal seizure, other an instruction from PM Anthony Eden to the Chiefs of Staff to prepare plans to recover the Canal and, hopefully, to overthrow President Nasser, and it is likely that this outloading was put in hand to provide later alternatives.

The Battle of Britain Flypast over London on 15 September 1954 startles the pigeons around Nelson's Column. Five 10 Sqn aircraft participated.

10 Squadron's Crewroom in Eskisehir, Turkey during 'Operation Thunderhead'.

As there was also no doubt a need to maintain the best-possible semblance of normality, a visit to the UK by Marshal Zhigarev, Chief of the Soviet Air Force, went ahead and a planned RAF visit to Honington went ahead on 6 September, when the Formation Team participated actively in the day's events. Flt Lt Ernie Lack's four-ship Demonstration Team had been rehearsing for this and Battle of Britain Day whilst other crews sharpened their bombing skills. Training continued and eventually, on 24 October, the first deployment in support of Operation MUSKETEER, the Suez operation. (Ground crews, equipment and spares had been flown out by Transport command Hastings aircraft in the preceding days.) The CO's crew left early that morning for what is recorded as 'Lone Ranger 145, as briefed." Two days later, a further seven crews departed in two waves of two and five, respectively. Routing via Malta, their destination was RAF Nicosia, in Cyprus, which was full to bursting. Accommodation was either in tents or the Station Squash Court, all concerned buckled down, and local familiarisation and Force training exercises were flown on 28 and 29 October.

Units of the Israeli Army moved into Sinai on 29 October and battle was joined with Egypt. The pre-agreed Anglo-French demand that fighting stop was made, with an ultimatum demanding a cease-fire, otherwise the UK and France would intervene. Israel agreed but, as expected,

Ernie Lack's Canberra WH 646

Egypt did not and the go-ahead was given. Sqn Ldr Sproates' crew with another squadron crew led the first wave in the first raid on the late afternoon of 31 October on Almaza airfield, with marking done by Canberras of 139 Squadron. The planned target had been the airfield at Cairo West but, that morning, the Prime Minister had been approached by the US Ambassador to be told that it was very close to a road being used to evacuate US citizens to Alexandria, and that the US hoped that their safety would not be hazarded. This meeting had the desired result and, whilst a signal was sent to the HQ in Cyprus to cancel the raid, it was not sent to the Air HQ controlling the air operation. When it was finally received, Valiants from Malta were well on their way and were recalled; the Cyprus-based Canberras were also airborne and were diverted by radio to attack Almaza. With hostilities now initiated, as the night continued a further six crews were involved in a second raid on Almaza, and in others on airfields at Kabrit and Abu Sueir. Air attacks on Egyptian airfields continued on 1 November and the Egyptian Air Force was effectively disabled – eight 10 Squadron crews flew on attacks at Cairo West and Luxor that day. With no effective Air opposition, other targets could be attacked and six crews flew in two separate raids on Cairo Radio on 2 November and two attacks on Almaza Barracks on 3 November. The final sorties flown were on the morning of 5 November, when six crews attacked an armour and ammunition depot at Huckstep Barracks. The airborne assault on Port Said occurred that day and advances on the ground were made. However, by then, the global reaction to the intervention – and, above all, the hostile US reaction – had put the British and French governments in a very difficult position. A ceasefire was declared at 2359 hours on 6 November – the military operation was proceeding very well indeed, but the entire venture had proved to be politically disastrous.

Events moved quickly in the hours after the ceasefire. The Squadron was given just four hours notice to return to Honington and all eight aircraft left on the night of 7 November to fly home via Luqa, arriving next morning. From a unit point of view, things had gone fairly well – in the absence of other navaids, visual bombing (sticks of six 1,000 lb bombs) had been reasonably accurate, and no sorties were lost to unserviceability.

Back at base, crews found that a Standby commitment was enforced – 4 hours readiness, relaxed to 12 hours every third day – and this limited the amount of training that could be achieved. On 8 December, the state was relaxed to three days readiness and the commitment was terminated on 12 December. However, instructions also came that nine Navigators were to be detached to No 44 Squadron, with

which they returned to Malta on 17 December. Sqn Ldr Sproates signed off the F540 for December in these terms:

> "The date for the squadron disbandment was given as 15 January 1957, and the packing up of the unit was begun. Aircraft were allocated away, a Board of Survey carried out on the files, and the property and silver packed away. By the end of the month No 10 Squadron had virtually ceased to exist as an operational unit, but there was still much to do before the disbandment would be complete."

The disbandment was duly completed, but it was not to last long. Having given place to the first V-Bomber, 10 Squadron would fairly soon find itself introducing the third into service.

THE VICTOR YEARS, 1958 – 1964

Having disbanded and made way for a Valiant element of Bomber Command's new 'V Force,' 10 Squadron re-emerged in the following year as the first unit to be equipped with the third aircraft of that trinity, the Handley Page Victor. (This would also be the last of the Squadron's several HP aircraft, a line that had started exactly 30 years earlier with the Hyderabad in 1928, and carried on through the Hinaidi, the Heyford, and the Halifax. The company resisted being merged into what became the British Aircraft Corporation in the 1960s, only to go into liquidation in 1970.)

The Victor had been developed as the HP80, with a first flight by a development aircraft on 24 December 1952. As with the two other V-bombers, its origins lie in the development of a UK nuclear weapon programme after WW2 and in the need for aircraft capable of delivering such weapons to replace the Lancaster and Lincoln. The Handley Page proposal resulted in an aircraft with a distinctive crescent-shaped wingplan and, in its B1 form, the variant that 10 Squadron and two others would receive, the aircraft was very much a basis for further development. Indeed, a considerably uprated B2 version was ordered in 1956, before deliveries of production B1s began. The B1 was powered by four Armstrong Siddeley Sapphire engines and, at altitude, had a superior performance to the Vulcan – a fact demonstrated during a test flight in 1957 when an aircraft exceeded Mach 1.0 in a shallow dive. The Victor was designed and built to carry the UK's *Blue Danube* weapon, a decision that gave it a very large bomb bay indeed. Although smaller nuclear weapons would follow, this gave it the capacity to carry either *Grand Slam* or *Tallboy* bombs of WW2 fame, or up to forty-eight 1000 lb bombs, a figure restricted in practice to thirty-five. Like the other two Vs, the aircraft had a crew of five – two Pilots up front, with a Navigator Plotter, Navigator Radar and Air Electronics Officer (AEO) seated side-by-side in a rear-facing compartment behind the cockpit. The Pilots had ejector seats but, after early thoughts that an escape module might be feasible, the

The smooth and futuristic lines of the Victor are well displayed on XA 928. The Victor's designer, Sir Frederick Page, attended a 10 Sqn Guest Night at Cottesmore on 10 July 1958.

QRA (Quick Readiness Alert) Crew.

QRA Scramble .

No 3 Victor Course Rear L - R : FLt Lt Mobberley, Flt Lt Nichol, Flt Lt Cole, Plt Off Firth,Plt Off Hollington, Plt Off Drew, Fg Off Featonby, Flt Lt Neville Front L - R Flt Lt Matthews, Flt Lt Kemp, Sqn Ldr Bartlett, Sqn Ldr Phillips, Flt Lt Peters, Flt Lt Mellor, Flt Lt Demmer.

XA 928 follows 2 other 10 Squadron Victors around the taxiway at Cottesmore in September 1958.

Flt Lt Rooney's crew and their Crew Chief ready to board.

Happy Hour Awaits.

Victor XA 934. *This aircraft, later operated by No 232 OCU at Gaydon, crashed one night in October 1962 after No 4 engine exploded and caught fire on takeoff. 3 crew members, the Captain, a Nav & AEO were killed.*

Prime Minister Harold Macmillan watches the Scramble aircraft land, next to ACM Sir Harry Broadhurst. Gp Cpt Johnnie Johnson (of Battle of Britain fame), RAF Cottesmore's Station Commander, is second from the right.

HM King Hussein of Jordan visited Cottesmore on 24 April 1959 and flew with OC10 Wg Cdr Owen to Farnborough.

Gear-down flypast.

The Shah of Iran visited Cottesmore in May 1959 – all part of the Government drive to obtain overseas sales for the Victor. Here capt Flt Lt P. Wilson introduces Fg Off C.S.Q. Drew and the rest of his crew.

A somewhat posed photo of the 1959 Battle of Britain display aircraft crew. Flt Lt Stan Kemp is seen holding the map. 25 years later his son Chris was also to join 10 Squadron as a VC10 pilot.

AOC 3 Gp carries out his annual inspection of Cottesmore in May 1963. AVM B.K. Burnett talks to nav plotter, Sqn Ldr R.H. Mullineaux, with captain Flt Lt K.W. Rooney and copilot Fg Off J.M. Rattenbury next in line. Flt Lt J.R. Morgan and Fg Off D. Beane complete the crew line-up. Regrettably the names of the Master Technician and groundcrew are unknown.

10 Squadron Victors lined up lined up for Press Day, September 1958.

Brake Chute Deploys

Flt Lt P H J Peters and crew under XA 928's tail, adorned with 10 Sqn's Winged Arrow.

three rear crew did not. As had been the practice in WW2, Bomber Command opted for 'constituted crews' throughout the V-Force, with the same five men remaining together for as long as possible to progress up the Command's crew classification ladder. Training was focussed upon achievement of the smallest practicable bombing errors, with bombing runs normally scored by one of a number Radar Bomb Scoring Units (RBSUs). Bombs were aimed visually or through radar fixing, often using offsets, and three navigational techniques were practised - Primary, when all radar and computer systems could be used; Secondary, when the H2S radar was not used, and fixing was done using astro or radio position lines; and Limited, when H2S and Doppler drift and groundspeed were unavailable, and the Navigators were in a position not unlike that of their predecessors in WW2. Astro techniques developed in the V-Force were capable of fixing with a degree of rapidity and accuracy that was not readily achievable elsewhere. In addition, airborne compass swings were regularly flown, with the sextant used to minimise deviations and maximise the accuracy of the heading fed to the Nav/Bombing System (NBS) by carrying out checks at altitude and normal operating speed, and with all equipment functioning. (The H2S radar was by then in its Mk 9 iteration, but Halifax Navigators from 1944 would have been familiar with its use – and they would have been delighted with the relative simplicity of operation of Gee Mk 3 compared to the original that they had known.)

RAF Cottesmore, in Rutland, was selected as the first Victor base so, once again, 10 Squadron found itself on an airfield by the A1, albeit a deal further south than RAF Leeming had been. (The Station Commander was Gp Capt (later AVM) J E 'Johnnie' Johnson, DSO, DFC, the RAF's highest scoring fighter Ace in WW2.) Wing Commander C B Owen DSO, DFC, AFC was nominated as Squadron Commander and the formation of the unit was authorised with effect from 15 April 1958, with an establishment of 8 Victor B1 aircraft. Conversion training was taking place at No 232 OCU, RAF Gaydon, and by the end of June 1958 there were 4 crews on strength and training sorties had begun. Two further crews arrived in July, when the Squadron was also formally welcomed by the Station at a dinner where, appropriately, Sir Frederick Handley Page CBE was Guest of Honour. Unfortunately, a setback occurred in August when the manufacturers found a crack in a Victor tailplane and all aircraft were grounded for inspection. A crack was found in a Squadron aircraft, and after it emerged that repairs could not be made at Cottesmore, the tailplanes were removed and returned to Handley Page. Four aircraft were serviceable by the end of the month on this score, but a further grounding had been imposed on 19 August after the partial seizure of an aileron control during pre-flight checks. A modification kit was provided and all aircraft were cleared by the end of the month. However, as a result of these technical issues barely 50 hours were flown in August, with half of them devoted to rehearsals for the SBAC Farnborough Show and for a demonstration to be part of an ITV documentary programme.

The main task in September was participation in formations and demonstrations of the new aircraft, with just three bombing sorties and four cross countries flown (albeit that one was to Madeira). An aircraft plus an airborne reserve for a V-Force formation at Farnborough were provided in the first week, immediately followed by flying on 8 September for the ITV programme 'The Heavyweights.' The Station and Squadron Commanders were involved in commentary for this whilst Sqn Ldr Burberry and Flt Lt Peters flew various demonstration patterns and showed off the streaming of the brake parachute. A formal national Press visit to show off the Squadron and its new aircraft was the next event on 10 and 11 September, and it resulted in considerable newspaper and magazine coverage. That was followed on 20 September by unit participation in Battle of Britain day displays at Cottesmore and in a touring V-Force formation, but that was by no means the end of the Squadron's ceremonial participation.

For the best part of two months, preparations had been underway for the presentation of the Squadron Standard that should have occurred in 1956, and these came to a head on 21 October when HRH Princess Margaret visited RAF Cottesmore. The Squadron paraded in two Flights, with the CO as Parade Commander, and Plt Off R B Nelson as Ensign of the Standard Party. After the presentation and a formal lunch, Her Royal Highness was conducted on a brief station visit, including an inspection of a Victor aircraft and,

Prime Minister Harold Macmillan visited 10 Sqn on 1 April 1959 - the 41st Anniversary of the formation of the RAF. Two aircraft, flown by OC10 Wg Cdr C.B. Owen and Flt Lt P. Wilson with their crews, carried out a demonstration Scramble.

V-Force advert for Aircrew in the 1950s featuring 10 Squadron's Victor B1s

just prior to her departure, the day's formalities concluded with a flypast of three Victors, three Vulcans, three Valiants and three Canberras. Most flying for the month had been in connection with the Royal visit, but some Range and Cross Country sorties were flown. Then, on 29 October, the CO took an aircraft to Honington to take part in a static display laid on for the RN Staff College. On landing, the brake chute failed to deploy and on application of the parking brake in dispersal, a system failure expelled hydraulic fluid under pressure on to the hot brakes. A brake fire ensued and, inevitably, a tyre change, thus extending over two days what should have been a brief, routine visit!

As the year ended, a training pattern involving a mix of range sorties, cross countries, and Rangers developed but flying was still subject to the weather, with December showing only 46 hours in all, thanks to fog and low cloud on over twenty days that month. However, 1959 would see a change of pace.

The 1958 Defence White Paper ('Britain's Contribution to Peace') had started from a position in which there was "no military reason why peace should not continue to be maintained for another generation or more through the balancing fears of mutual annihilation." Later, looking at the UK's national deterrent capability, it continued: "If the deterrent influence of the bomber force is to be fully effective, it must not be thought capable of being knocked out on the ground by a surprise attack. Measures are accordingly being taken to raise its state of readiness so as to reduce the minimum time needed for take-off." This was, of course, during the era of 'The Four Minute Warning,' the approximate warning time that the UK might have of a surprise Soviet ballistic missile attack and, if retaliation was to be assured, steps had to be taken to maximise the chances of survival of nuclear-capable assets. One approach was to work on reducing the time taken to get airborne after an Alert and in early March 1959, the Squadron participated for the first time in Exercise MICK, designed to familiarise crews and Ops staff with Alert and Readiness procedures, and during which two pairs were 'scrambled.' In addition, the crews of the CO and Flt Lt Wilson carried out practice scrambles in preparation for a demonstration during a visit by Prime Minister, Harold Macmillan, plus the BBC and Press, on 1 April.

Dispersal of the force was also taken into account and nominated airfields would become used to seeing Victor or Vulcan aircraft on regular training detachments. The Squadron carried out its first dispersal Exercise MAYFLIGHT in early May, when four aircraft and five crews deployed to Boscombe Down for five days. In outline, a Setting-up Party went ahead some two weeks earlier to erect tents and marquees and dig soakaways and sullage pits; an Advance Party was flown by transport aircraft as the exercise began and was followed by the Main party, with all concerned accommodated in caravans or station buildings. The four aircraft were scrambled on the morning of 5 May, and a fifth remained on Standby at Cottesmore throughout the exercise. Exercise attacks were carried out successfully and the post-Exercise report shows that many useful lessons were learned by the host station and the Squadron for what was a continuing relationship.

Despite the need to continue working-up as an operational squadron, the sheer novelty of the aircraft could not be denied and, on 24 April, the unit had a visit by HM King Hussein of Jordan, at the end of which the King joined the CO's crew for a demonstration flight. The plan had been for a short cross country leading into a bombing attack but, as the King had some sinus trouble, a low-level flight to Farnborough was substituted, starting with two practice radar approaches to Cottesmore, one flown by the King who expressed himself as being "most impressed" with the Victor. This was by no means the last time on which the aircraft was shown off. In June, there was a demonstration for the US Air Force Chief of Staff, General White, participation in a V Force display for MPs at Waddington, a one hour flight for Defence Minister Duncan Sandys, and two crews involved in support of a British Trade Fair in Lisbon.

With the Squadron now into its second year, crews were building experience through a variety of exercises, salted with overseas trips on Western and Lone Rangers. (Western Rangers normally routed via RCAF Goose Bay in Labrador to Offutt AFB in Nebraska, the home of HQ Strategic Air Command. There were RAF detachments established at both locations. Lone Rangers at that time were mainly to RAF Wildenrath in Germany, or to bases in the Mediterranean.) Eleven sorties were flown on MANIATE in July, that year's major UK Air Defence exercise involving simulated attacks on the Continent and the UK, and CinC Bomber Command went with the CO's crew on a Near East tour taking in Idris, Nairobi and Akrotiri. Two crews also began to practise for the Battle of Britain display season, and two displays at Toronto were included in that month's Western Ranger schedule. Training profiles were also flown for the Command Navigation and Bombing Competition. This took place over three nights in September, involving sixty-nine crews from eleven squadrons, six provided from 10 Squadron. Flt Lt Richardson's crew took third place in the navigation section, and the squadron was placed eleventh overall. Four crews deployed to Boscombe Down for MAYFLIGHT II, another dispersal exercise in November – this time involving loading of drill and inert weapons. That month also provided an example of the Service aircrew habit of making light of any accident from which the victim walks away. It appears that a crew was being recovered to base by road when their vehicle went off the road in fog. The occupants were treated for minor injuries in a hospital and the F540 compiler was unable to resist noting that the Navigator concerned *"is no doubt the only male member of the Squadron to have been stitched by a gynaecologist!"*

In January 1960 the weather limited flying for the month and it was noted that two aircraft had been fitted with the T4 bombsight, giving a visual bombing capability. In the following month, Wg Cdr Owen was posted to the Air Ministry and OC B Flight, Sqn Ldr R B Phillips DFC, AFC, was promoted to Wg Cdr to replace him as Squadron Commander from 15 February. The weather improved and the customary training exercises were flown, together with competition profiles - that year's Nav/Bomb Comp was in May when the results were "about average," but

better things were ahead for some. At the beginning of June, the CO's crew flew a proving flight to RAF Gan in the Maldives, via Akrotiri, Nairobi and Bahrain, to assess routes for a significant detachment to RAAF Butterworth in Malaya scheduled for July. Before that, however, there were two return visits to Boscombe Down – a KINSMAN in mid-June, during which four crews practised day-to-day operations from there in a total of twelve sorties, was followed by MAYFLIGHT III a month later on receipt of an Alert message, when the same crews deployed.

Although the UK had ceased to be the colonial power over the Federation of Malaya in 1957, the Communist insurgency 'Emergency' was just coming to an end. The UK continued to have interests and bases in the area and the Operation Order for Exercise PROFITEER had as its Mission:

> "To exercise medium bomber crews in the reinforcement of the Far East Air Force, and provide operating experience in the Far East theatre."

In that context, the plan was to deploy four Victor aircraft and crews, plus reserve aircrew and ground servicing personnel to RAAF Butterworth on the west coast of Malaya, for a period of eighteen days. (The station had been transferred to the Australians in 1957.) A Britannia of Transport Command was tasked as a support aircraft to the Victors and carried around half of the detachment, whilst the rest of the party left from RAF Lyneham on a scheduled flight to RAF Changi, Singapore, for onmove to Butterworth. A very new Airframe Mech, Peter Higgins, whose first overseas flight this was, recalled later:

> "There was I, not yet a 17-year old lad, working on aircraft that two years previously had featured in the 'Eagle' comic, and the Station Commander was Johnnie Johnson, whose book I'd had as a Christmas present……Soon we were off to Malaya. The first night stop at El Adem, Libya, was hot and the second at Aden, even hotter. Thence on to Gan in the Maldives, hot but lovely and, finally, to Changi, hot,sticky and humid. After two nights we were flown up country to RAAF Butterworth on a Beverley. The Aussies there treated us well, with plenty of beer and barbies – beefsteaks one inch thick and salted pineapple."

The Victors left on 19 July to fly via Akrotiri, Mauripur and Gan to Butterworth. Three arrived on schedule on 22 July, with the fourth a day late after an airframe change in Cyprus. A number of exercises had been anticipated, including some over Thailand, but these had to be reduced to give priority to preparations for participation by two crews in a flypast over Kuala Lumpur marking the official end of the 'Emergency.'

The Butterworth, Malaya Detachment of July – August 1960 was codenamed 'Exercise Profiteer'.

Three aircraft also paid a visit to Clark AB in the Philippines and two special flights were carried out, each requiring that the aircraft should reach 45000 feet within thirty minutes after takeoff. (The Navigators' logs for these were sent to HQ Bomber Command on return.) It was noted that high tropical temperatures degraded takeoff performance more than expected, whilst the very low temperatures at altitude improved performance, with 50000 feet easily reached towards the end of sorties, with the aircraft ready to go still higher. The four aircraft began the return flight on 9 August and all returned to base on 12 August as planned, with the servicing party following on some days later.

As the summer continued so, too, did the training routine and the aircraft was still being shown off at Battle of Britain Open Days around the UK in September. A slightly different NATO exercise punctuated the routine that month when simulated attacks on the US Sixth Fleet north of the Orkneys and on targets in Norway were planned. A first raid went well, a second was cancelled on account of weather and the third was "mainly unsuccessful due to the fleet dispersing." In October, unserviceability gave Flt Lt Peters' crew an extended stay at Offutt AFB and the next KINSMAN dispersal exercise to Boscombe Down took place, not helped by poor weather causing enforced cancellations of planned sorties. The 3 Group Standardisation Team visited during November, when four crews had air checks in bombing or navigation and two underwent ground checks. In December, Sqn Ldr Young's crew completed a Lone Ranger to Akrotiri but, a few days later, Flt Lt Wilson's crew had to abort theirs and return to base when a crashed Canberra blocked the runway at Wildenrath. A crew also went to Boscombe Down to check out the Telescramble installed there – this system was a refinement that enabled crews to receive instructions direct from the Controller at HQ Bomber Command when at cockpit readiness.

1961 began with the CO's crew taking a Western Ranger to check out the winter weather in Labrador and Nebraska, a trip that went to schedule. Then, in mid-January, five crews and four aircraft, together with servicing personnel, went out to Cyprus for a month's flying from RAF Akrotiri during the Squadron's first Exercise SUNSPOT. At that time, occasional augmentation of the RAF presence in Cyprus helped to underscore the UK's membership of the Central Treaty Organisation (CENTO). Other obligations were not forgotten as the Squadron record mentions a cocktail party at the end of the detachment organised by the officers to thank their colleagues at Akrotiri, at which "several RAF Nursing Officers from the Hospital were also entertained." In February, very little flying was possible at Cottesmore whilst the SUNSPOT detachment was away.

March saw a return to more normal business with another KINSMAN at Boscombe Down, a KINGPIN used to exercise UK Air Defences, and a GROUPEX. Johnnie Johnson was by now an Air Commodore and Senior Air Staff Officer (SASO) at HQ 3 Group and he flew five sorties with the Squadron during the month. He was clearly getting his eye in once again with the Victor in preparation for a flight to Pakistan planned for June. (An aircraft was also returned to Handley Page at Radlett to be resprayed for that flight.) Before that came Bomber Command's annual Nav/Bomb Competition from 15 to 19 April. Six crews took part and Sqn Ldr Young's crew had the best bomb score from the first night. The others also did well and the Squadron emerged as winner of the Armament Officers' Trophy, the main bombing award first won by the Squadron in 1955. Four-ship Scramble practices were also resumed in preparation for a demonstration as part of a 15 Squadron Standard Presentation programme on 3 May. Later in the month, an aircraft was flown to Le Bourget for static display at the Paris Air Show, where a three-Victor flypast was planned on 3 and 4 June. Another MAYFLIGHT to Boscombe Down was held from 10 May, when Scramble and Alert exercises were flown until 13 May. Four crews deployed and two others participated from Cottesmore.

In June, a number of events punctuated the training round, starting with Sqn ldr Williams' crew leading a formation of three Victors and a Vulcan in a flypast at the Paris Air Show, followed by two aircraft sent to Mildenhall and Sculthorpe for US Armed Forces Day displays. The CO flew with Air Cdre Johnson on his visit to Pakistan from 16 to 27 June where, amongst other things he and Wg Cdr Clare, OC Ops Wg at Cottesmore, gave a number of lectures to the Pakistan Air Force Staff College. The outbound route to Karachi was direct via Akrotiri, whereas

the return did a circuit via Gan, Nairobi and Akrotiri. July was a good month for training and Flt Lt Leinster's crew returned from a Western Ranger after a mix of hydraulic and starter motor problems had given them an extended stay at Offutt, whilst an aircraft was flown to Malta to assist with CinC Bomber Command's recovery from Australia to the UK. In September, the records include an interesting item, Exercise MACASSAR. Starting on 18 September, this was of indefinite duration, with an aircraft on permanent readiness and a crew at 15 Minutes readiness. The airframe was changed weekly, and the aircrew daily at 0900 hours. The crew was to remain together at all times, with a room in the Officers' Mess set aside for them, together with a MT vehicle. In retrospect, this can be clearly seen as a precursor trial of the permanent QRA arrangements for the V Force that came into force on 1 February 1962. As in past years, September brought the SBAC Show at Farnborough and Battle of Britain Open days. At Farnborough, it had been planned to have bomber aircraft overflying at heights at which prominent vapour trails – 'contrails' – would be visible, but weather conditions restricted this display to three days out of seven. In October the Squadron had a HQ 3 Group Standardisation Team visit, took part in a new Station level Alert exercise, TITTON, and sent four crews on a short detachment to RAF Finningley. The following month saw four crews return to Boscombe Down for another KINSMAN dispersal exercise with others participating in SPELLBOUND, a UK Air Defence exercise, whilst December was marked by a flight that was particular to its time. For five years from 1958, Bomber Command had a number of squadrons operating nuclear-armed US Thor Intermediate Range Ballistic Missiles (IRBMs) under a two-nation key system. These were deployed on twenty sites from Yorkshire to Cambridgeshire under the codename Project EMILY. A number of these were assigned to HQ 3 Group and, in early December, Wg Cdr Phillips crew flew to Vandenburg AFB in southern California to observe a launch and returned with SASO, Air Commodore Johnson, replacing the Co-pilot. A reserve aircraft had also been positioned at Offutt AFB, Nebraska. More prosaically, two aircraft conducted an exercise to test the defence systems of the Type 41 frigate, HMS Lynx and, on returning from a training sortie on 19 December, Flt Lt Bichard's crew was diverted to Wittering thanks to adverse weather at Cottesmore. On approach, the starboard main undercarriage failed to lower, resulting in an emergency landing and Category 3 damage.

1962

The new year began with another change of command – Wg Cdr Phillips left on 1 January and was replaced on 29 January by Wg Cdr T C Gledhill AFC, with Sqn Ldr Young taking charge over the gap period. With that done, the first weeks of the year were marked by crews participating in another Exercise PROFITEER to RAAF Butterworth in Malaya. On this occasion, 15 Squadron was also participating as the lead unit and there was no separate 10 Squadron Operation Order as in 1960. The exercise was scheduled from 11 January to 4 March and the Squadron sent two aircraft and crews initially, with a third crew positioned in the supporting Britannia transport aircraft. The aircraft remained in place but all three crews returned in February when Sqn Ldr Young's crew went out, completing the Jungle Survival Course as they passed through Singapore. They then brought one of the deployed aircraft home on 4 March, with a 15 Squadron crew flying the other.

On 1 February 1962, a system of 'Quick Reaction Alert" (QRA) was instituted throughout Bomber Command, requiring one aircraft with a crew from each squadron to be at 15 minutes Readiness at all times. This laudable aim was a dominant factor in each squadron's planning from then on but, to begin with at Cottesmore, it was agreed that there would be a single aircraft from the station until aircraft and crews had returned from Malaya. Indeed, the Squadron records show that only one aircraft was available for training through most of the month – Malaya, a Cold Weather Trial at Goose bay, a Major Inspection and an airframe with Cat 3 damage accounted for all of the others. Despite this significant change in operational posture, the future for 10 Squadron with the Victor B1 aircraft had become limited. Proposals had been made to upgrade the original B1 fleet to B2 standard, but these were judged to be unaffordable. A compromise of sorts was reached when it was agreed that a B1A version, with some of the Electronic

Countermeasures (ECM) kit destined for the B2, would be introduced. Modified aircraft began to appear with other squadrons from mid-1961 with the first B2s in early 1962, but 10 Squadron's older aircraft were not to be upgraded on cost grounds. Consequently, changes in the wind can be seen with four of the Squadron's AEOs sent on B1A courses in January, and the first postings-out for B2 conversion in March. And whilst the crew numbers would recover, the Squadron had barely seven crews on strength in April against an establishment of eight aircraft. The gloss of having been first with the aircraft was beginning to dim.

The annual Nav/Bomb Competition was held in April when, despite a very good start, the Squadron emerged as fifth out of sixteen overall. Another MAYFLIGHT dispersal exercise to Bosconbe Down was called on 7 May and was followed in mid-month by Exercise TITTON, when five sorties were flown on demanding mission profiles as part of the crew classification process. Flt Lt Cole's crew had the month's Western Ranger flight with aircraft XA 929 and, on landing at Offutt AFB, the home of HQ Strategic Air Command (SAC), the brakes locked on and the aircraft slid along the runway to a stop. In itself, this was well done, but it soon became clear that the Victor was blocking the takeoff of a SAC Airborne Command Post aircraft due to relieve another already airborne for some hours. A fraught situation, with threats that the Victor might be bulldozed to one side, was resolved by speedy work led by Chief Tech Bird, the Crew Chief, who was able to arrange for a very speedy wheel change that allowed the Victor to be towed off the runway. After a thorough check of wheels and brakes on the next day, the crew completed an uneventful return to base.

The same aircraft, XA 929, was used for a Lone Ranger to Aden in June when, on takeoff from Akrotiri for Khormaksar on 16 June, it crashed with all aboard killed. It is believed that an incorrect flap reading caused a wrong flap selection for takeoff and that, by the time the pilot realised what was happening, it was too late to avoid a crash. The aircraft was seen to lift its wheels briefly from the runway before the aircraft ploughed into the overshoot area where it rapidly broke up and burned. The Co-pilot ejected just before the crash but, as this happened well outside the design parameters of the seat, he did not survive. The crew comprised: Flt Lt G A Goatham, Captain, and Flt Lt A W Mitchell, Co-pilot; Flt Lt J Gray and Flt Lt D C Brown, Navigators; Fg Off A P Pace, AEO; and Master Tech D A Smith, Crew Chief. They were accorded a full military funeral and buried in the Dhekelia Military Cemetery, Cyprus, on 20 June, with a simultaneous service held in the Station Church at Cottesmore. Sqn Ldr Young, Flt Cdr (Air), and Plt Off Lawrence, the Squadron Engineering Officer, flew to Cyprus to participate in the Board of Inquiry.

On 16 June 1962 Victor XA 929 crashed on takeoff at Akrotiri, Cyprus killing all the crew. They were buried in the Dhekalia Military Cemetery and a Memorial Service was held at Cottesmore at exactly the same time.

A wooden plaque commemorating XA 929's crew was installed in the Cottesmore Church but was subsequently removed when the base closed. It is now held by No 10 Squadron.

Fg Off John Rattenbury joined the Squadron in June 1962 as a first tourist Copilot with Flt Lt Rooney's crew and later recalled those early days:

> "Like so many of my predecessors, I arrived in Bomber Command young and inexperienced. The sights and sensations of an active station at the height of the cold war were all-absorbing. In retrospect, I think my crew (especially Mick Rooney) were very tolerant and understanding of this callow youth. Within days of arriving at Cottesmore, much of this callowness was knocked off. XA929 crashed at the end of the runway at Akrotiri and all the crew were killed. My own time on the squadron was incident free apart from minor unserviceabilities. Only one dicey incident stands out. We were returning to land on a foul night with driving rain and a strong crosswind. I was doing the approach and visibility, normally not very good in the Victor, was hopeless. We swung around a bit getting lined up and about 100ft up Mick Rooney yelled 'I have control'. He too was almost blinded and opened the quarter light with a blast of wind and rain to squint through at the runway before putting us down. Stressful though it was for us pilots, how much worse for the rear crew, facing backwards with no bang seats wondering what the clowns up front were doing."

July began with another KINSMAN dispersal exercise to Boscombe Down for two crews. Western and Lone Rangers were flown – and 4-aircraft Scrambles were evidently still something of a 'Party Piece,' albeit with 15 Squadron sharing the load. A demonstration had been laid on for a visit to the station by Field Marshal Viscount Montgomery of Alalmein ('Monty') in June, and another was scheduled during a visit by the Imperial Defence College on 12 July. September saw the customary participation in Battle Of Britain Open days around the UK, followed by a MICKEY FINN Command dispersal exercise for four aircraft and crews, again to Boscombe Down. In turn, this was followed by a NATO ADEX, MATADOR II, for which the Station's engineering staff pulled out all available stops for, in addition to the three aircraft for Squadron crews, plus the QRA aircraft, two others were made available to Honington squadrons for their QRA commitments. They may have been trying to impress before a change from Centralised to Squadron servicing that took place on 1 October. Prior to that date, the Squadron had only its First Line engineering personnel on strength.

As things turned out, October 1962 would go down in history for more than a change in unit servicing arrangements. The most threatening confrontation between the USA and the Soviet Union during the Cold War years occurred over a week later that month and, as a NATO ally, the UK's bomber force became involved and prepared for a nuclear exchange. Whilst three crews were at Boscombe Down in mid-October on yet another KINSMAN dispersal exercise, President Kennedy was being briefed on the findings of recent U-2 overflights of Cuba. These showed that Soviet missile sites were under construction on the island and the President began to consider his response to this new threat to the USA. On the evening of Monday 22 October, President Kennedy ordered a naval 'quarantine' of Cuba and made a TV broadcast, carried in the UK, to inform the US public of what was afoot and of the potential global consequences should the crisis continue to escalate. The military readiness of US forces was increased and a succession of personal exchanges took place between President Kennedy and Soviet Premier Khrushchev, and between the President and Prime Minister Harold Macmillan. On 27 October, a U-2 aircraft was shot down over Cuba by a Soviet-supplied missile and the pilot was killed, but the worst of the crisis passed the following day when Premier Khrushchev accepted a proposed solution and released a letter affirming that the missiles would be removed in exchange for a non-invasion pledge from the US. The Squadron ORB describes the period with immense restraint:

> "This situation created tension throughout the world and it was decided on Saturday October 27 to bring Bomber Command to Alert Condition 3. Squadron personnel were notified as soon as possible. Preparations were made on Sunday for the possibility of an increase in the Squadron Alert commitment. On Monday October 29 a second aircraft and crew was brought to 15 minutes readiness and other Squadron aircraft were held at 2 hours, 3 hours, 4 hours, 5 hours and 12 hours readiness. The normal flying programme continued uninterrupted where crew and aircraft availability allowed. This Alert condition was

held until Monday November 5, when Bomber Command reverted to Alert Condition 4 once more."

With the crisis past, life returned to normal – officers came and went on Courses, a new crew formed on completion of a former Co-pilot's Captain conversion, and the annual Standardisation Board visit took place. The Team's report "was most satisfactory, and those crew members who were examined gave a good account of themselves."

1963 – 1964

The UK winter of 1962/63 was long and harsh. In addition to an evening on which he was Station Duty Officer and had to deal with fifteen fuel tankers destined for Cottesmore snowbound on the A1, John Rattenbury recalled a method of runway clearance in use at many RAF stations, both then and later:

> Keeping the runway clear was a priority and the technical bods had mounted a jet engine sideways on a trolley towed by a tractor to blast the snow and ice away. If the blasting was too strong, asphalt would be removed as well so the ideal people to control the throttles were Copilots of course. One of my coldest duties ever!

Nonetheless, a good proportion of the Squadron's personnel managed to escape that winter at home as another detachment to RAAF Butterworth in Malaya took place between 21 January and 20 February, this time under a new nickname, Exercise CHAMFRON. The format was much as in 1960, with four aircraft and five crews involved but only one made it out on schedule. The others were variously delayed with technical problems at Akrotiri, Nairobi and Gan, and the fact that the allocated support Britannia (which had its own problems) could not be in two places at once did not help. However, flying was able to start on 28 January as planned, albeit with just three aircraft initially, and various bombing trials and air defence exercises were flown. Some 40,000 lbs of bombs were dropped and all of the crews completed the Jungle Survival Course at Changi. As with the deployment, only one aircraft returned home on schedule, whilst another took fourteen days to get back after incurring a variety of major problems whilst at Nairobi. The report on the exercise closed by saying that "Everyone is looking forward to next year's trip!" but, as events would show, that was not to be.

The first Lone Ranger exercises to Cyprus since the fatal accident in July 1962 were flown in March and these now included visual bombing details on the range at El Adem, in Libya. Flt Lt Kemish led the second crew, the one that had experienced the lengthy return flight from the Far East in February and, as chance would have it, they found themselves once again involved in an extended stay at RAF Akrotiri involving an engine change. Rangers in April were also badly affected by unserviceability, so much so that one to Karachi was cancelled, and technical issues also spoiled the Squadron's chances in the annual Command Nav/Bomb Competition (now called the Combat Proficiency Competition) that month. Flt Lt Kemish's crew had the best Nav score in the first phase, and others also did very well in later phases, but an undercarriage that would not retract forced a crew to abandon the competition and any chance of the Trophy was lost. Another MAYFLIGHT dispersal exercise for four crews to Boscombe Down took place in May, and an early sign of a decision not to upgrade the Victor B1 is revealed by an entry noting that Western Rangers could no longer proceed beyond Goose Bay "owing to lack of UHF radio equipment and identification systems." Without upgrades, the aircraft's future was clearly limited and, with it, that of 10 Squadron as a B1 unit.

To maintain a credible deterrent posture, the increasing deployment of Soviet SA-2 surface-to-air missiles had caused both the US and UK to investigate standoff missiles as a way of maximising the chances of successful nuclear attacks. Amongst other approaches, the UK had bought into the US Skybolt air-launched missile system and when that programme was cancelled by the US at the end of 1962, the decision precipitated the crisis that led to the 'Nassau Agreement' and the UK purchase of the US Polaris system for the Royal Navy. As a result, Bomber Command had to look at alternative attack profiles for the delivery of nuclear weapons by the V-Force to cover the years until the naval system would be in service. Trials proved that low-level attacks could be feasible although Pop Ups to altitude would still be required for bomb release. So, with the other V-Force

squadrons, the Squadron took on a new low-level role. Training for it was given priority in May, the first month, and twenty-three sorties were flown as crews began to adjust to a very different flight regime. Some training limitations were applied to reduce aircraft fatigue, with the majority of flights flown at 1000 ft AGL and every third sortie at 500 feet AGL. New navigation techniques had to be developed for, whilst the H2S radar could still be used, high ground along-route would limit its horizon. So, as the Nav Radar did what he could, the Plotter gave a running commentary from a prepared map on what the Pilots should be seeing and, if all was well, they, in turn, confirmed his predictions.

As in other years, the summer months brought displays and two crews were tasked for one at Chievres AB in Belgium, near the SHAPE HQ, one in the flying display and the other in the static park. Then, in early August, two aircraft were flown to RAF Scampton to take part in a Command review by US General Lyman L Lemnitzer, the Supreme Allied Commander Europe (SACEUR), when six other Squadron crews participated in the Victor contingent, commanded by Wg Cdr Gledhill. At the same time, another KINSMAN exercise was held from 8 – 12 July and, for the first time, a new dispersal base at RAF St Mawgan was used. Seven crews and five aircraft were involved. Over the autumn, the unit was as busy as it had ever been, with eleven crews flying over 280 hours per month in a full range of low-level, bombing and navigational sorties, interspersed with Rangers and exercises. By November, what would be the 'last time' for a number of events occurred – the last KINSMAN to St Mawgan, the last MICKEY FINN and the last Standardisation Team visit – for in December it was announced that the Squadron would disband on 1 March 1964.

However, before that, there was one more notable detachment as Wg Cdr Gledhill and Flt Lt Rooney took two aircraft to Nairobi from 4 – 17 December as part of celebrations to mark the grant of independence to both Kenya and the island of Zanzibar, now part of Tanzania. Unserviceabilities caused delays on the outbound flights but both aircraft reached Nairobi in time to prepare for the two planned flypasts, each of which was to involve eight Hunter aircraft of 208 Squadron, deployed from their base in Aden, RAF Khormaksar. A Pembroke comms aircraft flew the formation leaders several times around the Nairobi flypast route to permit them to choose timing and turning

Kenya & Zanzibar Independence Celebrations December 1963, Embakasi Airport, Nairobi. Two Victors from 10 Sqn, XA 940 Wg Cdr T.C. Gledhill & crew with Reserve XA 927 Flt Lt K.W. Rooney & crew, line up for a photograph with two RAF Hunters.

points. Flt Lt Rooney's crew flew a rehearsal for the Zanzibar route on 9 December and, on the following day, the official salute was flown past the Sultan's palace at Dar-es-Salaam. The Kenyan rehearsal day was 11 December, when the route was flown by the CO's crew. On 12 December, both crews took off at a minute's spacing, with Flt Lt Rooney's crew as Reserve. He followed behind the CO for as long as possible before peeling off to orbit out of sight of the ceremony at the Uhuru Stadium. The flypast was exactly on time and the formation then embarked on a tour of towns close to Nairobi, with the Reserve Victor following some fifteen minutes later. Both aircraft returned to Cottesmore on 14 December, but that was not quite the last recovery from an overseas trip. That distinction fell to Flt Lt Cole's crew, who flew the final Western Ranger to Goose Bay, with a diversion to Stephenville, from 23 to 25 January. John Rattenbury went to Nairobi with Flt Lt Rooney's crew:

> One of the overriding aims of a crew was to get on a 'jolly,' something other than the routine of training and practice. I remember being disappointed not to get the detachment to Butterworth. Those who had been before enjoyed telling us what we were missing. On the other hand, our crew did get Kenya and Zanzibar for the independence celebrations. We flew to Akrotiri and I had toothache! No way would I confess and be taken off the jolly. On the way over it got worse so I could hardly stand my oxygen mask. On landing, I had a swollen abcess and the tooth was pulled by the station dentist. I have a vivid memory of waking from the anaesthetic gazing through the window at the blue Med. We stayed a couple of days, I'm not sure whether that was my fault or part of the planning. At Nairobi, we practised the planned flypasts and crossed the country at low level. Mick did most of the flying so I was free to snap away with my (unauthorised?) camera. It was a memorable jolly, both the flying and the off-duty. We met officials and politicians, toured the game park and even tried to blag our way to a meeting with Jomo Kenyatta. We went to the markets and stocked up with souvenirs and ordered baskets of pineapples to be delivered to the airport on departure. They duly were, but when we looked at them, they were full of ants. A great heap of pineapples was left at the side of the dispersal. I did return with a Kenyan flag, snaffled from Brunners Hotel where we were staying by climbing through a toilet window.

Despite the imminent disbandment, Squadron personnel were kept fairly busy by a full training programme during both January and February 1964 and a second airframe was provided to cover 15 Squadron's QRA commitment whilst it carried out the annual Victor detachment in Malaya. On 28 February, the CO's crew and four others took off on a variety of training sorties before coming together to fly in formation over Cottesmore and the nearby towns of Oakham and Uppingham and, whilst an aircraft still stood QRA on 29 February, these were the Squadron's final sorties bringing to a close the Squadron's five and a half years or so with the Victor B1. Official Farewells had been said in both Officers' and Sergeants' Messes by then, and on 29 February the Squadron officers returned the compliment by throwing a party. At midnight, *"to the sound of the QRA alarm bells and Auld Lang Syne,"* the Squadron was finally disbanded and its long history as a Bomber Command unit had come to a close. Nonetheless, although denied the opportunity to fly the powerful Victor B2 with its Rolls Royce Conway engines, it would not be long before 10 Squadron would re-emerge in Transport Command, bringing another new aircraft into service … one with rather more powerful Conways.

CHAPTER 11

THE EARLY VC10 YEARS, 1966 – 1985

"Try a little VC10derness" – a popular BOAC slogan in the mid-1960s.

10 Squadron reformed on 1 July 1966 for what was to prove its longest continuous period of service to date – some 39 years before disbandment on 14 October 2005. The 'Shiny Ten' nickname that had stuck over the years was now more apt than ever, for the Squadron was to fly the new VC10 4-jet strategic transport aircraft – the heaviest aircraft in RAF service at the time, with a cruising speed at height matching many contemporary fighter aircraft and, in the minds of many, lines of unrivalled elegance! During those 39 years UK Defence policy would change almost continuously in response to shifting political realities and the demands of annual budget rounds, and 10 Squadron's operations would not be immune from these. But at the outset, things were upbeat.

After the Berlin Airlift, RAF Transport Command had been reduced in size as priority was given to the build-up of Canberra and strategic bomber squadrons in Bomber Command and RAF Germany in the 1950s. 10 Squadron had played a part in both areas, as we have seen. Now, in the mid-60s, it would emerge as the foremost squadron of a re-energised Air Transport Force (ATF). At the start of the 1960s, the UK still had substantial Defence commitments overseas, in Germany, the Mediterranean, and in the Middle and Far East and, after a decade of relative neglect, plans were put in place to provide a modern fleet capable of supporting this far-flung network of bases. Comets and Britannias had come into service in the mid/late 1950s to augment the limited capability of a Hastings fleet that dated back to the Berlin Airlift, but these were considered insufficient to meet the task. Operational Requirements were written to provide a high-speed passenger aircraft matching contemporary airliners and a long-range bulk freight capability. These would emerge as the VC10 and the Belfast, with both fleets to be based at RAF Brize Norton in Oxfordshire. (No 53 Squadron would operate the 10 Belfast aircraft, and the temporary co-location of the units first seen at Abeele in 1918 would become permanent.)

Deliveries of the 14 aircraft began in July 1966 (the first was XR 808) and, whilst the unit occupied former USAF accommodation at Brize Norton, these had to be made to RAF Fairford nearby and, for some time, services with a payload were flown from RAF Lyneham. These relatively short-term measures were necessary whilst facilities at RAF Brize Norton were prepared for the new era in RAF transport flying that began there in the Spring of 1967. From 1950 to 1965, the station had been a forward base for bombers of the USAF's Strategic Air Command. It therefore lacked the facilities that would be needed to provide a major air transport base and an extensive works programme was commissioned to provide these – a Terminal/Ops block, a transit hotel (Gateway House), a cargo handling facility, enlarged and floodlit apron space, squadron offices and, most prominently, the Base Hangar, at the time the largest cantilever structure in Europe, designed to accommodate six VC10 or Belfast aircraft. Work on the runway was also needed, something that would prove a great hindrance in the first months of operations.

The 14 Type 1106 Vickers VC10s built for the RAF were the product of a series of Operational Requirements over a number of years. In essence, they had the shorter fuselage of the Standard VC10 operated by the British Overseas Airways Corporation (BOAC), with the wings, 'wet fin' fuel tank, and uprated Rolls Royce Conway Mk 301 engines of BOAC's Super version. Each engine had a nominal rated

thrust of 22,500 lbs. In addition, for military use, they incorporated a large freight door on the forward port side with a strengthened floor. (The aircraft was a very capable load carrier for palletised ammunition, and that would have been a primary war role.) The first batch of five was ordered in September 1961, six more followed in 1962, with the final three in July 1964. A RAF Project Team was established at Weybridge from 1962 and deliveries were made from July 1966 to August 1968.

At the outset, the only VC10 training facilities available were with BOAC, which had the first aircraft. Squadron Leader Harry Liddell, a very experienced training captain in the RAF's Britannia force, was one of those selected for No 1 Course and later recalled those first weeks:

> "At the end of May 1966, along with about a dozen other pilots, I found myself attending the BOAC training complex at Cranebank for the 6-week ground school phase of type conversion. Besides myself on No 1 Course I recall Wg Cdr Mike Beavis, CO designate of 10 Squadron, George McCarthy the Wg Cdr Ops for Brize, Bill Somers (ex-Comet) and another two ex-Brit people, John Loveridge and Dave Ray. There were also some co-pilots and flight engineers, the latter with a syllabus of their own. The course was fairly intensive, its content the same as given to BOAC crews, with frequent tests as it progressed. However, we all passed, progressing onto the simulator phase. That was also at Cranebank as our own unit at Brize did not yet exist, and it was rather irregular as we had to be fitted into BOAC's own programme."

Moving on to flying training:

> "At this time (summer 1966) use of the Brize runway was very restricted owing to improvement works being carried out. As a consequence, both Belfast and VC10 aircraft were kept at Fairford, from where all local training flights were carried out. And as no movements facilities existed there, nearly all route flights had to be mounted from Lyneham, a highly inconvenient arrangement that continued well into the next year.
>
> My first conversion flight, captained by Alf Musgrove (one of the two project team pilots, the other being Brian Taylor), was made on 15 August and others continued sporadically up to my first solo a month later. Things necessarily happened rather slowly at first, with XR 808 being the sole available aircraft, but speeded up a bit after 809 was delivered. All my conversion training, including route, having been completed by the latter part of November, I was duly awarded a category, declared an instant expert, and set to work training the rest of No 1 Course."

The Initial Buildup

The bulk of the Squadron aircrew on No 1 Course, now including Navigators and Air Quartermasters (AQMs) who had completed most of their training locally, was posted-in with effect from 1 January 1967. However, as the planned office accommodation for both 10 and 53 Squadrons was part of the major station works programme, 10 Squadron was initially allocated some former USAF buildings on the south side of the airfield where, much as had happened at Binbrook in 1953, a degree of self-help was required in setting things out. The plan was that the crew establishment should grow to a total of 34, giving a crew to aircraft ratio of 2.4:1, a ratio higher than on any other RAF squadron at the time. Around 40 Corporal Air Stewards were to be posted in from April 1967 and, in all, the Squadron strength would rise to some 275 flying personnel. This number was calculated to support a planned core task of 27 scheduled services per month to the Gulf and Far East from April 1968 (3 to Hong Kong, 4 to Bahrain, and 20 to Singapore). These flights would be mounted on a 'Slip' pattern that would become second nature to 10 Squadron personnel in years to come, whereby crews were positioned at points along the route to the Far East. This enabled the aircraft and its passengers or freight to be kept moving, with crews changing and waiting at the staging posts in question to take over the next flight to pass through. Thus, a passenger would reach Hong Kong in around 22 hours, whereas the first crew to carry him (or her, as family moves on posting constituted a significant proportion of the scheduled task) might take about a week to go around the route and return to base.

A flight crew consisted of two pilots, navigator, and flight engineer. Cabin staff were a mix of SNCO AQMs and Corporal Air Stewards, the latter drawn from the RAF's Catering Trade Group 19 on limited flying tours.

Significantly, around half of these were members of the WRAF, something new in 10 Squadron's history. By the mid-1960s they had a well-established presence in Transport Command's Britannia and Comet squadrons, where Service families were regularly carried, albeit that the advent of 10 Squadron was about to cause a Command-wide shortage that would need a targeted recruitment campaign to increase numbers. (A first commissioned AQM was posted as AQM Leader in January 1969 and in 1970, AQMs were restyled as Air Loadmasters (ALMs). Female flight crew would only appear in the 1990s, after a significant change in Service policy that permitted women to fly.) In its first months, the Squadron flew with constituted crews but from 1 January 1969 adopted the policy common in the Air Transport Force whereby crew composition changed from task to task, relying on the quality assurance provided by its Categorisation system to maintain and, indeed, raise standards. When the Squadron was disbanded in 2005, there had been no accidents leading to injury or loss of life in 39 years – no mean achievement. As regards groundcrew, a centralised servicing policy pertained in the Strategic Transport force at that time and, for many years, no servicing personnel were held on squadron strength, although NCO Ground Engineers would join crews for flights away from normal RAF routes.

Whilst the core of 10 Squadron tasking was expected to consist of scheduled services, other tasks would include flights in support of Service exercises, and 'Specials.' The latter covered a range varying in character from month to month – for example, VIP tasks carrying Royal or senior Ministerial passengers, crew changes for RN vessels anywhere around the world, or short-notice Aeromedical call-outs. For these, the role fit on the aircraft would be varied as required. After allowing for supernumerary crew, some 130 seats would be normally be offered in the full Pax (ie passenger) fit; the mixed PCF (passenger cum freight) role would have a number of pallets in front of, perhaps, 65 pax; and the full freight role had all seats removed and the roller panels in the floor revealed to permit easy loading of pallets. In common with other RAF transport aircraft of the time, all seats were rearward facing. The aircraft carried stretcher posts and stretchers as a matter of course to permit Aeromed re-roling whenever and wherever required.

The Squadron's first operational task, Flight 2600, using aircraft XR 809, positioned at Lyneham from Fairford on Friday 20 January 1967. It left for RAF Changi, on Singapore Island, the next morning, with Wg Cdr Beavis and Sqn Ldr Taylor as designated captains, a co-pilot, four navigators, three flight engineers, and seven AQMs – in effect, all of the available and qualified squadron personnel! In February, four tasks were flown – one to Hong Kong and three to Bahrain – with another cancelled to give priority to the Air Training Squadron, now busy with No 2 Course. The members of that course joined the Squadron in March, increasing the crew number to eight, and ten operational tasks were flown. Four of these – to Cyprus and Bahrain – are recorded as the first to have carried full passenger loads, and were mounted primarily to exercise Movements staff in

4 April 1967 The first VC10 schedule to the Far East starts from RAF Lyneham; there being no passenger handling facilities yet available at Fairford or Brize Norton.

808 parked at Kai Tak with low cloud over Lion Rock

the handling of around 130 passengers in and out in a 90 minute turnround.

With five aircraft delivered, Squadron personnel numbers were now sufficient to position slip crews down-route and permit the start of scheduled services. The first of these, Flight 2300, left RAF Lyneham on 4 April 1967 for Hong Kong (RAF Kai Tak). The standard route at that time was: UK – Bahrain (RAF Muharraq) – RAF Gan (in the Maldives); Gan – Hong Kong or Singapore (RAF Tengah, a temporary arrangement pending the completion of work at RAF Changi in October) - Gan; Gan – Cyprus (RAF Akrotiri) – UK. There were 7 schedules tasked in April, carrying around 1500 passengers. A weekly schedule to Bahrain was started in May and numbers increased until after a year, in March 1968, some 6000 passengers would be carried on 26 flights to Bahrain and the Far East, plus 2 to New York. (By February 1968, the number of government and service personnel travelling between North America and the UK was deemed to warrant a weekly RAF schedule to New York. For families moving on Exchange postings, this supplanted a few days on the QEII. Initially this alternated between 10 Squadron operating in a mixed PCF (passenger cum freight) role and a Comet 4 of 216 Squadron in the purely passenger role until 10 Squadron had sufficient aircraft to support the weekly requirement. In November 1970, the destination was moved to Dulles International Airport near Washington DC, with Ottawa later added as an occasional stop.)

Work on the runway at Brize Norton ended in May 1967, so the Squadron was able to move fully into its HQ and the need for crews to be bussed to Fairford before taking an aircraft to Lyneham came to an end. (The move into the purpose-built office block on the main station site facing the Terminal/Ops Block, to be shared with 53 Squadron, occurred in April 1969.) However the new Terminal would not be in use before October 1968 so, in January that year, to increase aircraft utilisation by doing away with overnight positioning, clearance was given to operate scheduled departures from Brize Norton, with passengers brought by bus from Lyneham. That said, passenger arrivals inbound to the UK continued to be made at Lyneham, with the aircraft then recovering to Brize Norton.

Some personnel arrivals in this early period are noteworthy. In March, Flt Lt J L Bowmer rejoined the Squadron from No 2 Course as a captain, having survived more than the customary 30-Op tour in 1944. Major Bobby J Massingill joined in August 1967 from No 4 Course as the first USAF exchange captain and, by the early 1970s, there were four exchange positions filled by USAF and Canadian captains and navigators. Their RAF counterparts were serving with squadrons of the USAF's Military Airlift Command and with No 437 Squadron at CFB Trenton, Ontario. Then, in December, Flt Lt J L Walker also joined as a captain from No 5 Course. As Flt Sgt Walker he had last left 10 Squadron by parachute over Germany on the night

of 19/20 February 1944, when his Halifax was shot down and he and his crew became POWs. He and John Bowmer are the only individuals of the VC10 era so far identified as having served on 10 Squadron in WW2.

> On 1 August 1967, Transport Command became Air Support Command (ASC), when it absorbed Offensive Air squadrons likely to be tasked with Army support. On 1 September 1972, HQ ASC was merged into HQ Strike Command to create a single RAF operational command, but command of AT forces remained at Upavon, then designated HQ 46 Group.

As the Squadron gradually increased in size in terms of both aircraft and crews, it was able to take on an increasing number of Exercise and Special Flights, notably exploiting the VC10's speed and range to deliver replacement engines to other transport aircraft or V-bombers stranded around the world. It was also able to increase the number of slip crews positioned along the Gulf/Far East route and adjust the slip pattern, with shorter crew duty days permitting much improved flexibility in coping with delays or re-routings. By November 1967, it was able to participate in Operation JACOBIN, the evacuation of British forces from Aden. To maintain the concurrent Far East schedules, crews were positioned in Teheran and aircraft staged there rather than at Bahrain or Akrotiri for the duration of the operation. Six sorties were flown from Aden and the Squadron record notes dryly that "the aircraft were not full." In December, a task was flown to deploy slip crews of other transport squadrons as part of Exercise Travelling Causeway, an annual Air Transport Force exercise of its Westabout route to the Far East. (The fragility of the normal route through the Middle East had been re-emphasised by the 6-day Arab/Israeli war of June 1967 when some overflying rights were denied. Staging posts were established at bases in Canada, the USA and over the Pacific to create what became the 'Westabout Route.') In later years, the VC10 would regularly be used to deploy Britannia and Hercules crews ahead of large-scale transport operations or exercises.

Also in December 1967, the Squadron's first Royal flight was flown. The Australian Prime minister, Mr Harold Holt, went missing and after 2 days was presumed to have drowned. A Memorial Service was held on 22 December and HRH Prince Charles, Prime Minister Harold Wilson and Leader of the Opposition Ted Heath flew from Heathrow to Melbourne on a VVIP task laid-on at very short notice. Outwardly all went very well, but the flight showed up shortcomings in the stowage available for additional VIP catering equipment that made things very awkward for the cabin crew, a matter that had already been raised by the unit.

Flt Lt John Walker receives the keys to the new slip-crew accomodation in 1968, at The Changi Creek Hotel, from the Station Commander Gp Capt E. W. Merriman. AQM Sgt (W) Jacqui 'Teddy' Gibby looks on.

In late February 1968, word came of a significant development that marked the next 12 years of VC10 operation. The Squadron's captains who were not substantive Squadron Leaders were granted acting rank, and would be able to retain it for as long as they were associated with the VC10, for example as Instructors or Examiners. RAF crewroom history has rather lost sight of the fact that this was not a provision unique to 10 Squadron. It came about as a consequence of an Air Force Board decision of January 1968 to provide retention incentives in the face of a longterm undermanning situation, and applied equally

to the captains of 216 Squadron's Comets. These, however, were many fewer in number, and the type was withdrawn from service in the mid-70s so, for a long time, this appeared to be a discriminatory move in favour of 10 Squadron. It occasioned all manner of ribald comment over the years, did the Squadron's image no favour in the eyes of the RAF at large, had probably no noticeable effect on retention, and was finally reversed in late 1980.

Naming of Aircraft

Most unusually for a RAF aircraft, the VC10 was not given a service name. (It did, however, attract nicknames amongst overseas-based servicemen, particularly at those bases where unaccompanied service was common in the late 1960s: "Moonrocket" in Bahrain, a tribute to the many late night departures from Muharraq, and "Gozome Bird" almost anywhere.)

In November 1967, Wg Cdr Beavis made a case for its VC designation to be used by naming individual airframes to commemorate RAF or Royal Flying Corps recipients of the Victoria Cross. His submission was accepted, but as there were 51 VC holders but only 14 aircraft, a decision was made to restrict the choice to 21 holders from the UK who had lost their lives on flying operations. This reduced list had 7 from WWI and 14 from WWII. A final list was produced and the project was implemented by having the selected name on a blue scroll, under a Squadron badge, placed to the left of the forward passenger door, in a position that could not be missed by anyone using it. A formal ceremony to name the aircraft was held at Brize Norton on 11 November 1968, the 50th Anniversary of the end of WW1, using XR 810, which was to carry the name of David Lord VC. He had been posthumously awarded the Transport Force's sole VC after being shot down over the Arnhem drop zones in 1944, and the naming ceremony was performed by his brother, Wg Cdr (Retd) F E Lord.

The 14 aircraft were named as follows:

XR 806	George Thompson VC
XR 807	Donald Garland VC & Thomas Gray VC
XR 808	Kenneth Campbell VC
XR 809	Hugh Malcolm VC
XR 810	David Lord VC
XV 101	Lanoe Hawker VC
XV 102	Guy Gibson VC
XV 103	Edward Mannock VC
XV 104	James McCudden VC
XV 105	Albert Ball VC
XV 106	Thomas Mottershead VC
XV 107	James Nicolson VC
XV 108	William Rhodes-Moorhouse VC
XV 109	Arthur Scarf VC

When named after Victoria Cross holders the VC10 naming ceremony took place in the Base Hangar at Brize Norton on 11 November 1968 with XR 810, David Lord VC as the backdrop.

(To begin with, Squadron aircraft also carried the words 'Royal Air Force Transport Command' on each side of the fuselage, above the blue cheat line. Reflecting changes in the RAF's higher command, this duly changed to 'Royal Air Force Air Support Command' after August 1967, and then to 'Royal Air Force' after September 1972. In the mid-1970s, at an overseas stop, Wg Cdr A J Richards, then CO, was taken aback to be asked to which Air Force the aircraft belonged. Nonplussed at this, he made a successful case to have a Union Flag added on the forward (port side) passenger door normally used by VIPs, thereby providing a substantial hint as to which nation's Royal Air Force the aircraft belonged!)

The Long Haul

By mid-1968, the pattern of operations was well established – a core of some 27 monthly schedules to the Gulf and the Far East, with others to the USA, together with Exercise and Special Flights. Civilian and Service VIPs were regularly carried on schedules and designated VVIP 1000-series flights began to be tasked. Nonetheless, with a body of new-to-type aircrew, it would take a little longer to build up a select band with the A Category required by the Captain of The Queen's Flight for carrying HM The Queen. In due course, that would cease to be an issue but it did mean that when the first, formally tasked flight for an overseas State Visit occurred, most of the crew had to be supplied by the Air Training Squadron but with a number of Squadron cabin crew. The flight left Heathrow on the morning of 1 November 1968 carrying HM and HRH Prince Philip on the first leg of a visit to Brazil and Chile that would also involve Comet 4, Hercules and Andover aircraft over a lasting until 19 November. The captain was Sqn Ldr Alf Musgrove and his copilot, Sqn Ldr Tony Richards, who later returned to the Squadron as CO in 1975, recalled the departure:

> "The aircraft was hardly airborne when the cabin door clicked open onto the flight deck. I turned against the tension of my seat belt to see who was entering so soon after take-off and there stood the Queen! Over the intercom I said to Alf, "Alf, it's the Queen". He handed over the flying to me, and invited Her Majesty to come and sit on the spare 'jump seat', where she stayed with us four, the two pilots, the navigator and the engineer, for a short while. Our Royal visitor told us that she had always thought that the continual flying in banked turns over Windsor Castle encouraged people to look straight down into her garden! Was this right? By the time we understood what Her Majesty was commenting upon, I think we were over Cornwall!"

And some days later:

> "En-route from Rio de Janeiro to Santiago de Chile we were in thick cloud at altitude, when, without warning, three fighter aircraft appeared in close formation on our starboard side. At the same time, HM came onto the flight deck as three further Hunter aircraft appeared on our port side. Here we were, with six Hunter aircraft presumably belonging to the Chilean Air Force, being tossed around in the heavy cloud, all very close to one another – all unplanned. Horror of horrors! "Tell them to go away" – but sadly we had no communications with them. Fortunately no accident occurred – and we lived to serve another day."

(It might come as a surprise to hear that the Royal Fit for such flights was in no way lavish when compared with the interiors of many privately-owned aircraft. The royal passengers had a small lounge, dining and sleeping areas with plain furnishing and upholstery in a forward compartment, behind which the entourage would be seated.)

The initial flight crews sent to 10 Squadron were almost without exception very experienced, and the posting-in of first tourists came later. However experience alone was not always enough to eliminate errors. Despite having a very good navigation equipment fit for its time, in late 1968 a Squadron crew found itself seriously off track on a North Atlantic crossing when the navigator elected to use a Gyro/Grid steering technique but applied the gyro rate correction in the wrong sense. The error became apparent when a hard return on the radar was finally identified as the coast of Greenland, some hundreds of miles north of the intended track. An overhead fix was obtained on the Prince Christian NDB on the southern tip of Greenland and the aircraft was able to continue to land safely at Gander. (A first visit by the ASC Examining Unit was already scheduled for December. It appears to have been quite searching, and amongst its recommendations was that action be taken to expedite the fitting of LORAN to the aircraft. And in January the Nav Leader gave Squadron pilots a lecture on Grid/Gyro techniques.)

An unexpected reduction in the fleet size occurred in the Spring of 1969 when an aircraft was released to Rolls Royce for use as an airborne testbed for the RB211 turbofan engine, then under development. The US Lockheed company had chosen the engine for its new L-1011 Tristar airliner and, at that early stage in the design of high-bypass engines, the technological challenges involved were considerable. Suffice it to say here that Rolls Royce later went into bankruptcy and was then nationalised in 1971 on account of its perceived

national importance. (It was privatised in 1987, with HMG retaining a 'golden share' giving it a say in any future takeover attempt.) Despite all this, the RB 211-series engines went on to be highly successful. XR 809 was the aircraft chosen for release. Its last recorded route flight with 10 Squadron was to Hong Kong, between 8 and 10 April 1969, and after delivery to Hucknall on 17 April it was registered by Rolls Royce as G-AXLR on 30 July 1969. Although the intention had been that the aircraft would revert to RAF service on completion of RB 211 testing, the costs of restoration would have been considerable and, as the 1974/75 Defence Review had reduced the aircraft establishment to 11, there was no viable argument for doing so.

In 1969, flying achieved on all tasks, including training, averaged around 1600 hours/month, and there were a number of aspects of operations still settling down. A trial of frozen passenger meals was carried out in April and these became standard from August that year. In July, after a shortage of Air Stewards had limited the number of passengers carried to 99, a further 15 were posted-in for accelerated training, permitting normal service to resume from 1 August. (The RAF used the civilian safety standard of the time, and that required a cabin crew member for every 33 passengers carried.) A trial of continuous Double Crew operation was also carried out in July. An aircraft left Brize Norton on 29 July to fly a global route to Elmendorf in Alaska, Wake Island in the mid-Pacific, Changi and Bahrain, before returning to base 45 hours and 15 minutes later, on 31 July. This went well, with the medical monitoring staff reported as looking the most tired by the end! The technique was used for the rapid deployment of slip crews for the annual ATF Westabout exercise in early 1970, but it was never to become anything but an exceptional procedure. (Similar trials were flown on the Belfast freighter – but it had the advantage of a compartment with bunks beneath the flight deck.) There were also operational support flights of note that year. In March, four tasks were flown to position personnel of 2 Parachute Battalion and the Metropolitan Police on Antigua before they moved on to participate in Operation SHEEPSKIN, an intervention on the Caribbean island of Anguilla. Of much more longterm significance were the handful of flights flown in August to Ballykelly and Aldergrove in Northern Island – the start of what would become Operation BANNER, the movement of Army personnel from/to the UK and BAOR throughout the 3 decades of 'The Troubles,' an operation that would consume many hundreds of transport hours over the years. And on 1 September, an unexpected opportunity arose for 10 Squadron to demonstrate the range and speed capabilities of its new aircraft. HRH Princess Alexandra was due to leave Heathrow that morning by RAF Comet for an official visit to South Africa when it emerged that a young Libyan Army captain, Muammar al-Qaddafi, had staged a successful coup that ousted King Idris of Libya. Much would follow from that act, but the immediate effect was that routing through the country, as was planned for the Comet at RAF El Adem, was not possible. At very short notice, a VC10 was tasked to take over and, despite its having to fly around the longer CENTO route over Turkey and Iran to Nairobi, the Princess was delivered to Johannesburg on time.

A highlight of 1969 was the Daily Mail Trans-Atlantic

Seen here shortly after clocking in at the Post Office Tower, London, on her return, 10 Sqn's entrant in the 1969 Daily Mail Air Race was Air Quartermaster, Sgt (W) Heather Robinson. Although having won the Race in the fastest sub-sonic time of 6 hrs 29 mins 11 secs, she was to be disqualified later on a technicality.

Heather Robinson speeds through New York from the Manhattan Heliport to the Empire State Building, in the car which was not available for the return leg. A motorcycle was used instead and, having won the Race, it was this changing of vehicles which caused Heather's subsequent 'official' disqualification. Flt Lt Olly Tarran is on the back of the car.

Air Race, sponsored to commemorate the 50th Anniversary of Alcock and Brown's first crossing, between the top of the GPO Tower in London and the top of the Empire State Building in New York in either direction. As the official RAF entry was a Harrier, supported by Victor tankers and an Air Race Flight, 10 Squadron decided to compete independently using North America training task 2046 and entered WRAF ALM, Sgt Heather Robinson. Forty-five years later, Heather recalled:

"I had been selected to act as "the runner" with Pat Howard as my stand in, if required. On the morning of 5 May, I clocked out of the Post Office Tower and rode pillion on an RAF police motorbike to a goods yard at St Pancras station where an RAF Wessex helicopter was waiting to fly me to Wisley Airfield. VC10 XR 810 was waiting on the tarmac with engines running ready to whisk me to JFK Airport in New York. On arrival another Wessex flew me to a heliport in Manhattan and thereafter a sports car painted in the "Stars and Stripes" took me to the bottom of the Empire State Building. I clocked in at the top in an overall time of 7hrs 17 mins 52 seconds. We knew that that this was not a wining time and that the return flight with tail winds across the Atlantic would give us a better chance of success. As a crew we flew on the next day to Chicago and San Francisco, thereby continuing the training flight. We returned to New York on 8 May in preparation for the return attempt on 9 May. Race rules had stipulated that all modes of transport used on the outward attempt must be replicated on the return but, at the last minute the "stars and stripes" car was not available to us and I was told I would be travelling by motorbike. This was to be the critical factor in the final result!

I clocked out of The Empire State and rode pillion to the Heliport. The bike rider looked like a member of the Hell's Angels and I hung on for grim death as we sped along the New York streets. One memory is that every set of traffic lights was at red so we raced between them and braked violently at each one. The Wessex again took me from Manhattan to JFK. I entered the VC10 via the forward freight door and was hauled inside by George Sperring, the reserve engineer. We braced ourselves in the hold whilst the aircraft took off – no such thing as Health and Safety in those days!

Once airborne I went onto the flight deck to find the crew ecstatic at the short time it had taken from my leaving The Empire State to top of climb. The Nav's calculation of

the flight time led us all to believe that we could win our category for subsonic aircraft. We landed at Wisley after a flight time of 6 hours. The Wessex flight and ride on the police motorcycle followed and I clocked in at The Post Office Tower after a time of 6hrs 29mins 11 secs!!! It was a winning time.

However the broadcaster Clement Freud realised that we had changed the mode of transport in New York without notifying the officials. This had been a complete oversight, but rules are rules. Mr Freud was unhappy at the number of prizes already won by the military and he was key to our entry being disqualified. It was a bitter disappointment, but Rothmans of Pall Mall who had sponsored the subsonic category recognised that the VC10 had indeed completed the crossing in a record time. So, at a party held at 10 Squadron headquarters at Brize Norton, they presented the Squadron with a silver salver engraved "For meritorious achievement and good sportsmanship for the fastest subsonic time from the centre of New York to the heart of London". That salver remains with the Squadron silver and is a fitting reminder of the wonderful VC10."

November 1969 saw a significant change in the pattern of continuation training. After a successful trial on No 99 Squadron at RAF Lyneham, what became known as Periodic Refresher Training (PRT) became the norm for strategic AT squadrons. Whereas monthly continuation training (MCT) had essentially been aimed at pilots and flight engineers, with a navigator and ALM on board to make up a minimum crew, entire crews would now go to the Air Training Squadron for a programme of exercises involving all trades, at intervals determined by the operating categories held. Squadrons would retain a very much reduced number of local flying hours for their own use.

Squadron participation in Exercise Bersatu Padu was a significant feature in 1970. This was a major multinational and Tri-Service exercise, in which a UK aim was to demonstrate a capability to reinforce Malaysia and Singapore once a rundown of forces and bases announced in 1968 had been completed. Twenty-three flights to Changi were flown in the Deployment phase – five on 11 April, with two each day till 20 Apr. Additional flights were flown in May and June, with another twenty required during the recovery in July. All in all, some 1600-1700 hours would have been flown, with a substantial overlay required on the normal slip arrangements to cope with the peak periods. (A General Election held on 18 June, during the exercise, resulted in a change of government. Edward Heath became PM and the first VIP task for the new administration took Lord Carrington, as Defence Secretary, to Singapore, Malysia, Australia and New Zealand between 24 July and 6 August to discuss Defence arrangements in the region. The plan to withdraw all UK forces was partially reversed and a new Five Power Arrangement was concluded in April 1971.) Other notable tasks in 1970 included a task in early June to convey relief supplies to Lima after a very serious earthquake in Peru; flights to bring VC holders to London from India, Australia, and New Zealand; and the transfer of an Army battalion from Malta to Northern Ireland in late June.

The planned reduction of UK forces in the Far East and the Gulf became a significant feature in 1971 and would reduce monthly scheduled flying by some 300 hours. Additional Singapore schedules were tasked from April onwards for the withdrawals from the island and the last VC10 to leave RAF Changi (Flight 2141) flew out on 28 November, when RAF Tengah became the transport staging post for a much reduced retained presence. The rundown of forces in Bahrain had also begun in the summer and both tasks were completed just before Christmas. However, hardly had that happened than, on 28 December, the new Maltese government of Mr Mintoff, having failed to agree with HMG on a number of financial demands, announced that all British forces and families had to leave Malta by 15 January 1972, less than three weeks later. 10 Squadron figured chiefly in the dependants' airlift, with six aircraft per day tasked from 8 to 13 January, and the ATF brought some 5,000 passengers in all home by the deadline. (A limited Service presence was later resumed in Malta.)

In December 1971, 10 Squadron took over a significant and, at the time, classified commitment. Following the creation of the Berlin Wall in 1961, a number of actions were taken to ensure access to Berlin in time of tension. Amongst these was a plan to maintain civil air access by probing Russian intentions if civilian air services had to be

Top: In September 1973 XV 106 visited the Dallas/Fort Worth Airport, taking the British Ambassador, Lord Cromer to its opening celebrations. A 'British Airways' liveried Concorde renewed an earlier friendship made when a VC10 had supported a Far East and Australian Concorde sales promotion tour in June 1972.

Right: Landing alongside US Navy aircraft carriers at Mayport, Virginia was not-uncommon in the 70s and early 80s, particularly on 'States Trainers'.

A Vickers aircraft visits McDonnell Douglas at Lambert, St Louis, Missouri but little else is known as to why or when, although it was possibly during a USA 'States Trainer' flight , which is known to have called in there during April 1971. Note though, the very 'British' gentleman with his suit and brolly !

suspended. This required US, UK and French authorities to be ready to man a number of their national civilian aircraft with military aircrews and to continue civil air services accordingly. British European Airways (BEA) was the national UK carrier into Berlin at that time, using Vickers Viscount aircraft. The aircraft had RR Dart turboprop engines, as had the RAF's Argosy transport aircraft and, for some years, Nos 267 and 114 Squadrons from RAF Benson provided designated pilots who received periodic 'Dart Handling' training on the Viscount and who accompanied BEA crews along the corridors on Berlin services during those periods. After BEA began to use the BAC 111 on its Berlin services in the late 1960s, and with the RAF Argosies being withdrawn from service, it made sense to use pilots flying a near-equivalent aircraft, the VC10. Again, only a handful of designated personnel were used, and the scheme was continued into the British Airways era, on the Boeing 737, until the requirement ceased.

A little-mentioned task in the 1970s and 80s was the 'Spey' (BAC1-11) and 'Rivers' (B737) Training that a handful of pilots carried out with British Airways. Here Dick King takes a quick break outside via the rear stairs, to have his photo taken at Gatow, Berlin in 1985. The BA Fleet Manager, Cpt Mike Butterworth waves from the flight-deck window.

May 1972 saw the Squadron fly three notable VIP tasks. The first, on 15 May, took HM The Queen to Paris for a State Visit. On 30 May, the Under Secretary of State at the FCO, Mr Anthony Royle, was flown from Hong Kong to Peking (now Beijing) on an extended scheduled flight – the first official visit to Communist China since its recognition by the UK in 1949, and the first landing in China by a RAF aircraft in well over twenty years. Meanwhile, back in the UK, a most an unusual VVIP task was being mounted – to recover the body of the Duke of Windsor to the UK for burial, following his death in Paris on 28 May. First written around use of the Britannia, an outline plan for Operation VAGUE had been in existence since the mid-1960s for "the collection and conveyance of a very senior officer or his lady if he or she dies overseas," a coded reference to the Duke and Duchess of Windsor, living in Paris. (The plan also included the carriage of staff from an Undertaker's firm to prepare the body to be flown back. These were listed in the standard Operation Order section headed 'Friendly Forces.') On 31 May, with an escort party of senior military and other officials, the body was flown from Le Bourget to RAF Benson. (The crew captain was Sqn Ldr Dennis Lowery, who had flown the flight to Paris with Her Majesty earlier in the month. The plan remained in force as Operation HAZE and, in April 1986, similar flights were tasked for the recovery of the body of the Duchess of Windsor for burial next to her husband at Frogmore, near Windsor Castle.) Later, after Mr Royle's successful visit to China, he was followed in October by the Foreign Secretary, Sir Alex Douglas-Home. On this occasion, a full VIP flight was tasked involving three crews and, leaving on 27 October, it took a polar route via Elmendorf AB in Alaska and Tokyo to Shanghai and Peking. Sqn Ldr Ward's crew took the aircraft into China and were accorded VIP treatment themselves over the days they were there, including having a meeting with the Premier, Chou EnLai.

The closure of RAF Gan for runway resurfacing from mid-February to late March 1973 meant that flights to Singapore and Hong Kong were temporarily routed through Colombo, in Sri Lanka, prefiguring a more permanent arrangement a few years in the future. The number of ALMs and air stewards on strength was an issue for much of the year and a trial with just three cabin staff was held. Nonetheless, tasking held up across the year with, for example, additional flights for an Army unit move to/from Hong Kong, support of Phantom and Vulcan exercise deployments to the Far East, two aircraft detailed for flood relief in Pakistan in August and, in October, the deployment of a UN Emergency Force from Cyprus to Cairo to supervise the ceasefire between Egypt and Israel after the Yom Kippur War. There was also a degree of internal reorganisation during May as five flights were created (A – Captains, B - Copilots, C - Cabin staff,

D - Navigators, and E - Flight Engineers) with the Flight Commander (Operations) and Ops Staff remaining at the centre of things.

For 10 Squadron and all of the long-range transport squadrons, July and August 1974 were dominated by developments in Cyprus. In mid-July the Cypriot National Guard, allied with the EOKA movement, staged a coup and overthrew Archbishop Makarios, then the Cypriot President. Turkey intervened within days by invading northern Cyprus and, by mid-August, had advanced to the present 'Green Line' that divides the island. With open warfare breaking out in the peak tourist season, evacuation of holidaymakers became a priority and the RAF played a significant part. Starting on 20 July, the Squadron flew some 35 sorties that month, carrying tourists who had flooded into the British Sovereign Base Areas seeking safety and shelter. Then, in mid-August it was decided that British Service families should be brought home from Limassol and another concentrated airlift using all available air transport assets was laid on. HQ 46 Group was instructed to plan for a departure every 30 minutes and the record for 15 August shows twelve VC10 return flights mounted on that day alone. Inevitably, these events played havoc with previously tasked flights – and it cannot have helped that 10 Squadron was operating once again from Fairford from early June till the end of July as the runway at Brize Norton was relaid.

The Squadron brought in 1975 with a number of events marking the sixtieth anniversary of its formation in 1915 and the year would prove to be a significant one for the RAF's Air Transport Force as a whole – more on that in a moment, after a brief look at the Squadron's participation in world events. In June, a Standby crew had a weekend callout to go to The Gambia to collect a rabies victim, a flight that attracted some Press interest and, in the runup to that year's Referendum on whether the UK should remain in the EEC, one of the Far East schedules brought ballot boxes from Hong Kong, Tengah and Gan. Meanwhile, Angola was wracked by civil war as factions struggled for the upper hand as Portugal prepared to leave its colony. In late July, an aircraft was sent at short notice to Luanda to evacuate refugees to the UK. Matters continued to deteriorate until, in September, an airlift was mounted to carry Portuguese refugees to Lisbon. The UK Government offered RAF assistance and the Squadron flew forty-two flights from Lisbon to Luanda and back between 8 September and 30 October, with a single aircraft rotated from base at intervals, and 5,660 passengers flown out.

Defence Review, 1974-75

On coming into office in March 1974, Defence Secretary Roy Mason announced a radical Defence Review aimed at placing the UK's focus predominantly in NATO Europe and the North Atlantic. His subsequent statement to the House on 3 December (after a second General Election in October) included retention of forces in Hong Kong, reductions of forces in Cyprus, a withdrawal from Malta in 1979, withdrawal of virtually all forces stationed in SE Asia under the Five Power Defence Agreement, plus the closure of the Indian Ocean staging base at RAF Gan and the RN facility on Mauritius. A consequence of these decisions and general force reductions in all three Services meant that the RAF Air Transport Force (ATF) "would be progressively reduced by one half." Staffing of the ramifications had been carried out throughout 1974, and when a detailed rundown plan was announced, the effect on 10 Squadron was a reduction in established aircraft (AE) from 13 to 11, with a concomitant reduction in crews and funded flying hours. However, all 13 aircraft were retained in service, with 2 now designated as unfunded In-Use Reserves (IUR), and a similar approach was used with the Hercules fleet. The ATF's Britannia and Comet forces were withdrawn from service as, after a short delay, was the Belfast fleet. Some of the short-range Andover C1 aircraft were retained for use in Airfield Radio Aid Calibration, with the remainder sold to the Royal New Zealand Air Force.

The Review's decisions and, in particular, the progressive withdrawal from service of the Britannia fleet over a year from April 1975, led to significant changes in the Squadron's operating pattern. Masirah, an island off the eastern coast of Oman, joined the schedule pattern in August 1975, and the increase in Specials to/from Northwest Europe began to occasion comment in the monthly records. The VC10 had also to take on the majority of the Army's operational Operation Banner unit rotations in and out of Northern

HM The Queen & HRH Prince Phillip board a VC10 at Heathrow prior to a Far East tour in 1972. 10 Squadron VC10s were frequently in the news when VVIP trip were flown all over the world.

Sqn Ldr John Redding introduces his crew to HM The Queen in July 1976 when 10 Sqn flew Her Majesty & the Duke of Edinburgh to the American Bicentenary celebrations. Only those VIP crews qualified with the covetted 'A' Category could fly Her Majesty.

HM The Queen seen here leaving a VC10 at RAF Gütersloh, Germany.

June 1972 - The body of HRH The Duke of Windsor is flown back to the RAF Benson from Le Bourget, Paris, by RAF VC10 in 'Operation Vague'. The Squadron was later to return the body of his widow, the Duchess of Windsor after her death in 1986 in 'Operation Haze'.

One of 10 Sqn's well-loved characters, MALM Ken Goreham seen here with the UK Air Attaché in Pakistan, Gp Capt D. J. Green, during the 1973 flood relief flights to Lahore.

British subject evacuees from strife-torn Iran return to the UK in February 1979. Similar scenes had been repeated with evacuees of all nationalities, earlier in 1974 following the invasion of Cyprus by Turkish forces.

Royal Air Force

NEWS

The only newspaper of the whole Service
WEEK ENDING June 24, 1972
No. 294 Fortnightly Price 4p

June 1972 - Operation Vague : the Duke & Duchess of Kent met the VC10 carrying the late Duke of Windsor from Le Bouget, Paris to RAF Benson.

XR 809 crew deliver the aircraft to Hucknall, handing it over to Rolls Royce for the RB 211 trials. In uniform: Sqn Ldr Geoff Bowles - nav, Sqn Ldr Alfie Musgrove – capt, & Flt Lt Brian Lingard - flt eng.

Re-painted, 809 becomes G – AXLR, RB211 Flying Test Bed at Rolls Royce, Hucknall

Gan Island 1975 HRH Prince Charles and crew

A striking view of a VC10 tail at Ascension Island . The white stones, just visible on the hillside above the left wing, were a form of grafitti made by those who had transitted the Island by VC10s or Hercules.

Ireland. Then, from the first flight after Christmas 1975, the new route for a reduced frequency schedule to Hong Kong used civilian facilities at Bahrain and Colombo, with periodic 'extensions' from Hong Kong to Brunei and Korea. The last flight through RAF Tengah occurred on 10 February 1976 and there were only five more flights to RAF Gan before the staging post closed on 28 March, when CO, Wg Cdr Tony Richards, flew the last VC10 departure to end "a number of years of happy association with the RAF's Coral Command station." Some indication of the change in the Squadron's operating pattern can be seen by comparing the flying hours achieved in December 1975 and January 1976:

December 1975: 941 Schedules, 238 Specials, 120 Exercises.

January 1976: 536 Schedules, 568 Specials, 282 Exercises.

10 Squadron had also to take over the periodic 'Gurkha Trooping' task between Hong Kong and Kathmandu, via Calcutta, that had been flown by the Britannia. This task was mounted using a single aircraft, two crews, and some technical and ground support based on detachment in Hong Kong. The first 'TenDet' was mounted from the end of January to early May 1976, with the aircraft changed using the Hong Kong schedules, when thirty-four flights to Kathmandu carried 4,204 passengers and 635,811 lbs of baggage and freight. As experience grew, the itinerary was altered to avoid a probable morning fog period in Kathmandu, and a maximum gradient climb out technique was developed for departures from Kathmandu over the foothills of the Himalayas. The crews also carried out some preliminary night flying at Hong Kong, and those turning approaches from the famous checkerboards to land on Runway 13 were felt to be easier than the "eye-squinting into-sun early morning arrivals with which the Squadron is so familiar."

The changes also led to an increase in westbound flying. After a proving flight in December 1975, support of the UK garrison in Belize became a regular task from January 1976 and a scheduled task from June. (The airfield there was far from ideal for VC10 operations. The limited runway width, with no turning loop at the end, meant that great care was needed in turning after landing to avoid going off the edge; and a combination of high temperature and short runway length meant that outbound aircraft had to stage at Nassau to refuel to permit a worthwhile payload to be carried.) The North America schedules to Dulles and Ottawa increased to two per week from June 1977 and, to maximise airframe usage, a crew slip in Washington DC (the 'Dulles Slip') was established. Later, when MOD was persuaded that there could be scheduling advantages by running the Belize schedules through Dulles rather than Gander, the advantage of having a slip crew in Washington was reinforced.

The demise of 216 Squadron, whose Comet aircraft had a specialised VIP role, also had an impact on 10 Squadron. Whilst a dedicated VC10 could still be tasked for official visits by HM The Queen and senior members of the Royal Family, and by the Prime Minister and Foreign Secretary, senior Service officers of 2-star rank and above would normally now travel on scheduled tasks, using a small, curtained VIP compartment at the front of the passenger cabin. This arrangement was not without its difficulties for the Squadron cabin staff who often had to placate other slightly less senior officers travelling in the 'normal' passenger seating behind the compartment.

Another incidental effect was that the Squadron was directed to take over 216 Squadron's association with a City of London Livery Company, The Worshipful Company of Coachmakers and Coach Harness Makers. Granted its Charter by Charles II in 1677, the Company today promotes the interests of the automotive and aerospace industries in the UK, and maintains links with all three Service. The Squadron maintained an active association by hosting regular visits until it was disbanded in 2005, when the task passed to RAF Brize Norton.

As 1976 progressed the Squadron settle into its new operating pattern, but the airframe and crew effort implied in supporting the increasing number of sorties per month was often a matter of comment in the monthly records. Multiple sectors per day/per task became much more common – a task to deploy the groundrew of a Coltishall squadron to Decimomannu and return the outgoing RAFG personnel to Bruggen was noted as needing 5 hours ground handling time for 6 hours 20 minutes flying. All in all, a great change from the predominantly long-range situation just ten years earlier. However, some things remained much as before and crews had still to deal with both weather and 'circumstances.' The record for May noted diversions to take the victim of a Coral Snake bite from Belize to Miami, Service personnel with 'Compassionate A' status were delivered to Losssiemouth and picked up from Masirah, and an aircraft diverted to Nassau when its nosewheel tread was found after takeoff on the runway at Gander. There were still loads a little out of the ordinary – for example, in late May one of the four original copies of Magna Carta was flown to Washington to be displayed for a year under the Rotunda of the Capitol as part of the US Bicentennial celebrations. (This was done in some secrecy. The crew was not told until shortly before departure and, on arrival at the Handling Agent's dispersal at Dulles, all movement in the area was stopped by the US authorities until the item was unloaded and moved onto one of a number of limousines due to take it into the city.) Then, in July, again as part of the US Bicentennial events, HM The Queen was flown to Bermuda to board *HMY Britannia.* She then sailed along the Eastern Seaboard and rejoined the aircraft to fly from Philadelphia to Washington and later to Newark. (Her Majesty went on to Montreal to open the 1976 Olympic Games but the VC10 was not used for that part of her itinerary.) And even after ten years in service, there were still some 'firsts' to score – the first aircraft to land on Edinburgh Airport's new runway in May and the first VC10 task into Berlin, RAF Gatow, in July. (In fact, as the task that day required a 'double shuttle', the aircraft landed there twice.) There was also a bit of Search and Rescue in September when, shortly after takeoff for the UK, a crew was asked to search for a ship in distress off Cyprus. The vessel was located, its position was passed to the Cyprus Rescue Centre, and the flight back to Brize Norton was resumed. A second 'TenDet' left for Hong Kong at the end of September. Working to the pattern established earlier in the year, it flew thirty-eight round trips to Kathmandu before returning to Brize Norton shortly before Christmas. (The increasing incidence of 'off-route' flying and a resulting need for more visa applications than before had by this stage caused some practical problems. The time needed to obtain visas whilst passports were toured around Embassies in London was restricting crew availability to the extent that the Foreign Office agreed that 10 Squadron personnel could hold two passports during their tours on the Squadron.)

There was also by this time a well-established and impressive VC10 display routine that showed off the aircraft's lines to good effect, and over these years it was flown by Sqn Ldr John Redding and his Co-Pilot, Flt Lt Jimmy Jewell. Jimmy had a background in display flying, having flown Spitfires with the Battle of Britain Memorial Flight, and was an able cartoonist with a sharp sense of humour. The finale of John's display involved a 360 degree turn in front of the crowd, with flaps and undercarriage extended. Full power was then selected, the aircraft was cleaned up and a spiral climb to 3000 feet commenced. At that height, idle power was selected and the ear-splitting sound of four Conway engines dramatically gave way to near silence. This caught an ATC controller at a Biggin Hill Air Fair by surprise and he transmitted: "VC10, is that it?" to which Jimmy replied: "Yes, if we do any more than that, it spills our tea!"

The operating pattern continued into 1977, Her Majesty's Silver Jubilee Year, an event that involved considerable Service involvement. For the Squadron, an aircraft flown by Flight Commander Sqn Ldr 'Sid' Adcock took part in the flypast at the RAF Review at Finningley on 29 July. (Some weeks earlier there had been what may fairly be described as 'mild surprise' when it became known at Squadron level that a first rehearsal had been flown and that the Squadron was not mentioned in the HQ STC Operation Order. The CO, Wg Cdr Bob Peters, duly 'made representations' up the line and a slot was found immediately behind the Hercules of the four RAF Lyneham squadrons.) The Squadron Standard was paraded with those of other squadrons during the ceremonies earlier that day, and an aircraft was provided for the Static Display open to the public on the next day. Earlier in the month, two aircraft had been tasked to fly Heathrow – Gutersloh – Heathrow in the VIP role for the Royal Review of BAOR on 6 July – one for HM The Queen and a royal party, and another for the Secretary of State for Defence, Mr Mulley, the Army Board and other senior officers.

However, whilst planning for these Jubilee events was in hand, out in Central America, Guatemala was once again asserting its right to annexe Belize, this time in a particularly threatening manner. Although Belize was then

Unrecognisable today as Singapore's International Airport , RAF Changi looking south-westwards.

VC10 s, one of which is seen here at Hong Kong, were emblazoned with 'Royal Air Force Air Support Command ' on the fuselage, until it was shortened to just 'Royal Air Force' in 1972.

Wet night turnround at Gan. The VC10 was able to avoid many of the tropical storms on the ITCZ (inter-tropical convergence zone) line of weather, by flying much higher than its predecessors.

Changi : The Creek Hotel

Transitting passengers saw little of the Maldive Islands sunshine. Crews however were luckier and slipped here for a few days at a time.

VC10 flight-deck crews will recall that the Flight-Planning staff at Brize had the best view over the Terminal Building apron from their office windows.

self-governing, it was not yet independent, and the British Government decided that reinforcement was necessary. And so, at very short notice, the Squadron found itself making rapid plans to play its part in the deployment of Harrier aircraft. Thus on 6 July, just as two Squadron aircraft were engaged on Royal business, two others began to activate the main deployment route to Belize by flying out the VC10 and Hercules slip crews plus the Victor tanker and Harrier route support needed for Operation OUTWING. The VC10 was used to carry palletised Harrier weapons, and the use of Military Operating Standards to a maximum takeoff weight of 340,000 lbs was authorised and used for the long outbound legs from RAF Wittering to Bermuda. (The normal maximum was 323,000 lbs.) Squadron aircrew were reinforced by OCU and Examining Unit personnel in an attempt to meet the operational task with a minimal effect on other services, and postponements rather than cancellations were employed wherever possible. The airlift began on 7 July but was stopped on the following day as a result of political developments – eight sorties had been flown and another eight were planned. The Role Equipment Flight at Brize Norton did sterling work in stripping airframes down to the basic roller-ball floor needed for palletised loads – for example, the aircraft used on 6 July in full VIP fit for the Army Board was off to Belize on 8 July. (As it happened, a reporter from the RAF News was around the Squadron at the time, working up a feature on the Washington schedules. This led to his publishing of an over-excited and wholly embarrassing extra article that appeared to pitch the then Flt Cdr (Ops) against the Guatemalan Army.) The latter part of the year was marked by delays and re-routings arising from Air Traffic Control strikes or disputes – first in Canada in August, followed by the UK till October, and then in Italy in December. Notwithstanding all of that, the next TenDet set off for Hong Kong in October for the next round of Gurkha Trooping, and a rather special flight took Marshal of the RAF Lord Elworthy to Wellington, New Zealand, with a former RAF Colour to be laid up in the Cathedral there. Mid-November brought a national firemen's strike and the Squadron played its part in Operation Burberry, a joint-service response to a call for 'Aid to the Civil Power' that brought the 'Green Goddess' fire engines out of storage. Internal flights positioned troops around the country, Squadron personnel were included in the deployment, and an exercise recovery flight delivered RHA troops directly from Cyprus to fire fighting duty in Yorkshire. Then, in December, troops had to be flown to Bermuda from both Belize and the UK as serious civil unrest broke out after the hanging of two men found guilty of shooting a former Governor and his ADC in 1973. There was also a short internal task to RAE Farnborough that is worthy of mention. This was for investigative work in connection with the planned installation of the Omega navigation system in the VC10. For some time the Squadron had been flying in the North Atlantic Track system after MOD had been able to satisfy the relevant authorities that a Navigator using Doppler, Ground Position Indicator (GPI) Mk 7 and Decca ADL 21 Loran C could maintain the required track accuracy. (Loss of one of these equipments meant that a crossing could not be continued.) Omega would provide greater accuracy than Loran C and would give coverage in oceanic areas where there was no Loran.

On another front, by 1977 the two ATF stations, RAF Lyneham and RAF Brize Norton, had become liable for periodic Tactical Evaluations (TACEVALS) by HQ STC in its NATO guise as HQ UKAIR, together with practise MAXEVALS and MINIVALS. The ATF had been formally declared to SACEUR in wartime (after national reinforcement) since 1974 and was the last Strike Command force to become subject to such evaluation. (There had been a degree of resistance within MOD, largely arising from reluctance to have exercise play interrupt the MOD-agreed task.) The first of these was called at Brize Norton in early May 1977 when, as it happened, the Squadron was involved in a special operation lifting troops into Northern Ireland – a circumstance that meant that the unit's operational performance was easily assessed. However, doing its normal job and generating crews for exercise tasks was not a problem but, at the time, the station and squadrons at Brize Norton were ill-equipped for the NBC and other incidents introduced by Directing Staff. A good deal of imagination had to be applied to the unreality that was inescapable with these and the patience of unit personnel was sorely tried as a result. Nonetheless, "lessons were learned" and gradually,

over time, matters did improve.

When the task for January 1978 was set by the MOD Air Transport Allocation Committee (ATAC), a 21% overtask was agreed to cope with additional flights to Hong Kong for the rotation of the resident Army battalion, support of a RN Task Group in Australia, and a major exercise in the Caribbean. In the event, the latter was cancelled as the units involved were still involved in the UK firemen's strike, so the impact was reduced. Nonetheless, additional crews were temporarily added to the Far East slip pattern to increase flexibility and minimise delays. In addition, as the Gurkha Trooping task resumed (16 January to 9 April), its crewing was also partially integrated with the HK slip, a reorganisation needed as a result of the imminent closure of RAF Kai Tak that would see the RAF presence centred inland at RAF Sekong, in the new Territories. Crews would now operate outbound along the slip, stay in theatre for some two to three weeks and then join the slip back to the UK, a change that gave an increased number of squadron personnel experience of operating in and out of Kathmandu. Ground crew were to be accommodated with the Army in Hong Kong and controlled by the OC Airport Unit, rather than a 10 Squadron Detachment Commander. As matters turned out, this was the Squadron's final Gurkha Trooping season as the task was subsequently put to charter.

It would be a rare year during which the ATF was not used as a tool of HMG's desire to participate in world events, and 1978 saw 10 Squadron tasked in support of several such. A number of VIP flights were mounted to carry Foreign Secretary, Dr David Owen, to destinations in Africa in his attempt to find a resolution of the many problems that followed the Rhodesian declaration of Unilateral Independence in 1965; in May a flight was tasked as part of an international response after the massacre at Kolwezi in Zaire; and in June, two aircraft and eight crews were used in an operation to fly Fijian troops into Tel Aviv following UN resolutions that led to the creation of the UN Interim Force in Lebanon (UNIFIL). The planned route for the nine flights involved went through Bahrain, from where it was not possible to fly directly into Israel. The solution was to file a flight plan for RAF Akrotiri and re-file for Tel Aviv overhead Cyprus. French Air Traffic strikes disrupted flight over Europe for much of the summer and, using military routings, the Squadron had to take on the trooping flights to Germany that Britannia Airways, the MOD charter company, was unable to fly.

Furthering the unit's war role, the Squadron's second TACEVAL was called on 24/25 July during this period, with some aircraft flown as part of the exercise and, in September, the Institute of Aviation Medicine (IAM) oversaw four flights to Northern Ireland undertaken as a trial of new aircrew NBC flying clothing. September also saw the Squadron's most intense participation to date in NATO's annual AUTUMN FORGE reinforcement exercises when eighty-five sorties were flown in support of deployments for Exercise Bold Guard, and similar peaks of short-distance activity followed in subsequent Autumn months.

The Hong Kong slip continued to operate in the background to all of this activity, with a notable disruption occurring in November. A crew flying the customary double leg from Brize Norton via Bahrain to Colombo was forced to divert to Madras, the nominated alternate airfield, by exceptional crosswinds at Colombo. On landing, the crew was out of duty time and a night stop was unavoidable. However, as there was no diplomatic clearance to use Madras other than for diversion, a bureaucratic logjam ensued and, initially, neither crew nor passengers were permitted to leave the aircraft. Eventually the crew were allowed to leave for a hotel and minimum rest time, but the passengers were forced to stay on the aircraft and were only allowed into the airport terminal for a meal.

World events, this time in Iran, came to the fore in December 1978. The Shah's position had become increasingly difficult and Ayatollah Khomeini, by then living outside Paris, had become a significant external force. On 9 December, ten crews were brought to 24-hour standby in preparation for the withdrawal of British nationals. After a few days, an apparent improvement in the situation permitted a relaxation to 96 hours but, meanwhile, there had been delays and a cancellation to the month's task. The Shah left the country 'on holiday' in early January 1979 and the Ayatollah flew into Teheran in February. As the regime finally collapsed, the FCO asked that transport aircraft be positioned in Cyprus against the likelihood of repatriations being required and a VC10 and a Hercules were duly despatched. The Squadron sent two crews and all personnel were requalified in pistol firing whilst at Akrotiri in case the use of personal weapons was needed. Both RAF aircraft flew to Teheran on 17, 18 and 19 February and, in very tense circumstances and surrounded by armed Khomeini supporters, they flew 656 people out to Cyprus for later repatriation to the UK. In the UK, that very cold winter, marked by public sector labour disputes, is now remembered as 'the winter of discontent' and it did not leave the Squadron's task untouched. A VIP departure from Heathrow was delayed as Prime Minister James Callaghan set to with a shovel to help union members clear some snow and ice for the press cameras.

1978 had also seen the arrival at Brize Norton of some other VC10s acquired from Gulf Air and East African Airways for conversion as tankers in the Air to Air Refuelling (AAR) role. This was not an issue affecting 10 Squadron at that time but it would do so in due course.

After a harsh winter marked by diversions and fuel shortages in the US, a cold and wet spring followed, as did the end of an era that had started with Flight 2300 on 4 April 1967. The Hong Kong slip was at an end! On 28 May 1979, Flight 2040 left Brize Norton for the final scheduled passenger flight as the task was moved to civil charter with British Airways. However, that was not to be the last time that the Squadron would visit Hong Kong. In June a slip had to be re-established to fly in troops to reinforce the Chinese frontier at the request of the Governor. The plan was cancelled and the crews brought home, only for it to be partially reinstated. In such circumstances, it is worth noting that the Squadron flew HRH The Duke of Kent from Hong Kong into Peking, Shanghai, Hangchow and back in connection with a British Trade Mission in the same month. Also in May, the first aircraft fitted with the Omega navigation system went into service and, most unusually, a passenger evacuation using the emergency chutes was carried out at Townsville in Australia when a takeoff was abandoned due to severe engine vibration after a turbine blade failure.

In the General Election held on 3 May 1979, the Conservative party gained a majority and Mrs Margaret Thatcher became Prime Minister, a post she would hold until the end of November 1980. During those years she amassed around 1000 hours as a 10 Squadron VIP passenger and forged a particular bond with the Squadron as a whole. Her first flight with the Squadron was in late June, to Tokyo, for a G7 Summit meeting. (A slip crew was positioned in Moscow for this, and it is likely that the onward flight over Siberia to Tokyo was a first for any RAF aircraft.) After the Summit, she flew to Canberra and returned via Singapore and Bahrain. In July, Mrs Thatcher made her second journey when two VIP flights were tasked – one for HM The Queen, the other for the PM and Foreign Secretary – for a Commonwealth Heads of Government Meeting in Lusaka.

In the mid/late 1970s the Squadron had been tasked with many flights as Foreign Secretary, Dr David Owen, pursued a resolution of the situation in Rhodesia post-UDI in 1965. Lord Carrington, Foreign Secretary in Mrs Thatcher's first Cabinet, was the beneficiary of those earlier negotiations that finally reached a conclusion in the Lancaster House talks held from September to December 1979. A ceasefire was agreed and, to oversee a short transition to elections, a 1500-strong Commonwealth Monitoring Force was raised and flown in. The UK element of the plan was codenamed Operation AGILA and, by the end of November, the Squadron had two crews positioned in Cyprus awaiting final agreement in London. In the event, on 11 December, the first flight (1175) took Lord Soames, the newly appointed Governor, to Salisbury, via Lajes in the Azores and Ascension Island. Deployment of the main UK force began a week later and, from 19 to 24 December, the Squadron flew fifteen sorties. These went via Akrotiri, Nairobi, Ndola and Lusaka, ending with a spiral descent from overhead Salisbury to reduce the threat from surface-to-air missiles known to be held by the Patriotic Front. The Squadron supplied personnel as Liaison Officers in Salisbury during the resupply period – twelve flights were tasked, with these officers making useful proposals to maximise payloads and the general efficiency of the lift. 'Free and fair' elections were held at the end of February and, between 2 and 10 March, eleven flights were tasked for withdrawal of the UK element of the Force. Two VIP flights were tasked to carry HRH The Prince of Wales and Lord Carrington to Salisbury for the Independence

celebrations in Salisbury on 18 April. Both shared the return flight and the second aircraft brought Lord Soames out on the next day. Rhodesia duly became Zimbabwe and Salisbury became Harare; Robert Mugabe became Prime Minister and from 1988, President.

So, as world events intruded once again to provide some unexpected long-distance flying, the essence of the monthly task remained much as it had become over the previous four or five years, with a higher proportion of shorter range work increasing the number of sectors flown but reducing average sector times. A continuing refrain each month was that the Squadron was being tasked below its authorised capacity in flying hours. That perception was not helped by an event like Exercise Crusader 80, the UK's biggest postwar exercise of the reinforcement of 1(BR) Corps in Germany, as part of NATO's Autumn Forge, for which the Squadron flew 163 sectors in September, each with an average length of less than an hour. Matters were not helped by the imposition of a Defence Spending moratorium in the second half of the year. Financial forecasts had suggested that MOD was heading for a substantial overspend in FY 80/81 and a reduction in Activity by all three Services was put in hand. The main impact for the RAF was a reduction in flying hours to reduce fuel costs and, in October, only 770 hours were flown. November and December averaged 680, around 50% of the authorised SD98 tasking rate. It can therefore have been no surprise when it was announced that the Crew Establishment was to reduce from thirty to twenty-four from 31 December 1980. The days of MOD tasking up to 1600 hours per month around a foundation of twenty-seven Far East schedules were now in the past. At a crew ratio of 2.2:1 for the eleven established aircraft, the Squadron was still established at a higher level than any other RAF squadron, but the longer-term implications of a reduced task could now be managed more confidently.

Falklands Conflict in 1982 – Operation CORPORATE

The Falklands War began on Friday 2 April 1982, when Argentine forces invaded and occupied the Falkland Islands and South Georgia, in the South Atlantic. The British government dispatched a naval task force to engage the Argentine Navy and Air Force and to retake the islands. The resulting conflict lasted 74 days and ended with an Argentinian surrender on 14 June 1982 that returned the islands to British control. 649 Argentinian military personnel, 255 British military personnel and three Falkland Islanders died during the conflict. The Squadron had an incidental involvement from the outset in that Lord Carrington, the Foreign Secretary, heard news of the invasion whilst returning by VC10 from Israel and, on the next day, 3 October, an aircraft was sent to Montevideo to collect Rex Hunt, the Falklands Governor, and the Royal Marines of Naval Party 8901, captured during the invasion. From then on, the Squadron was heavily engaged in the movement of personnel, stores and ammunition to Ascension Island; in the recovery of casualties from Ascension and Montevideo; and in repatriation flights when the war was over. As the situation developed from 2 April, there was a good deal of improvisation attempted and, in some cases, brought to fruition regarding military capabilities. This spirit even encroached upon 10 Squadron's routine airlift operations. One crew was taken aside and told to prepare for the possibility of a SAS insertion in Argentina, with no training or prior preparation, a near certain one-way contingency plan that was cancelled shortly before takeoff. Other less dramatic incidents developed and Dick King, a Copilot at the time, later recalled one trip in particular:

> "On the afternoon of 5 June 1982, the crew of which I was the co-pilot and which was captained by Flt Lt Brian Wheeler, arrived at Dakar Airport in Senegal to fly the round trip from Dakar to Ascension Island and back…. The VC10 taxied in and the crew which had brought the aircraft from the UK disembarked. The captain, Flt Lt Robbie Robinson, took Brian aside and was deep in conversation with him for a few minutes. The rest of our crew thought that he was merely briefing Brian on the cargo and other relatively mundane aspects of his flight out. This transpired not to be the case. In fact Brian was being told that we were not just going to Ascension and back but were now tasked with taking the aircraft further south to Montevideo in Uruguay in order to fly a "Red Cross Flight" back from there. Wounded sailors who had been injured by the sinking

Ascension Island Runway 13 at one mile from touchdown. After landing the aircraft was turned in the loop on the left of the runway and then parked on the apron to the right.

Concertina City' at Ascension 1982. Flt Lt Mike Emery and crew have a sundowner beer outside their night-stop accommodation, after their flight in from Brize via Dakar, Senegal. The resident VC10 Ground Engineers provided all the amenities.

The VC10 Ground Engineers' Steps at Ascension in 1982-84. Whatever accommodation the resident GEs were moved to, they took these concrete steps with them. Why? Only they knew ....It just seemed like a good idea.

The Zymotic Hospital at Ascension. After the night-stopping crews moved out of Concertina City they stayed at the long-disused isolation hospital near Georgetown. The Ground Engineers steps led the way to the sundowners platform on the cliff overlooking the beach.

Going Home on 6 July 1982: Troops boarding XV 109 at ASI after the Falklands Crisis was over

The Scots Guards Band welcomes the Regiment back from the Falklands, at Brize Norton in August 1982

of HMS Sheffield and HMS Coventry, together with Army casualties from Goose Green, had been evacuated from the area by hospital ship and taken to a military hospital in Montevideo. Robbie's crew left the airport for their hotel and Brian told us the news.

We had been told there would be charts and documents left in a green document bag on the flight deck, for our unanticipated crossing of the Atlantic. They were nowhere to be found. It later materialised that they had been left behind at Brize Norton. The Navigator and I went across to the Air Traffic Control Tower at Dakar where the flight plan would have to be filed but, with no charts, we would have a problem trying to ascertain which civilian ATC agency to contact. On the wall of the Flight Planning Office there was a fly-splattered US Air Force chart that must have been at least 10 years old. A quick trip for me back to the aircraft ensued, returning with a bag of the most vile-tasting coffee brand that was sometimes put into our in-flight rations. The age of bartering in Africa still lived. We now had a chart. Aircraft, even in the '80s, utilised a computer flight-planning system that produced a document from San Francisco sent to our handling agents and known as the Jetplan. It contained all the route tracks and other information we needed and our handling agent produced one for us sent from our HQ at Upavon. We now had enough paper-work to allow us to legally get airborne.

Unfortunately though there was the outstanding omission of two large red crosses painted on either side of the aircraft's nose. There hadn't been enough time in the UK to get this done before the flight had left. Since all of our flight from Ascension onwards would be in the dark, it had made sense to the planners for us to paint the red crosses on the nose once we had landed in Uruguay while we were resting before the flight home, and before our casualty load boarded. Our ground engineer would not be sightseeing when we arrived but would be wielding a paintbrush instead. We landed uneventfully at Ascension and off-loaded the outbound passengers and baggage from the UK. An hour and a half later we left Ascension and climbed up into the dark Atlantic night. As we neared the South American coast, some 500 miles or so out to sea, I called Porto Alegre once more to pass a position report. No answer. There was a great deal of static noise but no Recife to be heard. I repeated my call. Still no answer. Then, from out of the dark night, came a voice so loud and clear that I thought he was sitting next to me. "Go ahead, Ascot 2645. This is Buenos Aires. I will relay your position to Porto Alegre." Buenos Aires is in Argentina. We were fighting a war against Argentina. I pretended I had not heard him and we ignored the generosity of our enemy's friendly civilian air traffic controller. Although the equipment was not really designed to pick up other aircraft, after that we kept a very close watch on our weather radar in case we were approached by any unwanted visitors. Eventually we came within range and into contact with Montevideo on the clearer and shorter-range VHF radio for our position reporting. Our troubles had only just started.

It must have been about midnight, local time in Uruguay. The lady on the radio frequency advised us quite simply that we could not land in Montevideo because visibility there was down to about 50 metres in thick fog and the only solution was to divert. Where to? There are relatively few airfields in Uruguay suitable for long-range airliners such as the VC10 and, during our rudimentary flight planning back in Dakar with the 'bartered' chart, we had chosen an airfield called Maldonado. It was situated on the coast of Uruguay about 100 miles south-east of Montevideo, so we notified our charming friend that we would go there. She seemed amused as she replied, "I don't think so for two reasons. One is that they have thick fog also and the second is that they have no runway lights. You will divert to Durazno". Having never heard of Durazno before, we embarrassingly asked where it was. It was approximately 150 kilometres north-east of Montevideo, she gave us all the details we needed, and couldn't thank her enough. We later landed without incident but as we turned off the runway, the fog came down there as well. The only other suitable airfield within our fuel reserves had been Buenos Aires, in Argentina: not a good idea, we thought.

By now it was about 0130 in the morning and it transpired that Durazno was a relatively small training airfield for the Uruguayan Air Force, smaller than our own RAF Cranwell. A hangar we walked through did appear to have a considerable number of Pucara aircraft, which we knew were also flown by the Argentineans. In spite of the

Temporary Red Crosses were painted onto VC10 fuselages when medical evacuation flights carried the Falklands' wounded home to the UK from Montevideo, Uruguay.

Copilot Mike O'Donovan at Montevideo on one of the medical evacuation flights in June 1982. Flight Engineer Cliff Hall can be seen just inside the door. (Mike later became Chairman of the 10 Squadron Association from 2005–2014)

An entourage of vehicles arrives at a VC10 in Montevideo with passengers and patients on a medical repatriation flight during the Falklands Conflict in 1982.

Guild of Aviation Artist painter Penelope Douglas receives a Squadron plaque from OC10, Wg Cdr Gerry Bunn, after handing over her painting of a Red Cross VC10 at Ascension Island in 1982.

unearthly hour and our un-announced arrival, we were met by the only two officers on the base who spoke English. One was a Major and the other a junior Lieutenant. I forget all the minor details about the next hour or so but the outcome was that we were despatched to a small hotel in the nearby town of Durazno for a minimum rest, whilst our poor ground engineer remained at the airfield to paint the missing Red Crosses on the aircraft.

A few hours later I awoke in a basic hotel room to find an armed guard on the door, but I was allowed to go to the room that passed as the hotel restaurant and in dribs and drabs the rest of the crew arrived for breakfast. An hour or so later we were on the bus back to the airfield and were met there by our Ground Engineer. He told us that during the night he had been frequently interrupted in his painting by messengers telling him to go to the Operations Room where there had been telephone calls for him from the British Embassy in Montevideo, but they kept being cut off. One of them had asked him to check the aircraft's manifests - the documents that list all passengers' names and details of the freight being carried on board. We had no passengers but loads of stretchers and medical equipment in our holds. Now there is no need for a Ground Engineer to understand the finer details of aircraft documentation, and he had spent much of his valuable time searching the Loadmaster's paperwork left on board to find what the Embassy meant: all to no avail.

We then found that Durazno had no pressure-refuelling truck, so how were we to refuel in order to reach Montevideo? The Vickers designers had foreseen this possibility and had positioned over-wing refuelling points above each tank in the upper surface of each wing. The procedure should have

been simple: undo the twenty or so large screws holding down each point's cover and just pour fuel into each tank by a hose from any refuelling tanker, even if it didn't have pressure assist to help. Unfortunately this system had never, in the twenty or so years of aircraft in service, been used. The screws in each tank cover were totally encrusted with paint and each tank's cover-screws except one, the outer transfer tank in our right wing, refused to budge. It took forever to fill this small tank and even longer to then pump that fuel into the appropriate main tank from where we could use it, maintaining a balance in both wing's tanks for safe flight.

By four in the afternoon the task was completed and we took off for the short flight to Montevideo, and our arrival there was met by what seemed to be a very large number of people. Most significant were armed personnel and guard dogs. Nevertheless we started to off-load seats from upstairs and replace them with stretchers from the holds, in readiness for our wounded passengers, and I believe that some of the British Embassy staff were also present to assist us. After a while the attitude of the assembled throng seemed to change and we were advised to refrain from unloading the rear hold. The Daily Mail later reported that "In spite of vigorous protests from the VC10 crew, the Uruguayans insisted on off-loading the hold." In reality, faced with armed guards and Doberman or Alsatian dogs, there was no protestation at all and we readily agreed to allow others to continue the back-breaking work of unloading.

A box containing the personal effects of an RAF officer passenger who had left the aircraft at Ascension Island was found in the rear hold. In the box were certain items of RAF spares that were required for Harrier aircraft. (IFF spares) Un-manifested, they should not have been in his personal baggage in the first place and should, in any case, have been off-loaded at Ascension. How had anyone in Uruguay got to hear of this? Was the Embassy communications system tapped or hacked into? Had anyone been listening-in to the calls to our Ground Engineer when he had been cut off so many times at Durazno whilst trying to paint the Red Crosses? We never did find out.

By 1800 hours local time we were ready to accept our passengers and all 56 stretchers were occupied by the wounded who had been brought from the military hospital. As the sun set, we took off for the six hour night-flight to Ascension where, on arrival, our crew then became passengers for the final sector home to Brize Norton. During the flight to Ascension I walked down the back of the aircraft and spoke to some of our passengers who had previously been told that we had had a few problems during the previous 24 hours. One young sailor on a stretcher, with wounds to his head, even offered for me to use his stretcher for a quick nap! I declined his kind offer. Sat in some of the normal aircraft seats were three walking-wounded soldiers who had been injured at Goose Green. One had his arm in a sling and a dressing on his shoulder. I asked what had caused his wounds. He appeared most embarrassed and wouldn't tell. His mates next to him however insisted that I pull rank and order him to tell - he had been making use of a temporary toilet facility that had been rigged up near his camp. It consisted of two forked sticks and a pole stretched between them to use as a seat. Below the pole was a ten foot drop where the excrement fell. In the middle of the night, perched precariously on the pole and minding his own business, he had been the target of an Argentinean sniper and had been hit in the shoulder. Although not very seriously hurt, the impact had sent him reeling down the drop into 'you know what'. His mates had then refused to pull him out even though he claimed he was bleeding to death!

A walking-wounded visitor to the flight–deck was a naval Lieutenant Commander, who I think was the First Lieutenant from HMS Sheffield. Hit by an Exocet missile the ship had been caught up in flames. His face was a mass of newly-formed scarring where his anti-flash mask had not covered him, and yet he was quite normal and open in his recall of the event and a thoroughly nice chap. There was no inter-service rivalry aboard Ascot 2645 that night. Everyone, including ourselves, was just grateful to be going home.

Back in the UK, stories abounded that Mrs Thatcher had been personally involved with our escapade and that the British Embassy communications system had been compromised. Who was the mystery RAF officer whose kit had contained bits for the Harriers? Would our friendly Argentinean ATC operator have really relayed our position to Montevideo or his own fighter aircraft instead? I can't say because we were never told."

Flt Lt Mike Emery was captain on two of the later aeromedical evacuation flights from Ascension to Montevideo and back. The first was on 24/25 June, when all went well. His crew returned on 28 June for a departure on 29 June, a day, he recalls, "that is forever engraved on my memory." His story:

"It took a great number of ambulances to transfer all the casualties from the ship to the aircraft, and as we commenced takeoff it was dusk. At about 90 knots there was a bright flash and an explosion. The aircraft lurched and we abandoned the takeoff. I remember the Eng saying that the engines were OK, and my first thought was that a bomb had exploded. However as we stopped the No 1 engine ran down and, shortly afterwards, an inspection showed major damage on the engine – a turbine blade had sheared off. I can assure you that a Conway letting go at max power is quite a sobering experience, not at all like an engine failure in the Simulator. If it was loud for us on the flight deck, it must have been frightening for those at the rear and the poor fellow in the stretcher nearest that engine was apparently very distressed.

Returning to the Terminal, our problems were just beginning. I was fortunate to be able to get a HF relay to HQ 46 Group via Panama right away to alert them to the situation and get a relief underway – but there seemed to be nothing else that we could do immediately, apart from trying to tell the medical team and passengers what had happened. All the people we had dealt with earlier had departed. The airfield was semi-deserted and there was nowhere to offload the stretchers. Fortunately, the Military Attaché had seen something whilst driving off the airfield and he reappeared. As I remember it, he said that all the ambulances had dispersed, the Hospital ship had sailed and he doubted that it could dock again at night. The upshot was that after some five or six hours, ambulances were arranged and the patients were taken back to the ship – I'm not sure whether it had gone and returned, or whether someone managed to stop it sailing. The situation on the aircraft became very tense while we were waiting, with much swearing and cursing about 'the xxxx RAF,' and the Loadmaster and Medics had a difficult job in dealing with it. I felt very bad as I think we all took pride on being able to whisk these chaps back to the UK, and this had all happened at the very worst time, and we were helpless. I suppose we were very lucky that we had not had to evacuate the aircraft on the runway!"

How near they had been to an uncontained turbine failure and a full scale emergency was borne in on Mike and Dick King about a week later. They were on another stop at Ascension and were able to inspect the engine that had been brought there – the turbine casing was very badly damaged, but it had held and prevented fragments from being driven into the adjacent engine and tailplane. A replacement aircraft got to Montevideo on 30 June, and Mike's crew flew the patients back to Ascension on 1 July. Sqn Ldr Maurice Biggs and crew had delivered the replacement, and were left with the prospect of a long, over water, three-engined ferry flight to Ascension - not something that would have been authorised in 'normal' times.

As matters settled and the establishment of a Falklands garrison was put in hand, based around a new airfield to be constructed at Mount Pleasant, and capable of coping with the full range of RAF aircraft, a regular resupply system was needed. Accordingly, the Squadron was tasked with flying regular scheduled flights to and from Ascension, a task that would come to occupy some 45% of the monthly task for some years. However, the VC10 was not capable of reaching the Falklands within the accepted fuel planning regulations of the time. It would take some time, but the conflict would have implications for most RAF frontline aircraft and an in-flight refuelling capability was just over the horizon.

Squadron Commanders until 1985

Wing Commander M G Beavis AFC	July 1966
Wing Commander D E B Dowling AFC	November 1968
Wing Commander R L Lamb	March 1971
Wing Commander R I C Howden	January 1973
Wing Commander A J Richards	March 1975
Wing Commander R G Peters	February 1977
Wing Commander A M Wills	December 1978
Wing Commander O G Bunn	June 1981
Wing Commander L J Marshall	June 1984

'Dad's come home from the Falklands.' June 1982

Daily Mail 28p

Friday's People PAGES 19 - 30

Freed hostage tells the world: Terry Waite is safe

IT'S GREAT TO BE HOME

HE smiled his widest smile and his raised arms said: It's great to be home.

John McCarthy was back in Britain last night and revelling in freedom after 1,943 days' captivity in Lebanon.

Turn to Page 2, Col. 1

The Headline said it all. 8 August 1991 John McCarthy returns to RAF Lyneham in a 10 Sqn VC10 after more than five years in captivity in the Lebanon. Terry Waite also returned home by VC10 after his release later in November.

Taceval exercises in the 1980s required crews to 'don' and 'doff' the cumbersome NBC kit.

Nairobi, Embakasi 1985 : Army exercises in Kenya took place during the UK winter months and continue to do so today but troops are now carried in 10 Squadron's Voyagers.

Flt Lt Brian Johnston and crew pose with 'Mrs T' in Guam in December 1984, en route to Washington, having just signed the Hong Kong [handover]Accord.

In spite of the Westlands Affair causing her concern in February 1986, Mrs Thatcher visited Brize Norton after 10 years in No 10. Wg Cdr Len Marshall presented her with a jewellery box, made by co-pilot Chris Kemp. Mrs Jan Marshall, Wg Cdr Tony Webb (OC 241 OCU) and Denis Thatcher make up the group photo.

Prime Minister Margaret Thatcher was awarded a Log Book of her 10 Squadron flights having achieved 1000 hours as a passenger on the VC10. OC10 Wg Cdr Peter Bingham and Sqn Ldr Steve Baines made the presentation on board on 1 August 1990.

Converted Minuteman missile carriers were used as People Movers at Washington Dulles Airport to transport passengers to and from the VC10s on the Page Handling agent's apron.

No! The USAF never chose to operate the VC10 but XR 808 did become American and a film star in 1989-90. At RAF Gatow the film's makers allowed the RAF Benevolent Fund to benefit from the film fee.

Wg Cdr Peter Bingham, OC10 introduces an aeromedical team to HM The Queen, on board a VC10 when she visited Brize Norton in April 1989.

Copilot Flt Lt Chris Kemp, whose father had been a Victor pilot on 10 Sqn, hands over two chairs which he made for HM The Queen and HRH Prince Phillip when they visited Brize Norton in April 1989. It is believed that they were destined for a log cabin at Sandringham.

A de-fuelling incident at Brize Norton in December 1997, where the contents of the fin tank were ignored, caused XR 806 to sit on its tail and subsequently to be written off. Most of its fuselage remains as an aircraft-repair practice facility at the base.

Undergoing avionics mods at Wyton in 1983, XV 109 strikes an unfamiliar pose.

When 10 Squadron disbanded in 2005 its aircraft were taken over by 101 Squadron. Originally a 10 Squadron aircraft, XR 808 was the last VC10 ever to fly when it was flown to Bruntingthorpe, Leics. on 29 July 2013. After dismantling, it was later moved to the RAF Museum at Cosford where it has been rebuilt for display.

The sad remains of the flight-deck of XR 806

Falkland Islands : 109 taxiing off Mount Pleasant's runway after the first part of the Record-breaking flights.

XR 806 & XV 109 together at Mount Pleasant, Falklands Islands.

Who Makes a Transport Crew. . . . . ?

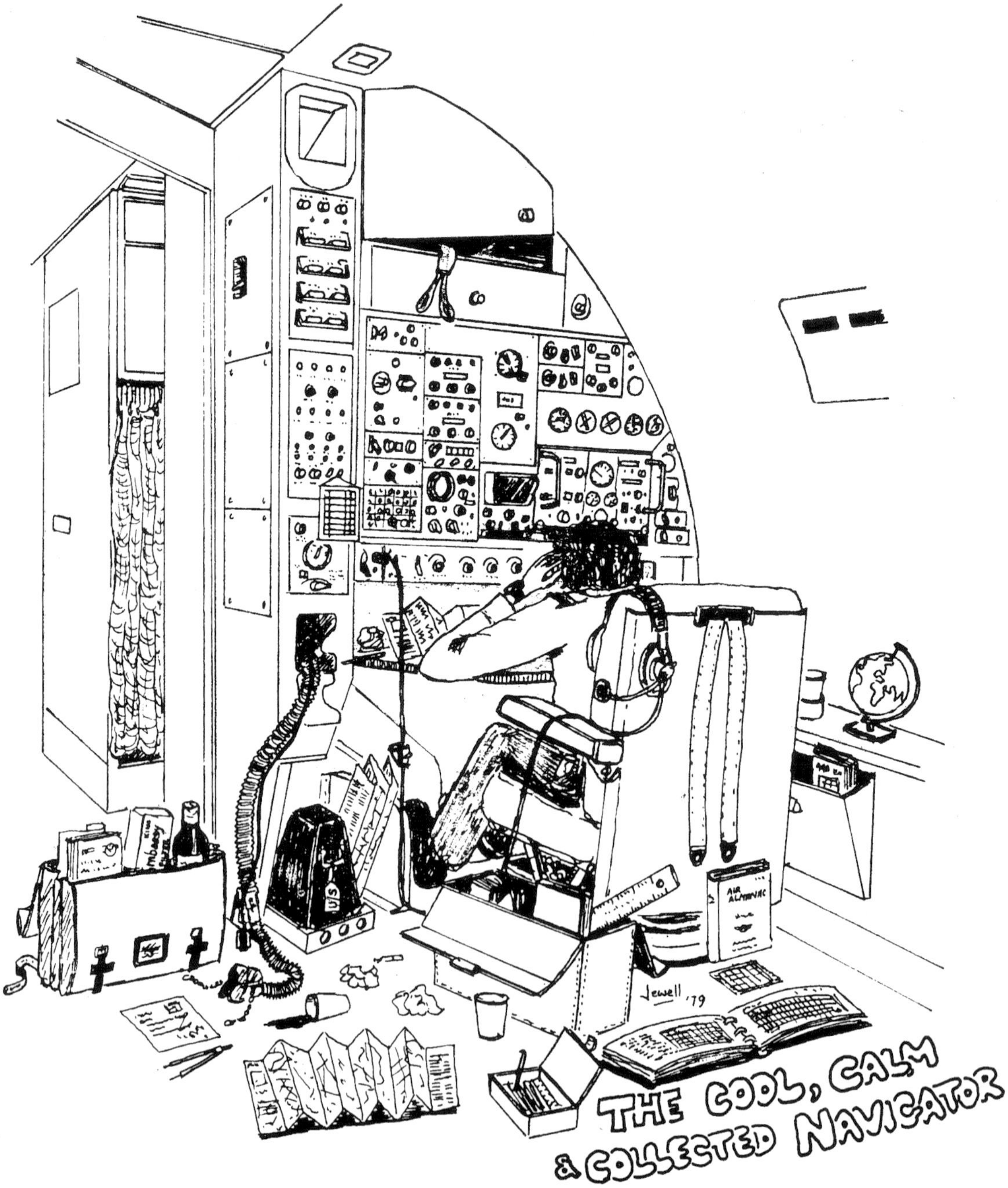

THE COOL, CALM & COLLECTED NAVIGATOR

Jewell '79

One of a series of cartoons drawn by Squadron pilot, Jim Jewell.

CHAPTER 12

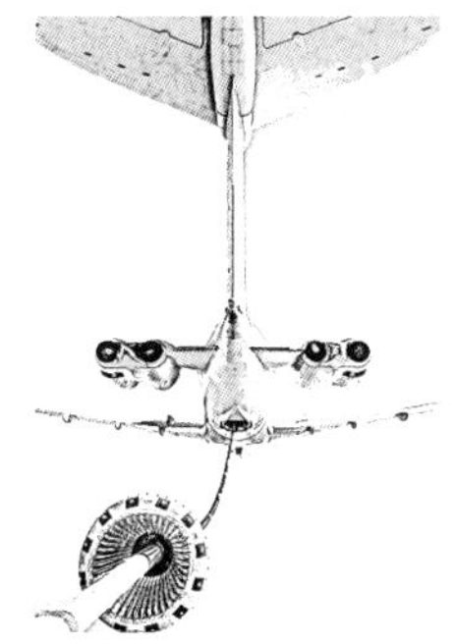

ADDING PROBES AND PODS, 1985-2005

The old order changeth, yielding place to new (Tennyson)

A significant change in the Squadron's task occurred in mid-May 1985 when the scheduled services to Ascension Island in support of the Falklands Garrison were transferred elsewhere. No 216 Squadron had re-formed at Brize Norton in late 1984 to operate nine Lockheed Tristar 500 aircraft acquired by the RAF from British Airways and Pan American Airways. These took on the Falklands resupply task and a proportion of other tasks that would otherwise have fallen to 10 Squadron. Exercise support then became an increasingly significant part of the Squadron task and, with the overall tasking level less than before, the incidence of VIP tasks became more noticeable. The Prime Minister and Foreign Secretary, in particular, had become very 'frequent flyers' in 10 Squadron's VC10s - Mrs Thatcher was flown on twenty-four flights between 1986 and 1988. Indeed, she had amassed some seventy-five flights by early January 1986 when, with her husband, she paid a private visit to the Squadron HQ. She spent considerable time with groundcrew of Role Equipment Flight, involved in the preparation of aircraft for VIP Tasks, and with other crew personnel before joining Squadron officers as Guest of Honour at a dinner that evening, when she was presented with a rosewood jewel box made by Flt Lt Chris Kemp, a Squadron Pilot. Then, on 30 September 1988, she visited Brize Norton to present a new Squadron Standard, thought to be the only occasion on which an incumbent Prime Minister has done so. She made some personal points in her speech that morning:

> "Over nearly ten years as Prime Minister I have come to know many of the Squadron's members and their aircraft. We have flown together to the most far-flung lands and continents. You have brought me safely to my destinations through ice and fog and sandstorms, with unfailing punctuality. Indeed I am living proof of your Squadron motto: 'Rem Acu Tangere' which might be loosely translated as 'we land even when the Prime Minister can't see the runway.'"

After the presentation, there was a short Reception in the Squadron HQ during which Mrs Thatcher was presented with a logbook recording details of the 810 hours and 406,861 miles flown with the Squadron since the election in 1979. The old standard, presented in 1958 by Princess Margaret, was handed over by a former CO, Air Cdre R G Peters, at a Laying Up ceremony in York Minster held on 24 November 1989.

The Prime Minister continued to fly by VC10 and a particularly unusual flight occurred a few months later, in early 1989. Mrs Thatcher had planned visits to Nigeria, Zimbabwe, and Malawi but there was an undisclosed hope in Downing Street that she could go on to Namibia on 1 April, the day on which a ceasefire in a long-standing conflict between South Africa and the South West Africa People's Organisation (SWAPO) was to be implemented under UN Resolution 435. Mr Robin Renwick, the British Ambassador to South Africa, was to go to Windhoek, the capital, to assess the situation there and a decision to go,

or not, would be taken just as the PM was leaving Malawi. It was agreed that she should go to Windhoek instead of flying back to London, and the accompanying Media contingent was only told of the change as the VC10 took off. On arrival, Mrs Thatcher paid a brief visit to a Uranium Mine but, by the time of her arrival back in Windhoek, news of incursions across the Angolan border had put the ceasefire at risk and she became involved in negotiations to keep the UN plan on track. After several hours, these were successful and, to quote Mr Renwick's words years later, "Mrs Thatcher boarded her plane extremely reluctantly. She was clearly attracted by the prospect of continuing to conduct the affairs of Namibia." A twelve hour return to London, plus a fuel stop in Lagos, was then in prospect, but with the VC10 crew well out of duty time. Anticipating this possibility and to ensure a rapid return to the UK, a second Squadron crew had been flown out on commercial flights. However, given the secrecy with which the possibility of a diversion to Windhoek was surrounded, only one member, a Squadron Flight Commander, was briefed on their true destination and a cover story based on the establishment of a slip pattern to support the UN mission was used. (Mrs Thatcher reached 1000 hours on the VC10 in August of that year, when another small presentation was made. She left 10 Downing Street in November 1990.)

By October 1987 the Squadron had become well-used to recurring Tacevals. By then these were of a very different character from the first in 1977, and the lack of realism and appropriate kit were in the past. That year's exercise was called between 9 and 11 October, when the Squadron was heavily tasked with seventy sorties in support of Exercise KEYSTONE, deploying troops into Central Europe. Despite exercise injects, all of these were flown successfully and an additional thirteen sorties were generated for simulated weapon outload destinations. Whilst these were disrupted, none was lost and a 'Satisfactory' rating was the well-earned outcome.

Introducing Air to Air Refuelling (AAR)

The RAF as a whole drew a particular lesson from the Falklands conflict, namely the importance of as many frontline and support aircraft as possible having an in-flight refuelling capability. By 1978 it had already been recognised that the existing Victor tanker force would have insufficient capability to meet the refuelling needs of protecting the UK Air Defence Region for which the Tornado Air Defence Variant (later designated F2, then F3) was about to be acquired. The RAF then acquired five ex-Gulf Air standard VC10s and four ex-East African Airways Super VC10s for conversion to the AAR role. After extensive conversion, the last of these entered service in May 1987 with No 101 Squadron, also based at Brize Norton, as K2 and K3 variants. In addition, as British Airways retired its remaining fleet of Super VC10s in the early 1980s, fourteen of the aircraft were also acquired by the RAF but, with the first conversions already in hand, these were stored at Brize Norton and Abingdon for some years until, in 1989, five were chosen for conversion to K4 standard as replacements for K2s, the oldest VC10s still flying but running out of airframe life. (The other aircraft were used as a source of useable spares then scrapped, used for fatigue testing, or for fire practise.)

Against that background, the transition of 10 Squadron to its final acquisition of a fully-fledged AT/AAR was a gradual one. The original C Mk1 aircraft had been built with much of the plumbing that would be needed to receive fuel in-flight and refuelling probes were supplied at the time, though never routinely fitted. However, as a consequence of the new policy, probes were gradually fitted to all Squadron aircraft from 1986 and, whilst most Squadron pilots did not start to gain AAR qualifications until 1988, some flights were made with the clear point of demonstrating the new capability in action. In November 1986, the first two air to air refuelled flights were carried out in support of Exercise SAIF SAREEA, a major UK exercise involving both the UK and Oman, where the destination was Masirah. Each flight was carrying 129 passengers. The refuelling bracket was over the Mediterranean, between Sicily and Greece, and each flight required four prods to transfer 46,000 and 42,000 lbs respectively, before the 101 Squadron tankers continued to Akrotiri.

There is no record of further transfers until, just over a year later, a further demonstration, Exercise MEDAL TRAIL, was laid on of the reinforcement capability now offered by the combined Brize Norton fleet of 10 and 101

Squadron aircraft. Using a 10 Squadron aircraft (XR 806), the Station Commander, Gp Capt Chris Lumb (with Squadron CO Wg Cdr Brian Symes), flew direct to the Falklands on 19/20 December, a distance of some 7350 nautical miles. The flight time was 15 hours 46 minutes, with fuel taken on twice from two 101 Squadron aircraft – an eighteen minute prod some three hours after takeoff and later, 200 miles south of Ascension, a thirty-eight minute transfer of 105,000 lbs of fuel. The flight constituted a FAI record for the route and its arrival at RAF Mount Pleasant was also the first by a 10 Squadron aircraft on the Falklands. Another FAI record time (14 hours 59 minutes) was established by a similar, direct return flight. That flight involved a continuous contact of 45 minutes for a transfer of 119,000 lbs of fuel, itself considered a record at the time. As part of the same demonstration, a second 10 Squadron aircraft (XV 109) established new time records (refuelled and unrefuelled) for the sectors Brize Norton – Ascension - Mount Pleasant, and again on the return. (Objectivity demands a mention of the fact that, only days after the VC10 effort, an urgent Aeromed requirement in the Falklands was answered by an empty 216 Squadron Tristar flying to Mount Pleasant. It took a shorter total airborne time of 14 hours 50 minutes but, with a one hour stop at Ascension, its overall time became 15 hours 50 minutes. An honourable draw could be claimed, but it remains the VC10 that is in the FAI record book!)

Despite these significant demonstrations of a new capability, the fact remains that whilst the aircraft were only receiver-capable, it was little needed by the general run of the Squadron's air transport task, where payload over most sector lengths was not limited by fuel capacity. At the same time, the dispensing capacity of the RAF's remaining Victor tankers and the growing number of 101 Squadron VC10Ks had no lack of customers from fighter squadrons at home. Thus, less than a handful of crews were trained in AAR techniques until there was some sign of a dual-role capability on the horizon. By late 1989, it was anticipated that the aircraft would acquire refuelling pods at some point in 1991 and crew conversion picked up as a result. By January 1990, three crews were to be AAR-receiver qualified and the Squadron had its own Instructor. Over this period, there are inferences in remarks made by Squadron COs in their contributions to Association Newsletters that there was a degree of resistance to the changing role: "Inevitably, there will be those who do not welcome these changes, but I am in no doubt that the move is in the right direction…" and later, "Like it or not, Tanking is coming." Indeed it was, and the capability was used to some effect from time to time. Flt Lt Steve Sansford, an experienced AAR navigator, had joined 10 Squadron from 101 Squadron in January 1990, and recalls:

> "On 27 July 1990, we flew XR 806 from Gutersloh to Calgary direct with a full load of passengers, taking on 58,000 lbs of fuel over the North Sea from a VC10K, and a further long-range fuel uplift occurred on 26 July 1991, when we took XV 102 from Brize Norton to San Francisco, taking on 60,000 lbs over Canada."

The Squadron had moved into the 1990s by marking its 75th Anniversary with an aircraft flying over all of the UK airfields used over the years. The weather was not good and the inertial navigation system was needed to ensure that this was accurately achieved - observers at Melbourne, the WWII base, confirmed that the aircraft was not seen but that it was very certainly heard! A little later that year, a substantial organisational change occurred when, after twenty-four years, First Line groundcrew were incorporated into the Squadron from RAF Brize Norton's Engineering Wing. The value of that change would quickly become apparent during Operation GRANBY.

Whilst these changes were occurring at Brize Norton, consideration of a fundamental shift in the UK's Defence posture had started in London as a response to the developments on the ground that had begun in Europe after the breaching of the Berlin Wall on 9 November 1989. Many were keen to reap the benefits of a 'Peace Dividend' and make savings in Defence spending, and the first round of such savings was announced in the 'Options for Change' paper of July 1990. This was followed by a series of Defence Costs Studies, aimed at streamlining the MOD's support functions, and concluding that many of these could be 'outsourced' to the Private Sector. These proved to be only the first of a series of reductions in the size and capability of the UK's Armed Forces that is almost certainly still not at an

end twenty years later. 10 Squadron would be affected in due course but, in the meantime, and within days of 'Options for Change' being announced, things were stirring in the Middle East. Twenty-five years later, they are very far from being put to rest.

Operation GRANBY

After the loss of the Ascension/Falklands resupply task in mid-1985, the Squadron can be said to have been waiting for some world event that would generate the levels of activity that had been customary before then. That impetus was provided when, on 2 August 1990, Saddam Hussein, President of Iraq, launched an invasion of Kuwait and took over the country. On 9 August the MOD announced that the UK was deploying Tornado F3 and Jaguar aircraft to the Gulf, the start of Operation GRANBY, its contribution to the international coalition effort that resulted in the liberation of Kuwait on 28 February 1991.

From the outset, 10 Squadron and the ATF as a whole was committed in support as Flt Cdr (Ops), Sqn Ldr Tony Jukes recalled:

> "During the spring and summer of 1990, the Squadron's VC10 aircraft had suffered from serviceability problems, and often only three or four aircraft were available for the world-wide task. It was against this background of airframe availability that, following a telephone call at about 2030 on 8 Aug 90, I reported to the Station Operations Room on the orders of the Stn Cdr to find an ASMA message ordering No 10 Sqn to generate six aircraft in full roller side guidance by 1000 on the next morning. I assured the Stn Cdr this would happen and departed for 10 Sqn Eng Flt (we had had our own Eng Flt for some time by now) and informed the duty JEngO of the requirement. I think there were just seven aircraft on base at the time and the response I received was not encouraging. However, I asked him to generate all the manpower available from both Squadron and Eng Wing resources and the shift FS came to his rescue and assured me it would happen.
>
> At 0600 the following morning, my Ops Sgt, Ben Mills, called forward the weapons and ammunition, AR5 NBC kits and war rations (4 cases of 10 man compo rations per aircraft). The Squadron moved to a full war footing during the day with the Ops Room being manned by a Flt Cdr and Ops staff on a 24 hour basis. Crew lodged their personal war equipment (NBC suits, respirator including war issue spare filter canisters and Kevlar helmet) at the Squadron HQ and admin out-briefings were being organised. Additionally the Squadron was allocated a minibus and became responsible for the transport of crews on base and within the MQ estate. The existing war/hostilities brief/aide memoir proved to be inappropriate for Op Granby and Sqn Ldr Derek Sharp rewrote and reissued an excellent update; and to ensure a coordinated management of airframe servicing and availability a Ground Engineer was located in the Squadron Operations Room.
>
> By 1000, six aircraft were available for tasking, although some carried a significant number of red line entries in the F700. However, no tasking for the operation was received until 11 August when bomb out-load to the Gulf commenced. In three days the Squadron cleared the ready-use bomb stores at Brize, Marham and Lossiemouth, with the exception of JP233 airfield denial weapons that required modified pallets and Tristar up-lift. A Military Operations standard takeoff weight (340,000lb) was authorised for a full load direct to the Gulf (to Dhahran, if I remember correctly) and one of our junior Captains was authorised for this mission.
>
> As direct flights to the Gulf were considered too limiting, a UK/Cyprus whirler into a Cyprus/Gulf hub and spoke pattern was established. Crews outbound from the UK picked up loads in the UK or Germany and slipped the aircraft at Akrotiri to the next crew scheduled for the Gulf sector. They flew a round trip, returning to Akrotiri. and all crews operating into the Gulf did so from and to Akrotiri on the main supply lift.
>
> Within a few days, all of the ten available aircraft were airborne on Op Granby and routine tasks and this remained so for thirty days, with no aircraft spending more than a few hours on the ground for loading and replenishment. The first aircraft to fail was at Brize Norton when Flt Lt Paul Atherton had a serious engine failure on takeoff. There was virtually nothing left in the engine cowling and a long trail of FOD on the runway, but Rolls Royce Treforest provided

a replacement engine and the aircraft was quickly returned to service. Training on the Squadron ceased; check rides were cancelled; and the OCU staff were absorbed into the Squadron's pool of crews. Only Squadron Flt Cdrs could authorise flights during this operational period. This was to ensure that all crews (including any dead-heading to other destinations) received an outbound admin briefing prior to departure from Brize, and tasks were only authorised following this briefing. (The outbound briefing included weapons policy, medical requirements (numerous jabs), accommodation and procedures at Akrotiri, and a variety of other things I cannot recall. Returning crews were debriefed inbound by the Duty Flt Cdr and weapons were returned to the Squadron. Crews were also briefed by Station Intelligence, Station Operations and the Met Office outbound and inbound. On a night watch it was not unusual to brief three crews out bound and two or three inbound.

Crew composition for Op Granby was 2 Pilots, Nav, Flt Eng, ALM, a Steward and a Ground Engineer, with all crew members wearing flying suits and flying boots. Crews arriving at Akrotiri from the UK, Germany or the Gulf were met at the aircraft by a driver with a Landrover and trailer in order for the crew to load/store their war kit. Such was the volume of kit carried by each individual that the trailer was always full to overflowing, and it was labelled and driven to a secure storage to be called forward when that crew departed. Weapons and ammunition were carried in separate sealed ammo boxes and lodged with Akrotiri Armoury during the crew's residence. At Brize Norton, all weapons and ammunition not on issue to crews was stored in the Squadron Operations Room.

Having moved the bombs into theatre, the Squadron then moved ground crews for squadrons deployed to the Gulf. Once that was done, ammunition for the 7th Armoured Brigade was moved from Germany to the Gulf, followed by the 'tankies' themselves (from 16 Oct 90). In addition to the main Op Granby bomb/ammunition and personnel tasks, general stores were carried and a weekly aeromed flight collected the sick and injured from the Gulf. The destinations in the Gulf were mainly Jubail, Dhahran and Riyadh, plus Bahrain, Bateen, Minhad, Seeb and others.

On 17 January 1991, Operation GRANBY ended and the ground and air attacks of Operation DESERT STORM began. Later in January, the Squadron was tasked to provide six airframes in the full aeromed role, but only one complete set of kit could be found. Two aircraft were eventually equipped in the full aeromed fit when all the "missing" stretcher posts had been recovered from the various locations at which they had been dropped off. The remaining aeromed requirement was filled by chartering civilian aircraft that would only operate to Cyprus. (On at least one occasion a charter aircraft arrived at Brize and the crew, when briefed on the risk and issued with NBC kit etc, returned to their aircraft immediately and departed never to be seen again.) Luckily the aeromed task proved to be minimal and essentially the weekly VC10 aeromed coped with the task."

By March 1991, some 5000 hours had been flown on logistic and aeromed tasks in support of the forces deployed to the Gulf. Inevitably, this effort had an impact on other tasks – for example, it had been possible to maintain the Dulles schedules until, as Desert Storm began, the slip crew was withdrawn from Washington. Overall, the operation would not have been possible without the logistic support provided by all of the ATF squadrons. This was, to a degree, recognised by the award of a Battle Honour, "Gulf 1991" but, as its aircraft and crews were not engaged in combat operations, the Squadron is not entitled to emblazon this on its Standard.

Shortly after it was all over, Squadron CO, Wg Cdr Peter Bingham, summed things up in a Foreword to an Association Newsletter:

> "Here we are, two weeks later and the War is over and the Squadron resembles a ghost town. Gone are the mountains of Chemical Protection clothing and respirators that we all seemed to live with, gone are the guns we were obliged to fly with and, more importantly, people are actually looking relaxed ...... However, even at this stage I can say that the Squadron achieved results that many thought impossible. During the peak of effort our engineers were providing 10 aircraft a day on the line for tasking and our crews broke all records in flying hours achieved. August 1990 saw the highest monthly total ever achieved by the Squadron – we

moved everything from bombs to video recorders to the Gulf and did not cancel an operational task. Luckily, we did not lose any people or aircraft, although there were a few narrow escapes with crews landing during air raids, evacuating the aircraft and diving for cover. One crew swore a Patriot missile locked onto them before being redirected by ground control, although this story tends to expand with each telling in the bar! We are all very thankful that the result was so decisive and that our troops returned with minimal losses; the highlight for us has undoubtedly been the return flights with aircraft full of jubilant people." (The CO added later that events in the Gulf had rather overshadowed the occasion in September on which a 10 Squadron aircraft had led the main formation of the RAF's Battle of Britain 50th Anniversary Flypast over Central London.)

Fg Off Charlie Duffy had just joined the Squadron as a first tourist Copilot and on 20 January 1991 flew his first flight into the conflict zone as Iraq fired missiles into Saudi Arabia. This may have been the Patriot missile incident mentioned by Wg Cdr Bingham, as Charlie later recalled for another Association newsletter:

"The next leg was from Dubai to Al-Jubail on the east coast of Saudi Arabia. The initial part of the flight was fine, the whole leg being my night sector.....At 11,000 feet we were then told to contact Dhahran as we entered their radar coverage ..... The Captain transmitted to check-in. No reply. He tried again – No reply. We'd just started to try once more when we saw the first missile being launched - a pale blue streak appearing from the ground. We were now two minutes from overhead Dhahran and still not in radio contact.

The Patriot missile vanished up above the cloud layer that was around 13-14,000 feet. The sky then suddenly illuminated from behind the cloud with an orange glow, and we presumed that an incoming Scud missile had just been destroyed ..... another blue flash from the ground appeared and vanished above the cloud – but this time nothing happened. The next thing we saw was an explosion as a Scud hit part of Dhahran. We watched the pressure wave expand from the centre of the explosion, and the street-lit part of the town now went black. From the ground now appeared four "orange balls" that we presumed were more missiles being launched. All four moved left to right across the flight deck windows, then two changed their direction and started to come towards our aircraft but dropped out of sight beneath ..... We were now hearing reports from other aircraft that Riyadh had just gone to Alert State Red ..... we decided to return to Dubai and try the trip again the next day."

In a somewhat different context, there were two flights from the Middle East that gained very considerable media attention during 1991. British journalist John McCarthy was one of a number of western citizens, some of whom were eventually killed, held captive in Lebanon throughout the 1980s by Islamist groups. He had been a hostage for over five years by the time of his release and he was flown back to the UK on a 10 Squadron aircraft on 8 August. Terry Waite had gone to Lebanon on behalf of the Archbishop of Canterbury to negotiate the release of hostages when, in January 1987, he too disappeared. He spent almost four years in solitary confinement and was finally released with another hostage, Tom Sutherland, in November 1991. A Squadron crew flew both men were flown back to the UK from Damascus on 19 November.

Amongst all of the other changes affecting the Squadron over these years, it is worth noting that in December 1991, Wg Cdr Al Stuart, who had been Flt Cdr (Ops) some years beforehand, returned to 10 Squadron as CO – the noteworthy factor being that he was the first Navigator CO of the VC10 era. For the first twenty years or so, only Pilots had been accepted as candidates for the appointment.

The Transport/Tanker Era (C Mk 1 K)

With the operational capability of the VC10 as a tanker well established within 101 Squadron, the next substantial change in 10 Squadron's role followed logically. In 1990 a contract was let with Flight Refuelling Aviation to convert all thirteen aircraft to what would become known as 'C Mk 1K' standard – ie the aircraft would now give and receive fuel, besides retaining a transport capability. The modification involved the fitting of a Flight Refuelling (now Cobham) Mk 32 hosedrum pod under each wing, together with TV and related controls at the Flight Engineer's station. XV 103 was the first aircraft to be modified and it was handed back to the

Two 29 Sqn Tornadoes take fuel from a VC10 over the North Sea. VC10s were later re-painted 'Air Defence Grey' and 29 Sqn disbanded in 1998.

Squadron on 3 December 1992, with the others following through into 1996. A year later, the Squadron had three of the converted aircraft on strength. In the first years of this new era, only about half of the unit's crews were trained in AAR procedures for routine 'towline' use around the UK or elsewhere – the tactical callsign 'Madras' was allocated for use during these sorties - and, of these crews, only about half were trained in the 'Trail' procedures used to deploy fighter aircraft to overseas locations. (Three AAR experienced crews had been transferred from 101 Squadron.) In June 1993, an aircraft went north to RAF Kinloss to provide AAR support to aircraft participating in JMC 932, one of a series of significant maritime exercises involving NATO forces. Then, in November, the first operational AAR detachment began, when an aircraft and crew were deployed to Bahrain in support of Operation JURAL, where the task was to refuel RAF and US Navy fighter aircraft operating air patrols to monitor and control airspace over Southern Iraq following the 1991 Gulf War. This deployment involved a trail with two Tornados destined for the Dubai Air Show, whilst a 101 Squadron aircraft trailed two Harriers and the Squadron aircraft was, of course, also able to carry supporting groundcrew for both fighter types.

On 20 January 1986, the Squadron had flown Mrs Thatcher to Lille, to meet President Mitterand for the joint announcement of the Channel Tunnel project. The Tunnel was formally opened on 6 May 1994, with ceremonies in Calais and Folkestone. Appropriately to the occasion, The Queen and PM John Major travelled by Eurostar, but a Squadron crew led the formation flypast over the UK Terminal that afternoon. In July 1994, whilst classroom and simulator training was retained, the flying instructional element of the OCU (by then renumbered as 55(R) Squadron) was transferred to create a 10 Squadron Training Flight. In early October, Saddam Hussein moved Iraqi troops towards Kuwait. Tornados in-theatre had to fly additional recce sorties, causing a need for increased AAR support. An additional Squadron aircraft with two crews went out to join Operation JURAL, and three or four sorties per day were then flown to meet the need. In addition, a trail to deploy six Tornados from RAF Bruggen in Germany to Dhahran was mounted, using tankers from 101 and 216 Squadrons. In the event, and to get the second wave away on time, a Squadron aircraft and crew had to be generated at very short notice

when a 101 Squadron aircraft went unserviceable. The Iraqi threat abated in late October, and one crew returned to base. The recovery of the additional Tornado reinforcements to Germany took place in early November via a highly complex trail in which the Squadron aircraft demonstrated both AAR and AT capabilities. As a result, approval was given for an increase in AAR flying hours and for more crews to become AAR trained.

As the proportion of AAR tasking increased to around 30%, particularly in support of air operations in the Gulf and over Iraq – the Squadron passed 1000 hours in support of Operation JURAL in June 1995 – its air transport tasking began to reduce, with a particular landmark occurring on 28/29 August 1995 when, only weeks short of a 25th anniversary, the last Dulles schedule was flown. (Food for thought: The first North American schedules had been flown into JFK International, New York, in 1968 – a task flown by the Squadron for a little over a quarter of its 100 years!) Earlier, in April, the new AAR capability had been demonstrated to Mr Malcolm Rifkind, Secretary of State for Defence, when he was flown non-stop from Heathrow to Windhoek, Namibia, and back, refuelling from a 216 Squadron Tristar. Another change was also in the wind. With the Cold War at an end, the RAF was moving towards preparation for 'expeditionary' operations and future Tacevals were to be planned on the basis of a simulated deployment to some overseas Forward Operating Base (FOB), a marked departure from the home base as a 'Fortress' concept that had dominated such exercises for the previous twenty years or so. Training for the next evaluation in the following year began with a Station Exercise held at the end of October 1995.

Despite the demonstrable capability provided by the Squadron in its combined AT/AAR role, by 1996, with the VC10 aircraft now some thirty years old – indeed, older than the younger Squadron personnel – they were becoming expensive to maintain. (They were still used for Royal and Ministerial VIP tasks but a combination of age and noise from early-60s era engines was making them less acceptable in the role, and a Sunday Telegraph report on a VIP flight in 1995 had referred to the "rickety old VC10. "Speculation emerged about the desirability of acquiring a "RAF One" but there was never enough political will to fund such an aircraft whilst other reductions in public spending were being made and it never came to pass.) In the circumstances, there was no immediate possibility of acquiring a fleet replacement and a modernisation programme was funded to extend fleet life until 2012. That said, the Squadron could not be immune to the rundown of RAF manpower that followed the changes underway in the mid-1990s. So, although all thirteen aircraft were retained in-service, the crew establishment was reduced to sixteen, with the groundcrew establishment similarly reduced, to a combined total of some 350 personnel. At the same time, the scope of the unit's AAR activity was expanding. Whilst maintaining the detachment in Bahrain, January 1996 saw the Squadron's first participation with RAF fighter aircraft in Exercise RED FLAG, where highly realistic combat training was possible in the USAF ranges in the Nevada desert. Later, in September, an aircraft was deployed to Singapore to exercise the Integrated Air Defence System (IADS) established under the Five Power Defence Agreement (FPDA), working with aircraft from Malaysia, Singapore and Australia. It was also soon found that the ability to dispense or receive fuel, whilst delivering passenger service in the rear cabin, made the 10 Squadron tankers eminently suitable for demonstration flights. Flights for Staff Colleges had been laid on and, on 11 September 1996, another was mounted on which the Air Force Board hosted a group of Welsh dignitaries for RAF PR purposes. Briefings were given at RAF St Athan and, once airborne, the guests witnessed a full spectrum of AAR work involving a variety of fighters and two other VC10 tankers whilst receiving full VIP treatment.

On the transport side, by April 1996, the only scheduled service retained was a weekly resupply of the training base at Decimomannu in Sardinia, whilst exercise and operational support had become predominant – a far cry from the days of twenty-seven Far East schedules per month generating almost 1000 hours per month more flying. It had also been decreed that the 10 Squadron aircraft should adopt the Air Defence Grey all-over colour scheme used on most other large RAF aircraft, and the first delivery of a re-painted airframe occurred in August. (The work was carried out under contract at Chateauroux, in France.) Crews had by now also largely abandoned flying in blue or KD uniform, as had long been customary in the strategic air transport force. Instead they adopted flying suits and tee shirts, another sign of an eagerness to put some distance between their new combined role and the old. The Squadron had never flown any one aircraft type for anything like thirty years before and a substantial generational change was afoot. Clearly proud of the degree to which the challenge of adapting to new circumstances had been met, the crews of the mid-1990s - as far in time from the introduction of the VC10 as the crews then had been from the pre-WW2 Heyford - appeared content to see much of what went before as a 'Jurassic Park' era, in the words of one record. Elsewhere, in summing up a year, it was noted that "At last the 60s, 70s and 80s image of bespectacled, silver haired aircrew flying long sectors around the world has gone, to be replaced by youthful, vibrant aviators capable of operating an old aircraft very flexibly and to the best of its potential." (It is ever the province of the young to question what went before but, by any standards this is an odd remark, given that those thus slighted had been young when that writer was at school, and were meeting the global air transport requirements of a different time with at least equal élan, flexibility and professional competence.)

That things had, indeed, changed was manifest yet again in June 1997 with the last VC10 flight into Hong Kong, for so long central to Squadron operational life. XV 108 was used to take the new Prime Minister, Tony Blair, and the new Foreign Secretary, Robin Cook, there for the formal handover of the colony to China at midnight on 30 June. And whilst transport tasks still occupied some 70% of tasking, with a geographically well-spread range of exercise support, during that year six Squadron AAR crews underwent conversion to the K2 aircraft of 101 Squadron, with the aim of sharing the support of the air defence fighters based at Mount Pleasant in the Falklands. (The K aircraft dealt with fuel in kilos, whereas the C1 aircraft used pounds, until eventually being converted to the kilo.) In May, the detachment at Bahrain moved to Incirlik AB at Adana in Turkey to participate in Operation WARDEN, now supporting patrols over Northern Iraq and, over the year, crews supplied AAR support to RAF aircraft deploying to the USA, Canada, Norway, Cyprus, Turkey, Saudi Arabia and Oman. (Unfortunately, the year closed with XR 806 being damaged beyond repair in a defueling incident at Brize Norton that broke its back on 18 December. The fleet size was further reduced to eleven in December 2000 when XV 103 was retired, after flying some 34,268 hours.) Around this period, and as part of the fleet modernisation, the aircraft of both 10 and 101 Squadrons had a navigation system update. As international use of the American Global Positioning System (GPS) had increased, the US decided to switch off the Omega navigation system during 1997. The inertial systems fitted in the aircraft were retained and the Omega sets were replaced by Rockwell-Collins FMS 800 equipment. This kit was linked to the autopilot, took automatic GPS updates that largely eliminated the need for fix-based updates, and signalled the approaching end of a need for a human navigator. (New RAF navigators became Weapon System Operators in 2003, and the last of those to graduate did so in February 2011.)

From 1991, the Former Republic of Yugoslavia began to disintegrate as long-standing ethnic tensions, suppressed since Marshal Tito's death in 1980, came to the fore in a series of civil wars and general unrest. The UK's first military involvement, Operation GRAPPLE, began as part of the UN Protection Force in Bosnia in late 1992. In April 1993, on behalf of the UN, NATO undertook Operation DENY FLIGHT, the enforcement of a no-fly zone over Bosnia and Herzegovina. As these operations continued, the Squadron was tasked with regular support flights into airfields in the Balkans. In 1999, when NATO became involved in the war on Kosovo, the Squadron was held in reserve for support of a ground offensive that did not materialise but, by year's end, flights into Skopje and Pristina took up a significant proportion of overall tasking.

The Squadron was also to become busily involved in events much further east that year. In September 1999, an aircraft had deployed to participate in another IADS exercise, based at Kuantan in Malaysia and then Paya Lebar, Singapore. During the exercise, unrest in East Timor led to UN intervention to restore order and a participating RAAF tanker was recalled to base, leaving ten Australian F-18 Hornet aircraft without support. On 10 September, the Squadron crew was tasked with trailing the F-18s back to

RAAF Tindall in two groups, a task that was successfully accomplished. At the same time, a second 10 Squadron aircraft was heading westabout to Darwin to deliver a RM SBS unit. After that, and with two aircraft and three crews in-theatre, a contingent of Gurkhas was flown from Brunei to Darwin for onward movement by other aircraft. With that move complete, the crews stood by for the move of a New Zealand contingent and, once it was clear that this lift would not be required, both aircraft recovered eastabout across the Pacific and the USA to the UK

With change continually in the air, the mixing of Squadron tasks with 101 Squadron increased and, in 2000, it was decided that all 10 Squadron crews should become AAR qualified and that those of 101 Squadron should be AT qualified. Wing Commander Tony Gunby had command of 10 Squadron from August 1999 to February 2002 and explains how it all came to pass:

> "I think it is fair to say that the Squadron was at a crossroads in several ways. There were still short and long-haul air transport tasks, a reducing number of VIP tasks, and some air refuelling tasks. This latter role was relatively new for 10 Squadron but the air transport demands meant that there was only enough AAR flying to keep a small number of crews current – something which was divisive and which caused lots of programming problems. A significant new line of tasking saw the Squadron flying into places like Pristina, Skopje and Split, in the former Yugoslavia, on a fairly regular basis, as well as maintaining an operational AAR deployment in Turkey. Taken together, I felt that the Squadron needed to embrace these changes and adapt to its new circumstances. A similar shift in emphasis was happening on 216 Squadron (Tristars), and its boss and I decided that the time was right for us to switch out of blue uniform and into flying suits full time, reflecting the increase in 'operational' tasking we were doing. I knew that this was going to cause some angst among the die-hard air transport people on 10 Squadron, but the switch went ahead and I still believe it was the right move to make. The only negative comment I actually recall, which I acknowledge may be down to selective hearing and memory, was from a passenger on an Akrotiri trooping flight who suggested that the service no longer had the feel of a commercial airline. Welcome to the 21st Century!
>
> The frantic pace of tasking, and managing an aircraft which was demanding an increasing level of tender loving care, ensured that the weeks and months flew by. Month after month, the Squadron overflew its hours' allocation and yet there was always plenty of unfulfilled tasking beyond that. The Squadron planners and operations team did a fantastic job keeping all the moving parts moving in the right direction. Yet even here there was room for change and albeit with teething problems, we saw the progressive introduction of information technology (IT), to support the tasking and crewing process, and the retirement of those chinagraph boards that we all had loved to spend so many hours looking at. Change was happening on the flight line as well, as we revised shift patterns for the Squadron's engineers to increase their availability at peak times. Throughout the change processes, I was keenly aware of upsetting routines and customs that had been in place for many years in some instances. But change was necessary and I truly believe that the Squadron came out the other side of the change process more operationally-focused and better-able to meet increasing demands with reducing resources.
>
> I suspect that most ex-10 Squadron aircrew and flight crew will remember the more exotic trips to the Far East, the 'long Dulles', 'MedMan' and so on. Juggling the various 'home' demands, with the need to stay current myself, was at times a struggle, but I was blessed with an outstanding executive team on whom I could rely totally. So whenever I could, I got away, but try as I might this tended to be on the shorter AT trips alongside doing my share of AAR tasking. Meanwhile the Squadron was getting involved in all sorts of unusual tasking, still conducting VIP tasks with senior Government Ministers, responding to events in East Timor, running a tanker detachment in Turkey supporting No Fly Zone operations over northern Iraq, and a multitude of other one-off tasks.
>
> The pressure to squeeze yet another quart out of the proverbial pint pot remained throughout my tour. This, and a growing feeling that the AT and AAR divide could not be sustained, led Brize Norton, and OC 101 and myself in particular, to look hard at the way in which the two VC10 squadrons were delivering their respective outputs. Once

again, the result was a programme of change which ultimately saw both 10 and 101 Squadron flight deck and cabin crews cross-qualified on all VC10 variants in both AT and ARR roles. This created a monumental additional workload for both squadrons' training staff - a challenge to which they rose magnificently. The subsequent ability to use the most appropriate aircraft and crew mix for a given task provided us with far greater flexibility, and I think it broke down some unfounded prejudices between AT and AAR crews. Some of the tanker crews even acknowledged that an AT task did involve rather more than a tanker sortie without giving any fuel away. As a result of the cross-qualifying, 10 Squadron was able to take up some of the Falkland Islands' tanker detachment workload from 101 Squadron and we were able to fly mixed squadron crews where necessary. The Squadron flying executives even found themselves authorising the other squadron's crews/tasks and vice-versa: quite revolutionary."

On the morning of 11 September 2001, four airliners were hijacked after takeoff from airports on the East coast of the United States. Two were flown into the 'Twin Towers' of the World Trade Center that dominated the New York skyline, another was flown into a wing of the Pentagon military HQ in Washington DC, and the fourth crashed in a field in Pennsylvania after passengers took on the hijackers. Over 3,000 people died that day. What we now call '9/11,' the first successful attack on US territory in modern times, has echoed through every year since and, whilst it had been by no means unknown beforehand, the Islamist Al-Qaeda terrorist organisation gained a global reputation. In response, the USA then convened a coalition of nations to track down and destroy Al-Qaeda base facilities in Afghanistan. As it happened, a substantial British force was bound for Oman for Exercise SAIF SAREEA II, a Tri-Service exercise of Joint Rapid Reaction Force procedures with Omani forces, and the coincidence allowed the UK to contribute to concurrent operations in Afghanistan. Wg Cdr Gunby had been nominated as CO of a joint 10/101 Squadron detachment to support the air element of the exercise:

"As we prepared to deploy, the 9/11 attacks took place in September, sending plans into limbo and forcing several on-going tasks to grind to a halt as the world's airspace was temporarily closed down. Over the days that followed, and as things kicked back into action, it became increasingly clear that there would be coalition operations in Afghanistan. Eventually, we were told we would have to continue with the exercise while concurrently flying operational sorties over Afghanistan – both from Seeb in Oman. Within a couple of weeks of 9/11 I found myself commanding a joint 10/101 Squadron tanker detachment, which was later expanded to include a 216 Squadron (Tristar) element. Once again, Team Brize Norton rose to the challenge of operating from austere facilities, unfailingly meeting the operational demands placed upon it; a hugely satisfying period which added the icing to my command cake."

Al-Qaeda was effectively expelled from Afghanistan and that phase of operations ended in January 2002, but the Gulf War legacy of RAF patrols over Northern Iraq continued with AAR support still based at Incirlik AB in Turkey. In addition, the Squadron positioned a slip crew in Cyprus to fly weekly transport flights to bases in Saudi Arabia, Kuwait and Oman. As the year went on, President Bush began to make out a case for action against Iraq, one of the nations he saw as part of a terrorist 'Axis of Evil' and, by early 2003, war was all but inevitable. Under the national codename Operation TELIC, Prime Minister Tony Blair committed UK forces in support of an action that began with air strikes on Baghdad early on 20 March 2003. For the UK, this was the largest operation since the Gulf war of 1990-91, with a RAF contribution of some 7,000 personnel and 115 fixed-wing aircraft. The VC10 tanker detachment moved to the Prince Sultan airbase in Saudi Arabia, from where up to seven aircraft of 'VC10 Wing' flew AAR support missions for a variety of coalition aircraft that could use the hose and drogue system. 10 Squadron's aircraft were also used to fly aeromedical evacuation sorties from Kuwait, through Akrotiri, to the UK. Four crews were based in Cyprus as slip crews for the task, over 1000 casualties were brought home, with the final flight recorded on 16 May 2003. Kuwait remained under threat of missile attack for a time, and a Squadron aircraft was on the ground when an Iraqi missile penetrated the air defence system and hit the city on the morning of 28 March. (On at least two occasions in April,

Iraqi children were flown out to the UK for treatment of injuries or burns sustained in the war.) Once the war had ended, the Squadron began flying to Basra for troop rotations. Conditions for slip crews sleeping in the ATC tower were basic and crews were not sorry when the task ended in December.

By 2004, both 10 and 101 Squadrons had developed a solid conjoint operation. Since 2001, they had been co-located in the same building, sharing the same Ops facilities for joint tasking, albeit with distinct unit identities and history. Together they maintained a continuous AAR presence in the Falklands and in Iraq, whilst providing a Quick Reaction capability from Brize Norton and still responding to other AT and AAR tasking.

October 2005 - Disbandment

Although the 10 Squadron aircraft were given a refurbishment programme extending their life until 2012, that would not prove possible for all of the more heavily used ex-civilian airframes used by 101 Squadron, and the four oldest K2 variants were retired between 1998 and 2001. The Tristar aircraft of 216 Squadron had also been delivered with extensive prior civil airline use. Thus by the turn of the century, the RAF's need for a replacement tanker/transport aircraft was recognised. Whilst the staffing of this requirement went on in London, the RAF was not immune from the progressive reductions in all three Services that marked the years from 2000. In 2005, the Squadron establishment was reduced to fourteen crews, with 101 Squadron reducing in parallel and a nod towards rationalisation came with a decision that 10 Squadron should disband, with its eleven aircraft to be passed on for use by 101 Squadron during their remaining years.

The Squadron disbandment occurred on Friday 14 October 2005. It was marked with some ceremony, a flypast, and an emotional Guest Night at which Air Chief Marshal Sir Michael Beavis, the first CO of the VC10 era, was Guest of Honour. On the following Sunday morning, the Squadron Standard was paraded at a Service in the Station Church that was followed by a Brunch for all Squadron personnel in the Officers' Mess, the final disbandment event, and the following announcement was published in the Court and Social columns of the Daily Telegraph of 21 October:

"The No 10 Squadron Standard was lodged at RAF College Cranwell yesterday, the final commemorative event marking the disbandment of the Squadron, and will be held in College Hall Officers Mess within the Rotunda. The principal guest was Group Captain Tony Gunby, Group Captain Operations, HQ 2 Group. The Reverend (Wg Cdr) Jonathan Chaffey officiated and Wing Commander Mike Smart, the outgoing Officer Commanding 10 Squadron, presided."

Whilst the arrival of a new aircraft was somewhat delayed, the Standard was not to hang there for too long.

Squadron Commanders from 1985

Wing Commander L J Marshall	June 1984
Wing Commander D B Symes	December 1986
Wing Commander P C Bingham	June 1989
Wing Commander A F Stuart	December 1991
Wing Commander S Duffill	July 1994
Wing Commander A M Bray	January 1997
Wing Commander A D Gunby	July 1999
Wing Commander A S Deas	February 2002
Wing Commander M A Smart	October 2004

Postcript: In the event, delays to the introduction of service of the Future Strategic Tanker Aircraft meant that it became necessary to extend VC10 service life. The last ex-10 Squadron aircraft, XR 808, the first to be delivered in 1966, was finally flown into retirement at Bruntingthorpe on 29 July 2013, after 47 years and 43,865 hours of distinguished RAF service. The last RAF VC10, a K3 variant, joined it there on 25 September 2013. It had been hoped that 808 could be flown into RAF Cosford for the RAF Museum, but flight safety concerns prevented this. Nonetheless, negotiations over the future of the airframe, the RAF's first VC10, continued after its delivery to Bruntingthorpe and, by early 2015, it had been agreed that it should be dismantled and moved to Cosford by road for subsequent reassembly and display. The fuselage was successfully moved north during the morning of Sunday 21 June 2015 as, at the RAF Church, St Clement Danes, in London, the Standard presented by Mrs Thatcher in 1988 was laid up. By happy chance, a symbol of the Squadron's early VC10 era was reverently put away whilst the first of the fleet was taken for resuscitation!

A new era of AAR began for 10 Squadron in December 1992 when OC10, Wg Cdr Alastair Stuart took receipt of XV 103's F700 (tech log) from BAe. It was the first of the original VC10 C1s to be converted and fitted with two point hose units.

XV 109 offers its services to two Harriers.

Sqn Ldr John Wolley, probably the Sqn's longest serving member, is despatched from the Crewroom by wheelchair on his retirement in April 1993, having served 14 years on the Squadron and a total of 42 years in the RAF.

After the Air Defence Grey re-paint the names of Victoria Cross holders remained on the fuselage. As some aircraft were later scrapped, the names were then added to those of other VC10s of 101 Sqn.

A 101 Sqn aircraft, included to express the gratitude of the 10 Sqn Association to the members of 101 Squadron who hosted them and kept the 10 Sqn VC10s flying until 101 Sqn too was disbanded. Now side by side again both Squadrons are now flying the Voyager.

The centre hose on 101 Squadron aircraft was essential for VC10 to VC10 refuelling capability.

Viewed from above a chase aircraft, whose canopy reflections mottle the sky, a VC10 tanker shows the length of its refueling hoses.

On 14 October 2005 another era ended when 10 Sqn disbanded leaving its former aircraft in 101 Sqn's charge. 101 itself disbanded in 2013 when its VC10s were retired, only to reform immediately and join 10 Sqn in flying the Voyager.

CHAPTER 13

VOYAGING INTO THE FUTURE

On 1 June 2011, the 10 Squadron Standard was collected from the Rotunda at the RAF College, Cranwell, in a ceremony that saw the Standards of Nos 13, 14, and 120 Squadrons laid-up on their disbandment. There was, however, no Squadron in-being at that time. On 1 July 2011, the Standard was paraded with seven others at RAF Brize Norton in a ceremony to mark the move of the RAF's C-130 Hercules squadrons to the station as RAF Lyneham closed. More unit personnel were being posted-in and, with the arrival of a first CO, Wg Cdr Dan James, the Squadron was officially promulgated as having re-formed on 1 August 2011. As in 1937, 1958, and 1966, the Squadron found itself introducing a new aircraft into service, but in ground-breaking circumstances.

That new aircraft is the Voyager, a version of the Airbus A330-200, designed to operate in the joint Air Refuelling and Air Transport roles. (Similar, but not identical, A330 variants have been supplied to Australia, Saudi Arabia, and the UAE.) What is different in this case is that the aircraft – 14 in all – are being acquired via a Private Finance Initiative (PFI) contract, similar in concept to many insisted upon by the Treasury for the building of significant infrastructure projects in the Public Sector since the late 1990s. The project is the first instance of a front line capability being acquired

Enter the Airbus A330 Voyager

in this way. After a competition for a Future Strategic Tanker Aircraft (FSTA) project and some years of complex negotiations, the AirTanker Consortium was announced as Preferred Bidder in February 2005, and a FSTA contract with a 27-year duration was signed on 27 March 2008. The RAF does not own the aircraft in the traditional way, but has operational control over them and their deployment whilst the Ministry of Defence pays only for the capacity used, subject to an agreed minimum annual usage. Deliveries are planned into late 2016 and, from 1 October 2013, No 101 Squadron was confirmed as the second squadron to operate the fleet, having retired its last VC10 two days earlier. What is termed 'The Core Fleet' comprises eight aircraft on the Military Register, with AirTanker operating one on the Civil Register. That milestone was passed on 29 May 2014 as the ninth aircraft was delivered, and the delivery of full fleet capability was marked by a ceremony on 30 July that year. The twelfth aircraft arrived in June 2015 and deliveries will be completed during 2016, with the other five aircraft being used by AirTanker to earn 'Third Party Revenue' through

In September 2013 a Voyager bids farewell to its predecessor VC10.

Overflying the Union Flag at the Brize Norton Gate of Remembrance through which Afghanistan re-patriation casualties passed, after arriving in the UK.

Modern technology in a comfortable environment make the Voyager a delight to fly.

commercial or military leasing. However, in time of crisis, all fourteen aircraft will be available to the RAF.

The Voyager is powered by 2 RR Trent 772B high-bypass turbofans, each capable of producing 71000 lbs of thrust. With a maximum takeoff weight of 233 tonnes, it can carry 111 tonnes of fuel in wing and fuselage tanks, and can cover some 8000 nm empty or 3400 nm with a payload of 44 tonnes. The aircraft can all be fitted with two underwing refuelling pods, but seven will also be capable of being fitted with a third centreline fuselage refuelling unit, and the aircraft are designated KC Mk 2 and KC Mk 3, respectively. (The centreline pod will mainly be used to refuel larger aircraft, and all pods will be removed whilst aircraft are operating in the civilian transport role. There is no provision for boom refuelling.) The aircraft will normally carry 291 passenger seats, at a 34-inch seat pitch, a little more generous than the pitch in a normal VC10 passenger fit. The aircraft is also capable of being fitted for aeromedical evacuation, and can carry eight military pallets in the under-floor cargo hold.

A normal flightdeck crew will comprise two Pilots, with a Mission System Operator (MSO) on AAR sorties. For AT operations, eight cabin crew will be added, with the MSO acting as Purser. Since not all of the pilots will be from the RAF, AirTanker will employ fourteen who will be Sponsored Reservists capable of being deployed should the situation demand. (The company also retains a number of civilian cabin crew for its own revenue flights.) Both Voyager squadrons are operating once again much as they did in 10 Squadron's last years with the VC10, in that crews are drawn as required for tasks from either or both, with AirTanker providing the Ops facility that would hitherto have been a squadron responsibility. That said, it has been agreed that

10 Squadron's first Voyager arrives back at RAF Brize Norton on Easter Sunday 2011 after its first training flight.

To the south of the well-known Base Hangar (left) AirTanker built the 'Hub' which houses the Squadron and all the ancillary back-up facilities.

10 Squadron will lead on Training and Force Development, whilst 101 Squadron leads on Operational Capability.

Although the civilianisation or 'outsourcing' of support capabilities had become an integral part of daily operations in all three Services since the mid-1990s, the FSTA contract is the first occasion upon which a frontline capability has been funded in this way. The detail above provides sufficient insight to appreciate that the successful integration of RAF and civilian AirTanker resources will be essential to the provision of the required military capabilities. The company has built a large self-contained HQ-cum-servicing facility – 'The Hub' – at Brize Norton, shared by the RAF and the contractor, in which the two squadrons have the third floor. A Simulator building was also completed in October 2012 under the contract – prior to then Pilot simulator training had been carried out in Bahrain.

2014 saw the Voyagers engaged in the Afghanistan withdrawl from Camp Bastion.

The first aircraft was delivered to Brize Norton in late 2011, but it was not until 8 April 2012, Easter Sunday, that the first RAF flight took place - a general shakedown flight over southwest England. Training of both pilots and cabin crews then got underway, with the aircraft making familiarisation visits to airfields in the UK, Europe and North America that led to a first live task to Akrotiri with MOD passengers in May. Training in refuelling was delayed by compatibility issues affecting the specified drogues and, after a change to the 'Sgt Fletcher' basket type used elsewhere, a clearance was granted and the first training sortie with Tornado GR4 fighters was flown on 20 May 2013. Fleet build-up continued with the sixth aircraft being delivered in June 2013 and a Release to Service to tank with the Typhoon followed in August 2013, by when some 350 contacts with both types had been completed, with 840 tonnes of fuel transferred. In addition, over 5,400 flying hours had been flown by then,

In May 2013 the Voyager was cleared to refuel Tornadoes and Typhoons.

representing 1,500 sectors carrying more than 110,000 passengers and 6,300 tonnes of freight, and over 1,000 hours were flown for the first time in October. Thus, with the fleet beginning to demonstrate its capability in both designated roles, and with many of the early challenges and frustrations involved in the creation of the joint military-civil project that is the Voyager fleet put behind him, Wg Cdr Dan James left in November 2013, handing over command to Wg Cdr Jamie Osborne.

The change of command occurred as the Voyager was about to join the Operation HERRICK air bridge, in support of British operations in Afghanistan. A first flight was flown on 8 December 2013, after which regular schedules were flown between Brize Norton, Akrotiri and Camp Bastion until the drawdown of forces in late 2014, and the relative comfort and reliability of the new aircraft made it quickly popular with homebound soldiers. The aircraft is also used in support of tri-Service exercises, with once-familiar destinations from the VC10 era such as Calgary and Nairobi reappearing in crew logbooks. The aircraft's AAR capability has also been developed. The 'Madras' tactical call sign has been reintroduced, clearances to refuel a variety of allied Air Force aircraft have been obtained, and training sorties around the UK are now routine. Tanker support of the fighters based in the Falklands was taken over in March 2014, and the Voyager squadrons now provide a continuous Quick Reaction Alert (QRA) capability in support of the air defence of the UK - for example, supporting fighters once again shepherding inquisitive Russian Bear aircraft as was so frequently done in the Cold War era. And, looking to the capability of the fleet as a whole, a significant milestone was passed in March 2014, when a Squadron crew flew the first military Voyager trans-Atlantic sector using an Extended Operations (ETOPS) clearance. (This procedure was introduced with the advent of the large twin-engined airliners in the mid-1980s and permits direct routings at distances that can be three hours from suitable diversions. AirTanker had been cleared for civilian operations in 2013, and took over the Falklands air bridge in October that year.) In August 2014, the Squadron's support of active operations continued when the Voyager was tasked in support of Operation SHADER, the RAF effort against Islamic State over Iraq, based in Cyprus.

A Squadron crew was airborne over Iraq supporting US Navy F18 aircraft on 1 January 2015 as the unit passed its 100th birthday, and the centenary was marked by the presentation of a new Standard – its third - by HRH Princess Anne, The Princess Royal, on Friday 30 January. The performance of the developing Voyager Force as a whole was recognised in June 2015 by the award of an Air Officer Commanding's Commendation. And so, from Brooklands to Brize, from a time when the sixty miles from Chocques to Valenciennes was a reconnaissance at strategic depth, to the Voyager with its 8,000 nautical mile range, aviation has come a long way. With a notable history in both World Wars and in the development of military aviation, from BE2c to Voyager, the men and women of today's 10 Squadron have a substantial inheritance of which to be proud. Long may they continue 'To Hit the Mark.'

Wg Cdr Dan James became the Voyager Squadron's first OC10 in 2011.

....followed in November 2013 by Wg Cdr Jamie Osborne, pictured here with his executives: Sqn Ldr Sean MacFarland , Wg Cdr Jamie Osborne, Sqn Ldr Phil Astle and Sqn Ldr Tim Rushworth.

Prior to the Voyager receiving its clearance to refuel other aircraft it was used in its adjunct role as a passenger transport. Prince Harry arrives at a snowy RAF Brize Norton after his first tour in Afghanistan in January 2013.

Two familiar sights: a 216 Sqn Tristar which the Voyager replaces, and the Terminal Building at RAF Brize Norton

The core fleet is made up of 9 RAF aircraft plus a further 5 'surge capability' 'civilian-registered aircraft, available when required.

Civilian cabin crews engaged in the FSTA programme at RAF Akrotiri on the first flight there.

In theatre at Camp Bastion, Afghanistan

Going home from Camp Bastion, Afghanistan 2014

Russian Air Force recce flights around UK airspace are always intercepted and followed by Typhoons, refuelled by Voyagers.

No 3 Squadron Typhoons from RAF Coningsby fly in eschelon-left formation with a Voyager.

Tornadoes with Paveway IV precision guided bombs successfully attack ISIL targets in Iraq during 2015 thanks to 10 Sqn's refuelling Voyagers.

Open all hours……

"TEN'S GEN"

ADDITIONAL MATERIAL

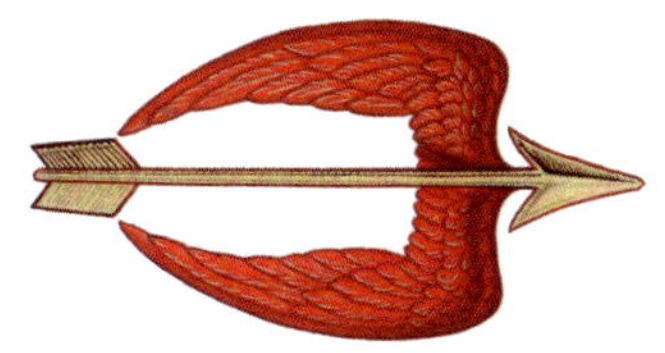

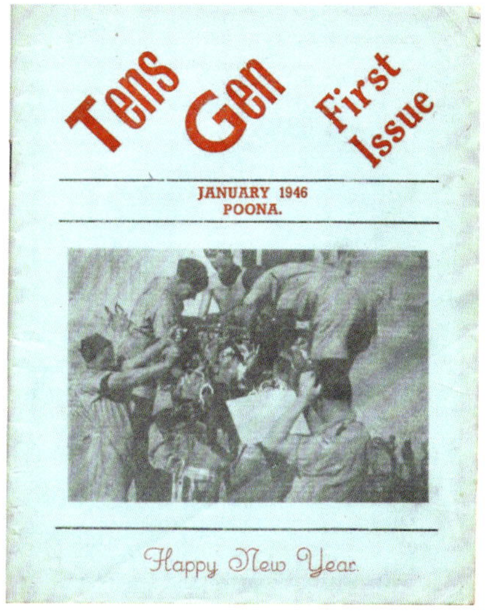

'Tens Gen': the Squadron magazine when based in Poona.

Captain Alfred Broomer's Story

Amongst the Squadron memorabilia there is a reminiscence, apparently written in the 1950s, by Captain Alfred Broomer. He had gone to France in October 1915 as a 2Lt with the Royal Lancashire Regiment, and served in the Line over the winter of that year before requesting a transfer to the RFC in 1916. His words paint a graphic picture of Squadron life in the latter part of that year. After his return to the UK in November, he remained with the RFC/RAF until leaving as an Honorary Captain on 28 April 1919. His story starts whilst on manoeuvres with 56 Division in April 1916:

I received a chit to report to the HQ of the RFC at St Omer. I do not recall any medical examination or colour vision test, although my memory may be at fault. A few days rest there in a small hotel were very welcome, but a dish of snails produced at one meal was certainly not.

By a stroke of good fortune, I was posted for training as an Observer to No 10 Squadron not far from Bethune, and quite near to the trenches where I had lived laborious days. On my first air lift over the static battle front it was a revelation to see how microscopic our infantry sector had been. The front lines were sometimes near together, but often far apart, and No Man's Land was riddled with thousands of pit holes, brim full of mud and water. Spanning back from each front were spidery zigzag communication trenches leading to blistered homes and battered landscape; yet down below – on scorched trees – the birds were gaily singing in spite of the echoing chatter of machine guns and the thunder of shell fire. Not far behind the lines, French women and children were busy working in the fields, under shot and shell, extracting the maximum from their shrinking farmlands, whilst their husbands or fathers were fighting in the French Armies to the South. In the few unbroken farms and houses near to temporary aerodromes the Flying Officers and observers of the RFC were billeted, and where a more substantial home or chateau was available this became the centre of administration and accommodation, as at the Chateau de Werppes, which housed the HQ Flight of No 10 Squadron.1852

When I arrived at C Flight as a Lieutenant of Infantry, I found an alert group of officers, three Captains and many subalterns, but I was impressed by their diversity in dress and nationality. There were English, Irish, Welsh, Scots, Canadian, and Australian and an ebb and flow of personnel like the changing tides. About half were pilots and the other half Flying Officer observers or trainee observers. After a satisfactory period of service as Flying Officer observers, they were posted to Blighty with the chance of training as pilots. I arrived at No 10 on 2 May and left for England on 23 November 1916 after completing 162 flying hours.

The duties of a new entrant were to master the Morse Code, learn how to use a Lewis gun, read a map in order to signal the results of artillery shooting to the batteries, to spot trains and enemy movements, and to help in dropping small 20 lb bombs just over the enemy lines, or 60 lb bombs on raids deeper into enemy territory. After a few busy weeks training, I was Gazetted from the Kings Own Royal Lancs Regiment to the General List as a Flying Officer observer on 26 July 1916, so I flew as a fully fledged observer for exactly four months. My first flight – the most pleasant one – was with Lt C B Bond, who took me over the lines at La Bassée and asked me how I felt. All was well. We donned the old fur-lined flying helmet, goggles, fleece-lined long leather flying coats, and long fleece-lined flying boots having thick red rubber soles. It was draughty in those open cockpits and particularly cold about 12,000 feet. The swish from the slipstream out side the fuselage was almost of gale force. On a bomb raid it was normal practice to assemble at a given rendezvous, say over Lillers, and to climb separately in long gentle spirals to about 12,000 feet beyond which height the nose of the plane fell gently but persistently back as the ceiling of the climb had been reached. Such rendezvous would take a full hour, so the enemy target had to be within the remaining fuel range. We had an escort of FE2Bs which appeared to float in the air as they passed beneath us. After two hours at this height a certain feeling of numbness used to creep up the legs from foot to knee, a certain sign of loss of circulation. Bomb sights were so crude as to be of negligible value, but it was a joy to release the load and the bombs appeared to drop straight down, having the momentum of the plane.

The German gunners generally held their Anti Aircraft fire until they had clearly sighted their plane and then burst HE all around, and fragments often punctured the fabric of the fuselage or wings. This was not a pleasant experience but with evasive flying little serious damage was done. When things 'were hot' the pilot looked grim and his observer felt pale because he was an inactive party. The only sad occasion of a direct hit in my time was when Lt G W Panter, an observer flying on patrol with Lt J E Evans had his arm shot off near the shoulder. Evans kept a cool head and took the plane down near to a field hospital and Panter lived to tell the tale. I remember that flight all too well because if Panter had not been available it would have been my turn to fly with Evans. I never did. Long after the war I saw a photograph of Panter in a national newspaper tobogganing in Switzerland. [The incident referred to occurred on 2 July 1916. Later awarded the MBE, Panter returned to the Army in 1919, advancing to the rank of Major.]

I did most of my flying with each of the three successive Flight Commanders, Capt J T Rodwell, Capt P W G A Harvey and Capt A W C V Parr; a few trips with Lt C B Bond and many a sortie with an Australian Lt W R Snow. He regarded each trip as 'a job of work,' was a qualified engineer and ended his career as a Lt Col with the DSO and MC and had charge of a squadron. Flying hours were normally between dawn and dusk, for good visibility was of prime importance; hence many hours passed away.

Each evening after Mess, the Flight Commander listed flying duties for the next day and you could make your plans accordingly. We indulged in tennis, bumble-puppy, shooting, punch ball and running, but a great deal of time had to be spent in the Mess for those on duty waiting for the weather to clear. In consequence many hours were consumed playing bridge for small stakes, the games usually ending by tossing 'double or quits' as an aid to Mess funds or Mess bills. For those on Dawn Patrol play frequently went on far into the night because it was easier to play cards than to indulge in broken sleep. The Mess was a friendly place for food and drinks, cards and conversation. We had a small bar in one corner, well stocked with sickly grenadines, wines and liqueurs. Pilots returning from home leave generously added to the stock after a visit to Fortnum and Mason's. There was no shortage of good food and plenty of time to enjoy it and we were fortunate to have better than ordinary kitchen staff. On one occasion we had a taste of haggis, thanks to that doughty Scot Lt D S G Turnbull.

After thirty five years it is hard to put into proper perspective the incidents which impressed me most, but I mention one or two from the misty past. On my second flight with Capt Harvey, on a new machine, when something went wrong with the petrol supply and we ditched with a bump in a ploughed field behind our front line. We were cut a bit and bloody but unbowed and lived to fight another day. The plane was shelled to bits before its appointed time. Returning from patrol one afternoon we were coming into land, over the road passing the aerodrome, when a horse drawing a beetroot cart suddenly bolted. On landing we found that the poor horse had been lashed with the aerial wire which the pilot had forgotten to wind in. For the same reason we often found the aerial wire and bob wrapped around both planes – an experience shared by more than one old pilot.

There was then the stir and commotion when the Father of the RFC, 'Boom Trenchard,' paid a visit to No 10. It was like the descent of a God from Parnassus. One evening a bombardment started so Parr and I went up to try to locate the enemy batteries. But all was black and nothing could be seen except flashes of light, some high, some low, on a black ground. When we turned to come home we could just see the La Bassée canal and the welcome oil flares on the

aerodrome. It was our first experience of flying in the dark and 'touchdown' was a great relief. I have memories too of the many hours spent swatting mosquitoes before getting into bed; they were small and they were musical but, by jingo, they could sting.

Finally, I think it should be said that we constantly owed our lives much more to the skill and care of our fitters and mechanics than we did to the lack of skill of German Anti Aircraft gunners. They kept the planes fit to fly, and did a quiet job outside the publicity of war. When I left No 10 somebody gave me an 18-pounder shellcase engraved with the names of the officers of C Flight. I conclude these personal anecdotes with the names it bears. Some were already known and established in other fields, some were on the threshold of achievement, and others failed to escape the dreadful penalty of war. All found a common denominator in the service of the Motherland in the First World War, in the 1st Wing of the 1st Corps of the 1st Army, BEF France.

Capt J T Rodwell
Lt H L Wallis
Lt G W Panter
Capt P W G A Harvey
Lt A Broomer
Lt W R Snow
Lt D S G Turnbull
Capt A W C V Parr
Lt H V Northam
Lt A M Mitchell

Lt C Fairbairn
Lt J E Evans
Lt C B Bond
Lt C W T Clifford
Lt C E W Foster
Lt A P D Hill
Lt Griffin
Lt J B E Langley
Lt F L Royle
Lt C M Crew

A Typical Day on a Bomber Squadron 1943/44

By Doug Evans

Doug volunteered to join the RAF in the summer of 1940 – his brother John had been called-up in the previous year and he, too, went on to become a Halifax pilot but with 158 Squadron.

Doug trained as a Rigger at RAF St Athan, qualified, and was posted to a Training Unit at RAF Kidlington, near Oxford, in January 1941. He was soon bitten by the flying bug, applied for aircrew training, and was accepted. He trained first in the UK, then in the USA and Canada, destined for heavy aircraft. He returned to the UK in January 1943 and, after training on both the Whitley and then conversion to the Halifax at 1658 Heavy Conversion Unit (HCU) where his crew formed before posting to 10 Squadron at RAF Melbourne in August 1943. He completed an Ops tour of 32 missions in May 1944, and was posted back as a Pilot Instructor at 1658 HCU. In April 1945 he was posted to 78 Squadron at RAF Breighton for another Ops tour, and that unit with the others in 4 Group transferred to Transport Command as the war in Europe ended in early May. He then converted to the Dakota and went out to the Middle East before returning to the UK for de-mob in December 1946. Post-war he became involved in civil airline flying, including a spell in the Berlin Airlift, and was eventually Fleet Manager for the British Airways' Tristar fleet.

Recalling as best he could after some 65 years, Doug describes a standard day on ops at RAF Melbourne around 1943/44:

0800: Breakfast in the Sergeants or Officers Mess.

0900: Report to the Squadron on station.

Pilots to the Flight Commander's office. Crew availability for the individual Flight and for the particular day was chalked up on a large notice board in this office, with each crew listed under the pilot's name.

Navigators, Air Gunners, Wireless Operators, Bomb Aimers, Flight Engineers would report to their respective Leader's office. These Leaders were usually men of long previous experience and seniority, having themselves completed tours of operations.

There would generally be informal discussion relating to ongoing operational experience. Sometimes, in the light of particular matters arising, the Flight Commander would hold a more formal discussion, with all aircrew members present. Quite often the Squadron Commander would call for all aircrew to report for a formal meeting or discussion.

1100 approximately, usually round mid-morning:

The Squadron would be advised "Ops ON tonight." This message would be received by Scrambler phone (probably by the Intelligence Branch) from Group HQ (HQ 4 Group at Heslington Hall, York). Group would have previously received its orders direct from HQ Bomber Command at High Wycombe.

Crews would be allocated from each Flight from the Availability list, and the Briefing Time would be passed to all rostered aircrew. Depending on the time of take-off, briefings would be held in mid-late afternoon or early evening.

Crews would try to fit in a visit to their aircraft for a general chat with their ground crew personnel. They would speculate on the 'target for tonight' – often quite accurately as they knew by then the fuel load required. Air Gunners would always proceed to the aircraft to inspect and harmonise their guns (4x .303 guns in both mid-upper and tail turrets).

Briefing – at the appointed time: All aircrew present, seated together at tables crew by crew, with mounting speculation as to the target. The Squadron Commander would open the briefing and the cover on the large wall map would be removed, revealing the target. Section Leaders might be called upon to speak, the Intelligence and Met officers would brief and, if the occasion demanded, the Station Commander or possibly the Base Commander would also speak. Post-briefing, the navigators would be busy for some time calculating their Flight Plans.

The Briefing would always include a Start Engines time at dispersal and, before then, crews would return to the

Mess to relax prior to a pre-flight meal, usually of bacon and eggs. At the appropriate time, they would than walk, dressed for the night's activity in a draughty Halifax, from the Domestic area onto the Station – about a quarter of a mile or so at Melbourne. The first call would be at the Parachute Section to collect their parachutes and a "Pandora's Box' containing a silk escape map of the target area, a compass, some Wakey-Wakey amphetamine pills, and other bits and pieces that could be useful if trying to evade capture. They would then catch a crew bus out to the aircraft dispersal.

On arrival at the aircraft, there would be more words with the ground crew and personal kit would be stowed on board. A cigarette before departure was essential for some, as was a last-minute 'pee', often over the tailwheel!

Start-up was done in absolute radio silence, as was movement onto the perimeter track to taxy towards the runway. The only illumination would be from a few goose-neck flares, so great care was needed in manoeuvering a heavily laden Halifax, sometimes in icy conditions and almost always in darkness. Take-off clearance was given by a green light from the Aircraft Control Pilot (ACP) situated in mobile hut by the end of the runway.

After take-off, course was set for a designated departure point on the coast, often Flamborough Head. (In planning large bomber streams to a target, HQ Bomber Command passed coasting-out and coasting-in times to its Groups and they, in turn, allocated departure times to their squadrons. There was no specified horizontal separation at these crossing points or, indeed, elsewhere.) Altitudes were allocated but only loosely adhered to, particularly when aircraft weight and ambient temperatures limited aircraft performance, as might often happen with the older Halifax MK II.

Crew Procedures: Intercom discussion would be kept to a minimum. To avoid potential collision, vigilance and alertness were maintained throughout a flight, and this was as important over the UK as over the North Sea or enemy territory. The aircraft had a very basic autopilot, but Doug Evans would not use it on operations, considering it too much of a risk should a sudden emergency arise.

On approaching and once over enemy territory, the Wireless Operator would be listening-out and transmitting to Group HQ as instructed (recalls were not unknown). He would also operate any anti-fighter devices installed, and would pass back upper wind velocities calculated by the Navigator. The Bomb Aimer (whose formal title was Air Bomber) could provide considerable assistance to the Navigator by reporting coastal pinpoints, towns and significant ground features, and the Morse idents of any ground beacons seen. Nearing the target, he would set up his bombsight with appropriate altitude and wind velocity information in advance of the bombing run to target, which he would control. After 'Bombs Gone' he would revert to assisting the Navigator for the return to base. The Air Gunners maintained a continuous lookout for enemy fighters and to report any overly close main force aircraft, whilst the Flight Engineer was concerned with monitoring fuel and other systems, and could assist with lookout from the astro-dome, near his position.

All crewembers' concentration had to be maintained throughout the return flight. That might involve another 2 hours flying over enemy-held territory before crossing the coast, where accurate AA fire could always be expected. And after crossing the North Sea and reaching base, crews would join a queue to land. After that, detailed de-briefing by the unit Intelligence Officers would follow accompanied by a welcome mug of tea or coffee with a tot of rum. And then off to bed, to await the next call to go and do it all again.

WW2 Bomber Raid Planning - The Tactics

By Douglas Newham

Squadron Association member, Douglas Newham LVO, DFC qualified as an Observer and joined 10 Squadron as an experienced navigator after tours flying Wellingtons with 156 and 150 Squadrons over Europe and North Africa, a tour at OTU as a Navigation Instructor followed by attendance on the Staff Navigator Course. He was Navigation Leader from 1943 to 1946 at Melbourne and in India, and he recalls:

No-one who was a member of Bomber Command's operational aircrew, or of the station Intelligence staff, will forget the half-million scale map of North-west Europe that covered the back wall of every Briefing Room. The red overlays showed known flak defences and searchlight areas, and attempts were always being made to keep these areas updated from operational reports. In early "Gardening operations" (to the uninitiated, this was dropping sea-mines) we used to try and creep between some of these red-marked areas, always hoping that a flak-battery had not been moved or added to the local defences.

In those same days (or more frequently - nights) our routeing to more distant bombing targets often crossed the English coast at a prominent landmark such as Flamborough Head, The Naze, Bradwell Bay, Dungeness or Beachy Head (the names still linger in the memory), maybe to a landmark on the continental coast such as Cap de la Hague, mouth of the Somme, or Terschelling, and thence almost direct to the target, but dog-legging as necessary to avoid known heavily defended areas. But in the latter days of the war the whole process of route-selection was significantly more sophisticated and, to my mind, has rarely been given the tribute it deserved. Yes! There were occasions when unforeseen circumstances caused "things to go wrong" – the tragic Nuremberg raid of 30 March 1944 was perhaps the worst – but, by and large, the overall tactics were well-thought-out. Most obvious were the choice of route, heights, and timings, and coordination between simultaneous operations on different targets. Additionally the planning involved deployment of numerous electronic-warfare counter-measures and "spoof raids". It was a complex operation, planned in a professional manner, and continuously trying to confuse and out-guess the German defences. At the time, we operational crews were largely unaware of the detail or depth of such planning.

Of primary consideration were the forecast weather conditions for selecting the appropriate target, and for the UK airfields for take-off and landing. The timing was also dictated by the duration of darkness over Europe - shorter during the summer months; and the phase of the moon – brilliant moonlight was advantageous for the German fighters – so operations deep into Germany avoided the long moonlit periods of the full moon, and at other times due attention was paid to the time of moon-rise or moon-set. We did not wish to be illuminated by moonlight against a lower layer of cloud like a train of ants crawling across a white table-cloth! Sometimes a small number of aircraft, deploying large quantities of "window", would simulate a major raid in order to tempt commitment of the opposing force of fighter aircraft towards a potential target, and the major raid would subsequently be directed elsewhere and at a time when the scrambled fighters would be needing to land and refuel before being directed against the main force. Multiple targets with similar timings could also be used to split the fighter defences into two or more individual operations.

The routes we were given were carefully selected to avoid the more heavily defended areas, not only around the major cities and industrial areas, but where practical to avoid the extensive flak and searchlight belts of the Kammhuber Line. The route was also generally defined with several changes of track, so that it gave the defenders no clear indication of the likely target. Where possible it also gave the widest berth from the German fighter assembly beacons. On some occasions the en-route turning points were marked by parachutes flares deployed by pathfinder aircraft. The altitudes we were required to fly generally stepped-up as

our weights decreased from fuel usage. Indeed there was normally a change of altitude part way along each leg of the route. In consequence these changes of track and altitude made it more difficult for the German controllers to position their fighters ahead of the bomber stream, and at the right altitude for interception. Our route and altitudes on the way home from the raid were similarly designed to afford us best protection from fighters and flak.

Nevertheless, it was the use of radar countermeasures that were the final touches to the complexity of planning the bomber raids. The fact that a major raid was imminent was obvious to the Germans from their radio monitoring stations along the coast of Europe. These could detect the transmissions of our H2S sets even whilst we were over England. Even use of H2S on pre-op test flights would alert the Germans to an impending operation. This was one reason why we were later instructed not to use this navigation aid until we were running out of cover of the navigational Gee system. We were again later told to be "abstemious" about use of H2S as some of the German fighters were equipped with Naxos, by which they could home onto an individual aircraft's H2S - even in the darkest of conditions. The British "Mandrell" radar-jamming equipment, in specially deployed aircraft, permitted an impenetrable anti-radar screen to be simulated down the North Sea or Channel, beyond which the German defensive Freya radar could not penetrate. In consequence the German fighter controllers had no indication of the direction from which a bomber raid might appear until it emerged from this screen. This was used to particular advantage if perhaps the first apparent raid might be one of the "spoof" raids of Window-droppers already referred to, and the fighters might be directed towards a non-existent bomber stream, or at least to a subsidiary raid.. Or indeed there might be no raid at all, and the German night fighters might in that case have been alerted or deployed unnecessarily. RAF "Moonshine" equipped aircraft could also give a radar image as if it were a substantial formation. It was a true game of cat and mouse, of hoaxing and feinting in order to confuse and detract from the German defensive efforts.

The complexity of the planning was still not exhausted, for aircraft of 100 Group would be incorporated in the bomber stream with further devices to disrupt the enemy's defences. This would include jammers for use against gun and searchlight-laying radar, and for jamming the radio channels used to issue directional instructions from German controllers to their fighters. Indeed, most of the main-force bombers also tried to drown-out these R/T orders by transmitting on the relevant radio frequencies, the engine noise from microphones within their own engine nacelles. German-speaking operators, in 100 Group aircraft or back in the U.K., were also used to countermand or confuse the controller's instructions, even telling the fighters to return to base because of fog-forecasts, or to proceed to a different city. The enemy tried to identify more easily their legitimate control directions by introducing women radio operators. It is a further tribute to the foresight of our planners that we were ready with German speaking WAAF's to continue the confusion. Yet more planning was involved with the operation of British night-fighters, equipped with not only their own radar, but also able to home onto the radar transmissions of their German counterparts. At a critical time these RAF night fighters would attack the airfields from which the German's operated, or over their assembly beacons.

I recall that on some occasions, at pre-ops briefings, we might be reassured by a scant outline of some of the defensive and diversionary tactics used by the planners and our colleagues in 100 Group on that, or previous raids. Amusingly, on one occasion a German night-fighter pilot had been heard to tell his controller that he had never been so confused and "b....d about" by conflicting instructions in his whole life. But it is only by more thorough study of such books as "The Most Secret War" (by RF Jones), and "Instruments of Darkness" (by A Price) that the contribution of our scientists truly comes to the fore. We are indebted to them and to those from Bomber Command HQ, and 8 and 100 Group HQ's, who did so much of the clever and complex planning which aided our proficiency, and increased our chances of survival. We thank them!

But the above still doesn't do adequate justice to the work of "the planners", at whatever level in Government or Command structure. The strategic choice of potential targets necessitated consideration of what would most help the overall and continually changing process of the war, and rested often at the level of the War Cabinet. The type of bombs to be used and the sequence of use in any one raid was at the tactical level of Bomber Command. Whilst on the Squadrons and Stations there were the innumerable tasks associated with planning the selection of crews and aircraft and the provision and timing of associated facilities. It never stopped! We crews were only at the "sharp end", and all too often unaware and unappreciative of the massive efforts that went on to facilitate our operations.

Douglas Newham LVO DFC
Air Observer (The flying "O") 156, 150 and 10 Squadron
10 Squadron Navigation Leader 1943 -1946

Ex-Sgt Ken Stewart, a 10 Squadron Halifax and later Dakota navigator, was one of the first recipients to receive the Bomber Command Clasp from Prime Minister David Cameron at No 10 Downing Street in March 2013.

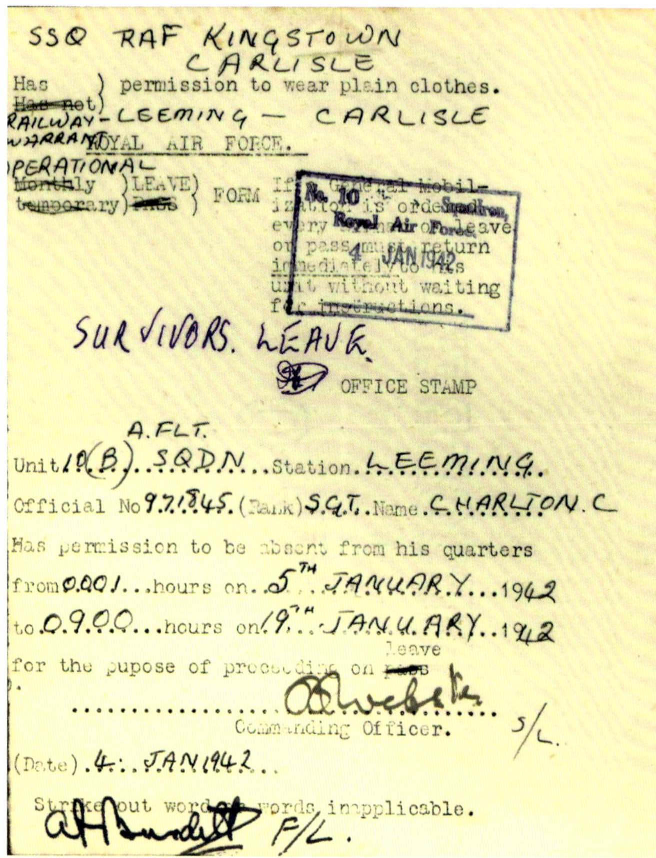

SSQ RAF KINGSTOWN
CARLISLE

Has) permission to wear plain clothes.
~~Has not~~)

RAILWAY- LEEMING — CARLISLE
WARRANT ROYAL AIR FORCE.

OPERATIONAL
~~Monthly~~)LEAVE) FORM
~~temporary~~) ~~PASS~~)

If General Mobilization is ordered every member on leave or pass must return immediately to his unit without waiting for instructions.

No. 10 Squadron Royal Air Force 4 JAN 1942

SURVIVORS. LEAVE.

OFFICE STAMP

A.FLT.
Unit 10(B) SQDN Station LEEMING.
Official No 971845 (Rank) SGT. Name CHARLTON. C
Has permission to be absent from his quarters
from 0001 hours on 5TH JANUARY 1942
to 0900 hours on 19TH JANUARY 1942
for the pupose of proceeding on ~~pass~~ leave.

Commanding Officer. S/L.

(Date) 4 JAN 1942

Strike out word or words inapplicable.

F/L.

FS Whyte's Halifax R9374, was seriously damaged by flak and a Me109, causing 3 engines to fail, after attacking the Scharnost & Gneisenau battleships in Brest on 30 December 1941. Ditching some 80 miles south of the Lizard they were picked up by an Air-sea Rescue launch after 5 hours in a dinghy. The tail-gunner Flt Lt Roach, killed before the ditching could not be retrieved. WOp Sgt Chris Charlton was given 2 weeks survivor's leave.

Wireless Operator Sgt Chris Charlton

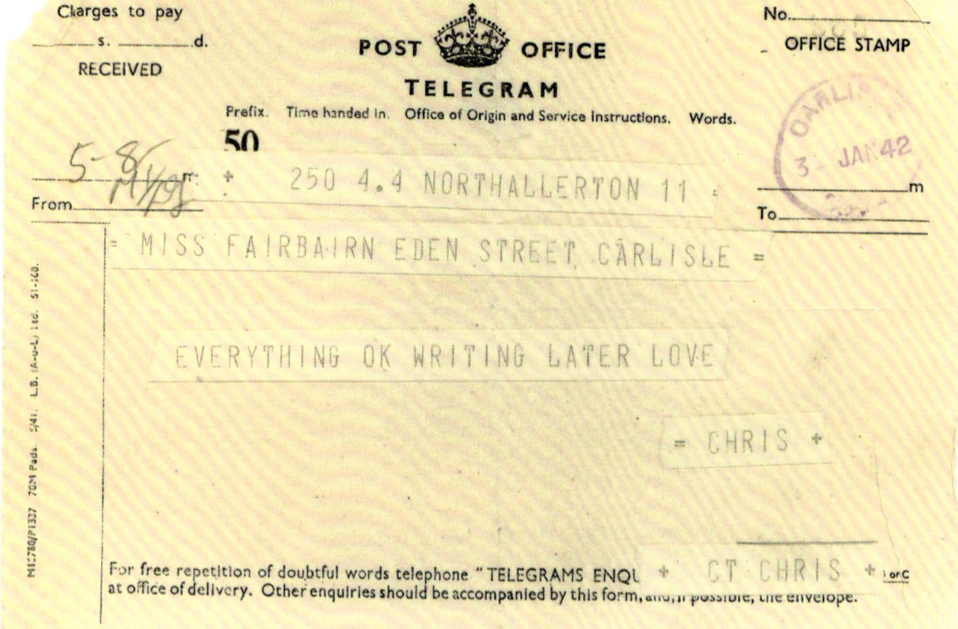

Charges to pay
s. d.
RECEIVED

POST OFFICE
TELEGRAM

No.
OFFICE STAMP

Prefix. Time handed in. Office of Origin and Service Instructions. Words.

50

CARLISLE 3 JAN 42

From
+ 250 4.4 NORTHALLERTON 11 +
To

= MISS FAIRBAIRN EDEN STREET CARLISLE =

EVERYTHING OK WRITING LATER LOVE

= CHRIS +

For free repetition of doubtful words telephone "TELEGRAMS ENQ + CT CHRIS +
at office of delivery. Other enquiries should be accompanied by this form, and, if possible, the envelope.

Telegrams were usually dreaded, yet occasionally brought good news: this time for Sgt Chris Charlton's girlfriend Mary Fairbairn, whom he later married in Carlisle in the spring of 1943.

'King's Wings'. DFC, 1939-45 Star, Air Crew Europe Star, Defence Medal, War Medal (1939-45). The medals of Flt Lt George Cartwright are similar to most bomber crews'.

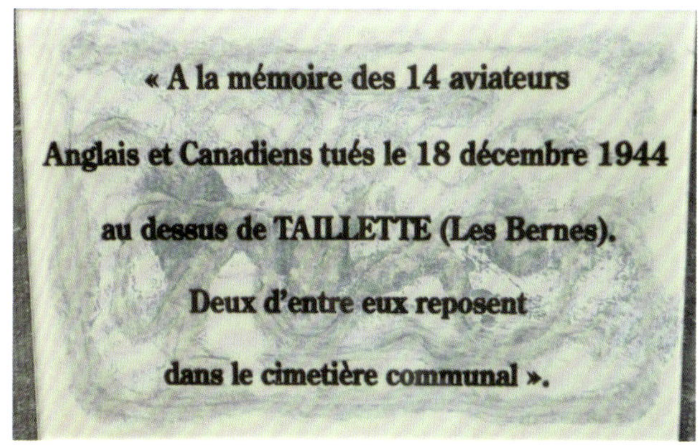

David Mole was but 7 days old when his father was killed in LV 818. After many years of research and administrative effort he succeeded in having this plaque erected in the village where his father is buried in the local cemetery.

A wooden board measuring some 15 x 9 ins which is now owned by the daughter of Flt Lt 'Homer' Lawson, navigator on the crew which flew in Halifax HK 357 ZA-J.

The graves of Fg Off John H. Waldron and Fg Off David J. Mole in Taillette. Never to be forgotten.

A Glimpse Inside The VC 10 VIP Compartment

The detailed planning and engineering work that went into preparing VVIP tasks for State or official visits of HM The Queen, the Prime Minister and senior Cabinet ministers might be thought to ensure that these would run smoothly. Usually they did – but Harold Macmillan's "Events, dear boy, events" would sometimes intervene!

In late January 1975, PM Harold Wilson, with other Ministers, was to make an official visit to President Ford in Washington DC, with a nightstop in Ottawa on the way. Flight 1121 had arrived there under clear skies, but the following day brought snow turning to freezing rain. The departure was late after a prolonged battle with ice that required repeated de-icing sessions but, eventually, the flight to Andrews AFB got underway. The approach there was to become subject to extremely violent windshear, a phenomenon much better appreciated today than then. The captain, Sqn Ldr Harry Liddell later recalled:

"My unease continued as we levelled off, for I noticed that the suburbs beneath us appeared to slide sideways as we passed overhead; evidently the wind at our 3,000ft altitude was unusually strong, yet there was no turbulence whatever. The navigator passed a Doppler drift value that I would have considered absurd had I not the evidence of my own eyes, and the tower came up with a revised surface wind that was now outside limits. I sensed an atmosphere of gloom; surely, after all that had gone before we would not now have to divert to Dulles? True, it offered an into-wind runway (more or less), but I could foresee tomorrow's British tabloids: "RAF incompetence, Harold at wrong airport etc etc", while it was probably best not even to think of possible passenger comments.

So, following the horizon director bars I lined up on the ILS; or better to say, that was my intention. What actually happened was that we settled level but displaced about two hundred yards left of centreline, the flight director giving an on course indication; another new experience for the day, apparently the drift was outside flight system limits. Resignedly I considered the situation; the tower was passing frequent wind updates, all indicating a strong westerly flow with rapid but erratic minor changes, so that there was a possibility that we might be presented with a lull at the crucial moment. Advising accordingly, I continued the approach; the glide slope bar moved down its vertical scale, DME crosschecked OK, further flap lowered with thrust and attitude adjusted to give the correct speed and descent rate, and we started down the last few miles to our arrival----or so I thought. However, my disquiet increased as the approach continued, for drift was such that the distant runway was seen through the clear vision panel rather than the windscreen proper. Moreover, we flying through completely calm air, nary a ripple never mind the bump & thump normal in such conditions, and my unease shaded into something close to fear.

We continued our smooth downward progress and then, passing the 1500 feet level, all hell broke loose. It was as if we had flown into a wall, with an instantaneous change from relative equanimity to sheer terror as the aircraft was racked and shaken by the most severe, sharp-edged turbulence I had ever experienced. Repeatedly we were forced violently into our seats and then against the straps, the instrument panel vibrated into an unreadable blur and, most frightening of all, the airspeed fell off to an alarmingly exiguous value This triggered the stall warning system, causing the control wheel to shake and hammer noisily in my hands, while crockery crashed distantly in the front galley to the accompaniment of a string of oaths. Calling for max continuous thrust I reached my hand toward the throttles, gratefully watching the speed recover as we quit that murderously disturbed air and were thus able to re-continue the approach as before. There was now some slight turbulence and the horrific drift was marginally less, indeed the latest wind was now just within limits. The point was near when I must make the final decision as to whether or not to land, only a short distance to go now.

The decision was made for me when we hit another zone of savagely disturbed air about two miles from touchdown, and again we were pressed rapidly up & down by "g" forces of elemental ferocity; the airspeed fell off the clock once more, the stall warning repeated its urgent, hideous racket and one wing dropped sharply. Delineated by our landing lights, a busy four-lane highway swept beneath the nose and I received a most distinct feeling that control of the aircraft, the situation, everything, was slipping from my grasp. Picking the wing up, I shoved the throttles forward while simultaneously yelling for full power, but the other wing dropped ominously in its turn as the engineer responded to my call; again taking remedial action, I attempted not to become locked into that cycle of aileron over-correction so easily achieved with the VC10's excessively sensitive lateral control. Mercifully the massive thrust from our four Conway engines powered us out of the mouth of disaster by sheer brute force and, watching the airspeed recover once more, I gave silent thanks to Rolls-Royce for our deliverance. Climbing away, we levelled off at the missed approach altitude where a stunned silence was broken by the tower requesting our intentions. Telling them to stand by I checked with the Cabin Supervisor, to be reassured that all was as well as could be expected----no need as yet to break out the smelling salts.

The tower then offered a precision radar approach for landing to the north, sugaring the pill by quoting a wind value that fell within company limits, if only by a minuscule margin. Accepting with a proviso that frequent wind updates were passed on the way down, we circled under the controller's guidance and lined up once more on the distant runway; not so bad this time, the drift angle merely very considerable as opposed to the previous near-impossible. Starting down the slope I mentally added a larger than normal increment to the bugged threshold speed; this time it was a rough ride all the way, but thankfully with none of that appalling, bone-crunching turbulence of minutes before. Nearing decision height our lights picked out the runway surface, conveying an impression that the wind was stronger than the stated value; were we being lured into another trap? Flaring for landing as the ground came closer yet, I called for idle power and pushed full right rudder to align us with the runway, simultaneously applying full left aileron and then easing the control column gently forward; indeed we were right on the limit. The bogies kissed softly, nosewheels following as spoilers were deployed and full reverse thrust applied. A feeling of profound relief swept over me, but it nevertheless remained necessary to hold full rudder and aileron almost to the end of the runway. As we slowed, I found that my right leg had the shakes; this I ascribed to the effort of sustained pushing on the rudder pedal, but it could equally well have been down to fright. Very little was said as we taxied to the ramp."

Less dramatic, but still awkward circumstances arose in early July 1976, when the first full Royal flight for some time was tasked. This was to start from Heathrow and proceed to Bermuda where the Queen was to board HMY Britannia and sail north into the US for a State visit as part of that year's US Bicentennial celebrations, using both the Royal Yacht and a VC10. With Her Majesty on board, the Copilot called on the radio for routine Start and Air Traffic clearances – to be told that there was no Fight Plan on file. Of all the plans filed with them that morning, the British Airways Ops staff had somehow overlooked this one! The captain, Sqn Ldr John Redding immediately intervened to make it clear that there was no way that, with his passenger now on board, he could wait; that he would start and proceed to the runway in use; and that by the time he got there he expected to have a departure clearance of some kind that would permit him to get airborne and sort out the rest of the route along the way – which is what happened. (In passing, the RAF engineering regulations for flights by HM The Queen involved very extensive preparation of the primary aircraft that took some two weeks to complete, with the provision of an in-place departure reserve aircraft that underwent a lesser preparation period. The Captain of The Queen's Flight also insisted that a proving flight be flown into Philadelphia and Newark to give the handling agents there hands-on experience of the VC10 and that was done using a training flight in June. Another trainer in July provided the reserve for the departure from Philadelphia.)

Sir Robert Menzies, Prime Minister of Australia for some 18 years in two terms, died on 15 May 1978. A VIP flight was laid on for a large party led by HRH Prince Charles, and including former Prime Ministers Lord Home and Sir Harold Wilson, who were to attend the funeral in

Melbourne. The flight went via Bahrain and Colombo, where Sqn Ldr Dave Dinmore's crew, waiting for the next Hong Kong slip, picked it up. The Governor, Air Chief Marshal Sir Wallace Kyle, was collected at Perth and the flight to Melbourne completed without incident. Some days later, on the return to Perth, as the aircraft prepared to descend, Air Traffic Control (ATC) advised that the airfield was closed as there was a wild bull loose on the airfield, but that the flight should continue its descent. With the aircraft at 5000 feet, the animal was still loose and the likelihood of a diversion to RAAF Pearce nearby was broached. The aircraft entered a holding pattern and, after some time, was vectored towards Pearce, with ATC arranging a routing that would permit the VIP Reception Party to get there from the international airport by road. Then, as the crew were ready for a final approach, came word: "Good news – we've shot the bugger! Continue your approach to Perth International now." So the aircraft headed back towards Perth, again on a circuitous route to give time for the Reception Party, now on the road, to turn back to greet its passengers. And, sure enough, as the aircraft finally taxied towards the Terminal, the police-escorted convoy could be seen speeding along an approach road, blue/red lights flashing. And other than an awkward approach avoiding thunderstorms on the way into Colombo, the remainder of the flight went as planned!

It ought to be remembered that, whilst the flight crews worked to get aircraft on and off chocks to the second, it was the cabin crews who were responsible for care of the VIPs on board, ensuring that they appeared at the top of the steps smiling, wide awake, just sufficiently well fed, and ready to face the official welcoming party. When a crew was away from standard RAF staging posts or established handling contracts, considerable initiative was needed to ensure that all ran smoothly and to the standard expected. ALM John Simlett first joined 10 Squadron from No 3 Course after a tour on Comets. He returned to 216 Squadron in December 1969, but found himself back at Brize Norton when the Comets were withdrawn from service in 1975. He looks back on some of his experiences in this area:

"VC10 VIP flights were the antithesis of the standard route flying we had experienced on our first tour. Now we relied totally on our initiative as there were no procedures; we were on our own. The difference lay in the type of passengers and the catering required. Generally, the main VIP party comprised four persons who required a five star silver service. Then there were the Private Secretaries, detectives, civil servants – usually a party of twelve – who required First Class tray service, as did the Press Corps, of about forty. Then of course there was the crew of fourteen, including ground engineers and RAF Police. We also provided a comprehensive bar and wine service, which the Press always attempted to drink dry. Without any real refrigeration, even ice-cubes could become a logistical nightmare.

My speciality was the catering, which for each VIP flight took many weeks of planning and ordering. I began by agreeing menus with the VIP party's representative. On a Royal Flight this would involve me having to go to Buckingham Palace. It was a matter of compromise between what they wanted and what I could get at the places we were to visit. This invariably came down to my knowledge because there was no published information available. I had gained the knowledge the hard way as there was no other method. For example, there was no way of getting the best caviar in the world out of the Shah of Iran's private warehouse, unless you knew a certain Ahmid Makhalian; it is illegal to buy trout in New Zealand so consequently there is no way to serve it onboard unless you know a chef who will give it to you for free and offset the charge for it on other items; QANTAS catered us in Australia, but only ANSETT would fly fresh oysters to where you needed them, if you knew who to ask. Every single item had to be onboard when you got airborne as there are no shops up there. Everything had to be planned to finest detail and ordered; from toothpicks to cheeseboards; ice-cubes to Lobster Thermidor; canapés to breakfast; mustard to soda water. All had to be kept at the right temperature by use of ice cubes or dry ice.

Having worked out the requirements for each leg, it became a matter of communicating it to the right person in the right department in the right country of the appropriate airline. Once again there was no guide or procedure and it was long before the internet. One had to write out all the details, as signals, onto the appropriate paperwork for the area using the correct six digit designators and then take them to the Commcen for transmission.

All this relied heavily on the Flight sticking to the agreed itinerary that the menus were planned upon. Any deviation and you could, for example, end up with all the raw ingredients for a cooked breakfast but find it was now lunch that was required. If you had major last minute changes, then not only would you have the next meal as a problem, but the catering you had ordered all around the world might now be compromised. Occasionally this happened ...take for example, Flight Number 1110..."

Flight 1110 in April 1977 was one of a number of flights tasked as David Owen, a new Foreign Secretary, endeavoured to find a political solution to end the standoff with Rhodesia that had existed since the declaration of UDI in 1965. They involved a good deal of shuttle diplomacy around a number of African countries and, for this flight, Mr Owen was joined by Andrew Young, the US Ambassador to the UN. John Simlett continues:

"We flew out of Heathrow on the 10 April 1977 bound for Cairo, and after a night-stop there, flew down to Dar-Es-Salaam in Tanzania. Thereafter we became part of shuttle-diplomacy between Johannesburg in South Africa, Lusaka in Zambia and Salisbury in the former South Rhodesia: Zimbabwe.

The main problem for the crew lay in the need for us to become 'Time Travellers'. For things inside Salisbury were in a very different decade to those outside of Salisbury. This came home to me during a stop there, when I was told that the itinerary was to change. I needed to get an instant signal to British Airways in Johannesburg to change the catering out of there. But British Airways didn't exist in Salisbury and with all the sanctions it was impossible for me to make contact with BA in Johannesburg. I checked the hotel telephone directory and there, to my great surprise, was an advert for British Overseas Airways Corporation. This is what British Airways had been called in a different era. So, armed with the address of BOAC, I boarded a taxi that took me into the city centre.

The BOAC office turned out to be a simple booking office for passenger tickets. A young lady sat behind the counter and after a chat, she said that there was no way she could help me. I asked how she communicated with 'BOAC' outside of Rhodesia. She said by Telex. Couldn't she Telex my catering requirements for me, I asked? I could see she was sympathetic, but she couldn't help as she didn't know how to operate the telex, and further, Heather the Telex operator was out to lunch. She didn't hold out much hope for me with Heather though – she sounded a bit of a dragon. I said I would call back. When I returned I could see Heather was at her desk. I went into full attack mode by going down on one knee and presenting each young lady with a bouquet of flowers and a box of chocolates. Including the time it took for them to stop laughing, it required over an hour for her to Telex my marathon of changes to the catering requirements."

The itinerary continued to Luanda, capital of Angola:

"There had been frantic communications between embassies to allow us to land at Luanda, but permission had still not been received as we approached the border. It seemed that there was a conference of Eastern bloc leaders going on there. The Foreign Office insisted we continue, and so on we went. At the last moment the Angolans relented and we landed in one piece.

The VIP party was only to be allowed into the airport lounge for a short meeting with an official because, just before we landed, a visiting head of state had arrived and he was getting all the attention. So we were not just an embarrassment to the Angolans but also a potential security risk. Things went from bad to worse when one of our crew chiefs gave a set of overalls to a crew chief of an adjacent civilian C130, who had recently had his flying-suit stolen. Within minutes a large squad of armed soldiers marched up. They were positioned around the aircraft and their leader marched up the steps to the centre door of the aircraft. He was very tall and as I'm pretty short I thought it prudent to stop him a few steps below me, so that I could look him in the eye.

'Where do you think you are going?' I asked, in my best Sergeant Major voice.

'I need to speak to the man who gave a parcel to that aircraft,' he replied, pointing.

'Well you can't, he's too busy.'

He took a step up.

'If you go through that door, you are illegally entering British territory and will be arrested!'

He paused, looked down and growled, turned on his heel

and stamped down the steps. He barked an order and all the troops formed up and marched away behind him. I think the two very large RAF Police behind me had more of an influence than I did!

We weren't sorry when the trip was over."

Mention has been made of the degree of engineering preparation required in advance of VVIP tasks. An inspection of the aircraft a day or so before departure was customary and would involve the designated captain, the Station Commander, OC Engineering Wing and others. Dick King recalls one of these occasions, when a senior engineering officer from HQ 38 Group also attended:

"As we entered the aircraft, the Group officer asked who had approved the attachment of a framed picture of a 10 Squadron Crest attached by four screws to the bulkhead on the left by the port mid door. The Brize senior officers looked somewhat embarrassed, confessed that they had no idea, and a junior officer from Engineering Wing piped up that they had been bought at Timms. "What on earth is Timms?" demanded the Gp Eng Off. "Are they official screws? If not take the thing off."

Now at this stage it must be said that all components of aircraft have to be of the highest engineering standards and metallic items are tested in the design phase to confirm that they do not encroach on the aircraft's magnetic signature. Any large metallic object could, it was felt, possibly interfere with an aircraft's compasses. This may or may not explain why defence budgets are so high, but the Timms screws had been purchased for about £2.50 from a local builders' merchants in nearby Brize Norton village, thus saving HMG a fortune. The 38 Group officer insisted that the crest be removed which was solemnly done there and then.

On boarding the aircraft two days later at Heathrow in order to take Mrs Thatcher to Milan for an EEC conference, I smiled as I boarded through the port mid door and saw the 10 Squadron Crest back on the bulkhead."

In 1983, earlier in his 10 Squadron time, Dick had been copilot to captain Taff Bevan for a VIP task taking the Chief of Defence Staff (CDS), Field Marshal Sir Edwin Bramall, on a westabout route to New Zealand:

"After official stops in Washington DC and Belize, we arrived at McLellan Air Force Base near Sacramento, California, at about 11 o'clock local time one morning. By then, CDS had become greatly taken by the service given him by one of our Corporal Stewards, Wendy Griffiths. Now, McLellan was due to be just a tech stop, and we were only scheduled to be there for ninety minutes. No official welcome had therefore been planned or anticipated, and our easy-to-please VIP was quite content to stay on board during the turnround. However, as we taxied in, we could see our planned parking area from some distance away – and lined up in ranks of three with a band was a large welcoming USAF guard of honour. Seeing this, we immediately rang the galley and told them to get 'his lordship' out of his casual gear and into a suit ASAP. We managed to find a longer taxiway route to the apron and executed the slowest taxi-in that a VC10 had ever done. Mclellan Air Traffic Tower was told of our dilemma but the waiting assembly were not; they just thought the British were mad to take so long to arrive. We duly stopped on the apron, the forward door was opened and, just as Sir Edwin, now suitably attired in a grey suit, was due to step outside, Wendy shrieked at him, "Wait a mo!" She dashed back to his clothes area, grabbed his trilby hat and rushed back to the vestibule. Plonking the hat on his head at a jaunty angle she said, "You'll do," at which he descended the steps with the dignity and aplomb befitting his position."

In June 1982, Dick flew his first VVIP trip as a copilot, taking Mrs Thatcher to New York and Washington DC. On the return flight to Heathrow, all was going well:

"From my point of view, as a newly qualified VIP co-pilot, my first trip with a VVIP seemed to have gone OK. The captain, Sqn Ldr Mick Dobson, had flown the first sector to JFK and, to give me the experience, he had allowed me the two last sectors, whilst he worked the radios. He had flown Mrs T a fair number of times prior to this trip. It was a sunny morning as we approached the top of descent point from our cruise altitude of 37,000 feet, at a position just south of Cardiff.

We had travelled a long way in the preceding two days. On these rapid, hard-working trips for our passenger, the aircraft, in its VIP fit, had a bed installed which was situated on the left side in the forward VIP compartment. Mrs T, whose programme for the day required her to go straight

from Heathrow to the House of Commons, was fast asleep in bed. She would be gently woken by the cabin crew for a light breakfast once we had started our descent.

The VC10 had two autopilots, only one of which was engaged at any one time. The autopilots could be locked in the Speed mode, the Height mode or a Mach number mode. We invariably only used the Speed lock in climbs or descents and the Height lock at our cruising altitude. It was both customary and necessary to glimpse a small dial on the facia panel, below and to the left of the co-pilot's flight instruments, before disengaging the Height lock, if it was engaged, before a descent was started. This dial was known as the 'Three Axis Trim Indicator'. In simple terms, with its huge tail-plane, the VC10 was unusual in that the whole tail-plane itself moved to adjust the aircraft's trim when the balance of the control column forces needed to be adjusted. In case the aircraft had become slightly out of trim, the indicator was diligently observed before the Height lock was swapped for the Speed lock at top of descent. At this point engine power would be reduced to idle and, if the speed lock were then engaged, the current speed would be held. Since there was now only idle thrust given by the engines, the aircraft would descend smoothly. That was the theory........

I glanced at the dial, noting that all the indicators were in their correct positions and that the aircraft was in perfect trim; the transition from Height to Speed should therefore be smooth. My left hand flicked the Height Lock switch back to the 'out' vertical position. With a sickening lurch the aircraft bunted upwards. Everything not tied down floated upwards for a second or so as we 'pushed' negative 'g'. Cries of alarm from the other three flight-deck crew members reduced me to an embarrassed frazzle. A rapid re-trim, reinstatement of the Speed Lock, and we descended to an uneventful and slightly more dignified landing in the June sunshine.

On later trips with her, I found that Mrs T always visited the flight-deck and thanked the crew for her flight before she left the aircraft in the UK. That day was no exception. After the engines had shut down the flight-deck door opened and she sat down on the jump seat between the two pilots. "Good Morning", she said to all of us and then, looking at Mick Dobson, the captain, she asked, "To whom do I owe my early-morning wake-up call, Captain?" With the aircraft's negative 'g' bunt earlier, she had apparently floated horizontally upwards off the bed. No sound came from the captain in reply.......... he merely pointed across the centre console at me, cringing in the right-hand seat. She smiled at me. "Well," she said, with a twinkle in her eyes, "at least you can now say, quite truthfully and literally, that you once got the Prime Minister of the United Kingdom out of bed." With that she thanked us for the trip and went on her way."

10 Squadron's UK Bases and Significant Airfields from 1915-2015

Lossiemouth
(1942 Tirpitz Det)
Leeming
Topcliffe
Dishforth
Melbourne
Scampton
Cottesmore
Oakington
Honington
Upper Heyford
Broadwell
Brize Norton
Fairford
Brooklands
Farnborough
Netheravon
Boscombe Down
Ford Junction
St Mawgan

The Battle Honours of No 10 Squadron

Western Front 1915 - 1918
Loos
Somme 1916
Arras
Somme 1918

Channel & North Sea 1940 - 1945
Norway 1940
Ruhr 1940 - 1945
Fortress Europe 1940 - 1944
German Ports 1940 - 1945
Biscay Ports 1940 - 1945
Berlin 1940 - 1945
Invasion Ports 1940
France & Germany 1944 - 1945
Norway 1944
Rhine

Gulf 1991
Iraq 2003

Periods Of 10 Squadron's Active Service

1	January	1915	-	31	December	1919
3	January	1928	-	20	December	1947
4	October	1948	-	20	February	1950 (Berlin Airlift Operations only)
15	January	1953	-	15	January	1957
15	April	1958	-	1	March	1964
1	July	1966	-	4	October	2005
1	July	2011	-		Currently active	(2015)

Commanding Officers Of No 10 Squadron

and Command Start Date

Wg Cdr	J. W. Osborne	Dec	2013
Wg Cdr	D. James	Jul	2011
Wg Cdr	M. A. Smart	Oct	2004
Wg Cdr	A. S. Deas	Feb	2002
Wg Cdr	A. D. Gunby	Jul	1999
Wg Cdr	A. M. Bray	Jan	1997
Wg Cdr	S. Duffel	Jul	1994
Wg Cdr	A. F. Stuart	Dec	1991
Wg Cdr	P.C. Bingham	Jun	1989
Wg Cdr	D. B. Symes	Dec	1986
Wg Cdr	L. J. Marshall	Jun	1984
Wg Cdr	O. G. Bunn	Jun	1981
Wg Cdr	A. M. Wills	Dec	1978
Wg Cdr	R. G. Peters	Feb	1977
Wg Cdr	A. J. Richards	Mar	1975
Wg Cdr	R. I. C. Howden	Jan	1973
Wg Cdr	R. L. Lamb	Mar	1971
Wg Cdr	D. E. B. Dowling	Nov	1968
Wg Cdr	M. G. Beavis	Jul	1966
Wg Cdr	T. C. Gledhill	Jan	1962
Wg Cdr	R. B. Phillips	Feb	1960
Wg Cdr	C. B. Owen	Mar	1958
Sqn Ldr	G. Sproates	Apr	1955
Sqn Ldr	D. R. Howard	Jan	1953
Sqn Ldr	I. D. N. Lawson	Jan	1950
Sqn Ldr	T. F. C. Churcher	Oct	1948
Wg Cdr	A. J. Ogilvie	Oct	1946
Wg Cdr	A. J. Chorlton	Jun	1946
Wg Cdr	A. C. Dowden	Sep	1945
Wg Cdr	U. Y. Shannon	Oct	1944
Wg Cdr	D. S. Radford	Jan	1944
Wg Cdr	J. F. Sutton	Oct	1943
Wg Cdr	D. W. Edmonds	Aug	1943
Wg Cdr	W. Carter	Oct	1942
Wg Cdr	K. Wildey	Jul	1942
Wg Cdr	D. C. T. Bennett	Jun	1942
Wg Cdr	J. B. Tait	***May	1942
Wg Cdr	D. C. T. Bennett	Apr	1942
Wg Cdr	J. A. H. Tuck	Sep	1941
Wg Cdr	V. B. Bennett	Apr	1941
Wg Cdr	N. C. Singer	Jul	1940
Wg Cdr	S. O. Bufton	Apr	1940
Wg Cdr	W. E. Staton	Jun	1938
Wg Cdr	S. Graham	Apr	1937
Wg Cdr	M. B. Frew	Aug	1934
Wg Cdr	C. B. Dalison	Feb	1934
Wg Cdr	H. K. Thorold	Feb	1933
Wg Cdr	P. C. Sherren	Aug	1930
Wg Cdr	A. T. Whitelock	Apr	1929
Wg Cdr	F. L. Robinson	Oct	1928
Wg Cdr	H. R. Bustead	Jan	1928
Maj	K.D.P. Murray	Sep	1917
Maj	G. B. Ward	Dec	1916
Maj	W. G. S. Mitchell	Jun	1916
Maj	U. J. D. Bourke	Mar	1915
Maj	G. S. Shephard	Jan	1915

***In an attack on the German battleship Tirpitz in Foetten fjord, Norway, on 27-28 April 1942, the Officer Commanding 10 Squadron, Wing Commander D.C.T. Bennett, who later formed the Pathfinder Force, was shot down. However, he escaped through Sweden and within five weeks was back in command of the Squadron.

10 Squadron's Crest, Standard and Badges

The original concept of the integral part of the 10 Squadron Crest came from an idea conceived by the Squadron Commander, Wg Cdr A.T. Whitelock in 1929. A medieval arrow was thought to depict a modern bomb and the wings to denote great speed. The Latin motto "Rem Acu Tangere" freely translates as, "Hit the Mark".

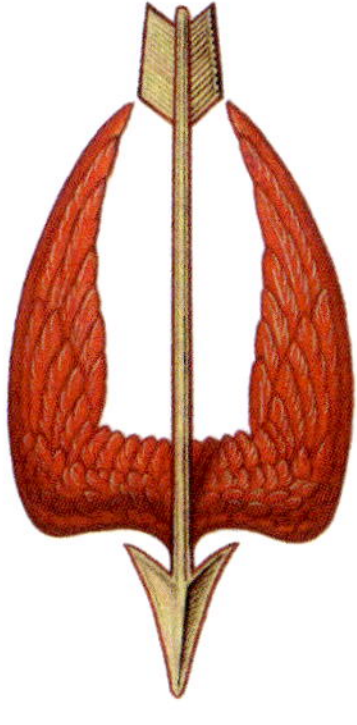

The Squadron Badge which adorned the Heyford aircraft in the 1930s.

A copy of the original design for the Squadron Crest is shown below

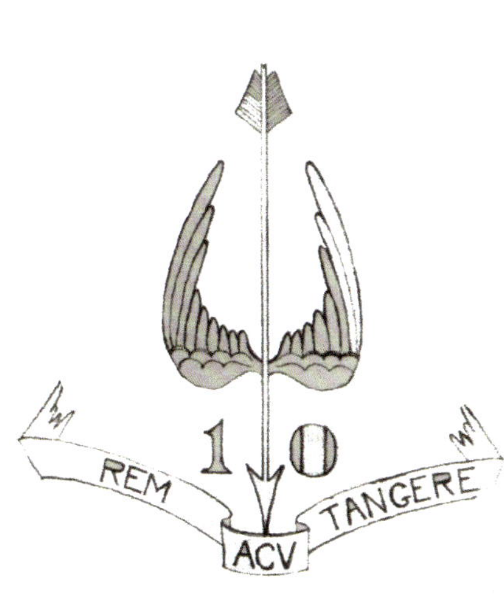

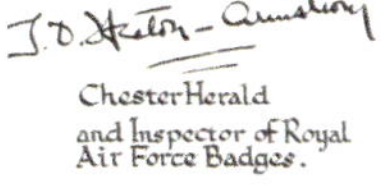

Chester Herald
and Inspector of Royal
Air Force Badges.

College of Arms.
September, 1937.

The original No 10 'Bomber Squadron' crest award that was approved by HM King George VI in 1937 and which now hangs in the RAF Club, London.

The King's Crown Crest – now with 'Bomber Squadron' removed since the Squadron was transferred into Transport Command on VE Day, 1945.

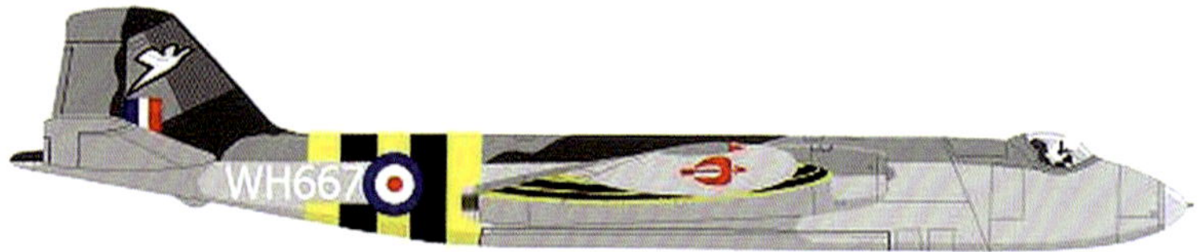

This print shows 10 Sqn, Honington's 'Suez Crisis' markings of 1956, In reality there was a shortage of yellow paint and most Suez aircraft were painted with white stripes instead. WH 667 was later to crash on take-off from Akrotiri , when with 100 Sqn in November 1980.

The Squadron Crest with the Queen's Crown of HM Queen Elizabeth II.

HRH Princess Margaret presents the Squadron with its first Standard on 21 October 1958 at RAF Cottesmore.

The First 10 Squadron Standard. A squadron must have completed 25 years service and earned the appreciation of the reigning monarch for especially outstanding operations before one is awarded.

The 10 Squadron Standard is paraded for the first time

30 September 1988 - Prime Minister Margaret Thatcher presents the Squadron with its second Standard at RAF Brize Norton.

A Voyager aircraft forms the backdrop for the 2015 Standard Presentation Ceremony.

Hats come off for the 3 Cheers.

The new Standard is unfurled for the first time.

At the start of the Centenary Year the third 10 Squadron Standard was awarded by HRH The Princess Royal on 30 January 2015.

HRH Princess Anne presents the new Standard.

Marchpast of the new Standard. Flt Lt Tom Fox is escorted by FS Kenny Murray and FS Mark Newberry with the Standard Warrant Officer Master Aircrew Darren McLean.

Many air force squadrons throughout the world create their own designs for badges that adorn aircrew flying suits. Today's 10 Squadron Voyager crews maintain this practice.

The third 10 Squadron Standard with its eight Battle Honours.

Blazer badge.

The 10 Squadron Association & Memorial

Throughout WW2 and until its transfer to Transport Command on 7 May 1945, 10 Squadron was a bomber squadron of No 4 Group, Bomber Command. In the early 1980s, a number of ex-members of the Group agreed to hold reunions in York every two years. These were initiated and organised by Bill Day, a former Flight Lieutenant and Air Gunnery Leader on 76 Squadron. In setting up a 4 Group Association he sought the administrative advice of a Norman Appleton, a local government work colleague of his wife, Marian. Norman, although never in a 4 Group squadron, had been an air gunner on Lancasters, just as the war ended. The first Reunion was held at Betty's Café in York. Subsequent meetings were held at York Racecourse's Gimcrack Rooms, a venue with the larger space needed to take the increasing numbers of those wishing to attend. They were always well supported by ex-members of 10 Squadron who attended in force, together with wives.

One of these was Doug Dent who had been an engine fitter, working on 10 Squadron's Halifax aircraft at RAF Melbourne. Doug, a most accomplished pianist, had originally planned to make his career in music until WW2 intervened. Nevertheless, having received his Royal School of Music qualification medal, uniquely annotated with King Edward VIII's name on the face in 1936, his talents were later put to good use in the wartime dance band that he formed at RAF Melbourne. Following that initial 4 Group meeting at Betty's Cafe, Doug and his wife Pam had become friends with Norman Appleton. Norman had become a member of the Guild of Aviation Artists and his paintings of Halifax aircraft were always popular when displayed at the Reunions. Other former aircrew colleagues from 10 Squadron's Melbourne days, included Doug Evans, a Halifax captain and Tom Thackray, a Halifax flight engineer. An honoured guest at the September 1981 Reunion, was Air Vice Marshal Don Bennett who had commanded 10 Squadron in 1942, before leaving to set up the Pathfinder Force.

By the late 1990s the number of those attending the 4 Group Association meetings was dwindling on account of the increasing age of its former members and the travelling distances involved, and a final Reunion was held in September 1999. However, during the early period of those reunions, Doug Dent had expressed an interest in forming an Association of past 10 Squadron colleagues and he asked for Norman Appleton's administrative help. Coincidentally, around that time in September 1983, Wg Cdr Gerry Bunn, then Officer Commanding 10 Squadron, based at RAF Brize Norton in Oxfordshire, visited Yorkshire answering an invitation to formally open a lounge dedicated to 10 Squadron, at the Blacksmith's Arms Inn at Seaton Ross, near the former RAF Melbourne airfield.

The pub had been the favourite watering hole of 10 Squadron's air and ground crews during the war, the Sunday scones-with-jam, afternoon teas that were served in the lounge of the pub being a particular favourite. The pub was nicknamed "The Bombers" by servicemen at Melbourne and was known locally as such until it closed its doors as a public house in 1996 to become private office accommodation. Indeed, before its closure, the pub even underwent a formal name change in 1993. Its tenants, Bob and Wendi Johnson had suggested that the name be changed from the Blacksmith's Arms to "The Bombers Inn," and their Tetley Brewery landlords kindly agreed. It was then an obvious choice to ask Norman Appleton to design a new inn sign.

Many original photographs of 'Gus Walker' and the Bomber Boys of Pocklington and Melbourne, adorned the walls of the Bombers and many can now be found in the nearby village of Allerthorpe, at the Plough Inn, where the Association enjoys a Remembrance Sunday Lunch every year, thanks to the generosity of its landlords Daniel and Harriet Morgan. They now carry on the old Bombers tradition of extending a welcome to anyone from 10 Squadron. Right up to the 1990s, monthly reunion lunches, organised by Pam Dent, were enjoyed at the Bombers at Seaton Ross, where

some of the older people in the village could still remember the WW2 airmen billeted in their houses. It was sad day when it served its last pint. Nevertheless, the monthly lunches continued to be held at the Plough until 2011, when the diminishing number of those able to attend resulted in their cessation. After the Bombers closed in 1996, the Inn's sign was purchased at auction by Mike Wood, one of John Rowbottom's employees at Melrose Farm. With a small team of volunteer helpers, known as the Melbourne Tower Preservation Group, Mike had plans to refurbish the old Air Traffic Control Tower at Melbourne and turn it into a 10 Squadron Memorial Museum, containing wartime memorabilia. In mentioning the Bombers, which all ex-Melbourne veterans will insist is essential to the plot, we have digressed somewhat from the main story about the creation of the Association, so let us return to 1983, and Gerry Bunn's visit to the Bombers.

By that year the Squadron had been equipped with the VC10 aircraft at RAF Brize Norton for some sixteen years. The day of The Bombers new lounge opening in September 1983 coincided with a 4 Group Association reunion at York Racecourse, to which Gerry Bunn was also invited and here he issued an invitation to all ex-10 Squadron personnel to visit RAF Brize Norton in the spring of the next year. This was enthusiastically accepted and over forty former Squadron members, plus wives, were guests of the Squadron in April 1984. A Halifax painting by Norman Appleton was presented to the Squadron at an informal dinner at the Bury Barn, Burford, near Brize Norton, on 13 April and it was at this event that Gerry Bunn suggested that a 10 Squadron Association should be formed for both its wartime and more recent peacetime, past members. The seed was sown; Doug Dent's initial enthusiasm was contagious; and on 1 July 1984 the Association was established. A committee was formed comprising Doug Dent as Chairman, his wife Pam as Treasurer, and with Tom Thackray and Doug Evans fulfilling the roles of Newsletter publishing, secretarial and past historical research. Serving as a flight commander on the Squadron at that time, Sqn Ldr Ian Neil agreed to be the first Association's Squadron Liaison Officer. Its President was Gerry Bunn who had recommended the formation and Air Marshal Gus Walker was appointed the Vice-President. He had been Station Commander of nearby RAF Pocklington, and Commander of the group of stations that included Melbourne when the Base concept was introduced in 1943. He went on to be the Senior Air Staff Officer for 4 Group in February 1945 and was immensely popular with all squadrons in the Group. Subsequent to his death in December 1986, his widow, Lady Brenda Walker became the Association's Vice-President until her death in July 2007 when former OC 10 Squadron, Wg Cdr Len Marshall, MBE took over the honour. In December of 1984, the first Association Newsletter was published by Tom Thackray. Its first article, on the first of its thirteen type-written foolscap pages, was an introduction by Gerry Bunn which began with the poem by John Gillespie Magee, "High Flight, " well-known to all flyers. It was a fitting way to begin reuniting past members of Shiny Ten.

Doug Dent's ambitions for the new Association now knew no bounds. Just three months after the Association formed in July 1984, he asked Norman Appleton if he would consider designing a Memorial for 10 Squadron to be situated at their former Melbourne base. Norman's account follows:

THE 10 SQUADRON MEMORIAL DESIGN

In my experience the people who usually design war memorials are either monumental masons – for the smaller type – or architects for those larger edifices which are the centre of activities in thousands of British villages and towns on the 11 November every year, most of these memorials having been constructed shortly after WW1 and of mainly Victorian style in design. Imagine my pleasure, therefore when in October 1984, as an Artist, I was invited by the 10 Squadron Association to design their memorial, which now stands at Melrose Farm, at the entrance to the former WW2 10 Squadron Airfield at Melbourne, 9 miles east of York. I was delighted to accept the invitation without reservation and without fee.

The mandate I was given by the Association's Chairman and Committee was in some ways quite tight and thereby a terrific challenge. The basic elements of the design were determined by a number of unusual and important factors, the most exciting of these being that it should reflect not only the honour and glory of those members of the Squadron who gave their lives in WW2 but also to commemorate the very close bond-ship between 10 Squadron's personnel and the residents of Seaton Ross and Melbourne and the willing hospitality offered by the two village communities during the wartime period, when 10 Squadron were based there from August 1942 to May 1945. It was clear that the Association wished the design to retain the dignity of a War Memorial without having the appearance of a tombstone and it was therefore obvious that a totally different approach to the design was necessary. Another unusual factor having a major influence on the design was the intention to make use of bricks taken from some of the derelict RAF buildings on the old Melbourne Airfield. To me this was another fascinating element, which would allow total freedom of design within the constraints of size and cost. The size would be partly affected by the planning considerations and the location of the site, which provisionally had been earmarked at the eastern side of the camp entrance. However when I showed my draft ideas to John Rowbottom, the owner of the airfield, the unsuitability of that site was obvious and John, showing typical local kindness and generosity, readily agreed to offer the westerly corner site, even though it was in fact part of his front lawn.

The cost of the Memorial was to be set at a maximum of £700 which was the amount that had been raised from Association members. The basic concepts of the design therefore had to reflect all the elements referred to above and depict, visually, a direct and immediate reference to (a) 'Shiny Ten', the nickname by which 10 Squadron was known, and (b) the RAF, so that present and future generations, upon seeing the memorial for the first time, would be left in no doubt, even at first glance, that the structure related to 10 Squadron, Royal Air Force. In addition, the structure had to be easy to maintain and also had to be vandal-proof, bearing in mind the semi-remote site, a matter referred to in Doug Dent's letter to me dated 15 November 1984.

The memorial is 7 feet high, 7 feet long, being 5 feet wide at the base and 2 feet at the top. The basic shape is a slim version of the balloon sheds used by the Royal Flying Corps from as early as 1910, a direct reference to the early heritage of 10 Squadron, originally formed on 1 January 1915. The front face of the structure is moulded in brick to form the number Ten, in relief. This was to allow recesses for placing flowers and the four-foot high numerals One and Zero are made from stainless steel to accentuate the 'shininess' of Shiny Ten. Doug Dent, the then Chairman of the Association was instrumental in obtaining the steel numerals and assisted in solving many other problems during construction work. The Royal Air Force element is represented by a Type 'A' Roundel inset centrally on the lower part of the front side, and the structure is topped with a shallow ridged coping stone. You can imagine my relief when the final design was approved without amendment by all the relevant authorities: the Squadron Association's Chairman and Committee, John Rowbottom, the site owner, the Parish Councils of Melbourne and Seaton Ross, the Local Planning Committee and the Manpower Services Commission (MSC) Community Programme Agency. At the time the design was being considered, the Government's MSC Community Programme scheme was just getting underway, and they were keen to take on new and challenging projects to prove the effectiveness of the scheme. In the interests of the financial constraints of our project, it was vital that we try and persuade the MSC to take on the construction of the memorial. When the scheme was presented to them, we were very pleased that they readily accepted the plan without alteration and designated the Humberside Community Programme Volunteer Group to carry out the work, which they willingly accepted. Mention must be made of the co-operation and assistance received from Mr T. G. Wilson, Clerk to Melbourne Parish Council who was responsible for obtaining all the relevant permits and who helped with the supervision of the construction work, John Rowbottom, who was a stalwart in smoothing things over behind the scenes and the Humberside Community Programme Volunteers who did most of the physical work on site.

May I say that I am greatly honoured to have designed 10 Squadron's Memorial and I can state without equivocation that its dedication on 15 September 1985 was one of the proudest events of my life.

Norman Appleton, GAvA

By the time the first Newsletter was published in December 1984, plans were well in hand for the building of the Memorial. Donations were received from far afield. By the early autumn of 1985 it was ready for its official unveiling and dedication. And so, on Sunday 15 September, in a formal military parade, on a rural country lane by a pig farm in East Yorkshire, the Officer Commanding No 10 Squadron, Wg Cdr Len Marshall, proudly unveiled the Memorial. The afternoon's honoured guest, Air Marshal Sir Augustus (Gus) Walker, GCB, CBE, DSO, DFC, AFC, took the salute as the Association's past Squadron members, serving personnel of 10 Squadron and cadets from No 110 Squadron (City of York) Air Training Corps marched past the Memorial. A moment after the Memorial's unveiling, a Squadron VC10 aircraft, XV104, slowly overflew the Memorial in salute, flown by Flt Lt Dick King and crew.

The 10 Squadron Association continued to flourish thanks to the enthusiasm of its founding members. Reunions were held initially in York each September, to coincide with the period of the bi-annual 4 Group Reunions. They then took place in May each year and were invariably held in the Oxfordshire area since their highlight was a flight in a 10 Squadron aircraft, if weather and service commitments allowed. One such reunion was held in the Witney Town Hall, with the local Town Band providing the musical accompaniment. Another, during the First Gulf War in 1991, was held at Stratford on Avon. Brize Norton reunions comprised a formal dinner in the Officers' Mess on the first day followed, on the subsequent day, by a planned flight in a VC10 or other day-time activity. A less-formal, buffet meal with dancing afterwards occurred later that evening in the Sergeants' Mess. This format remains today, with the organisation of the events being coordinated by the current Secretary and serving Association Liaison Officer.

The other important annual event is the Remembrance Day commemoration service is held at the Memorial at the entrance to Melrose Farm. Former 10 Squadron members and relatives attend and the Association Standard is paraded, whenever possible together with the Squadron Standard, escorted by a Colour Party. Those who gave their lives and the generosity of the local people who had helped the Squadron members in those far-off, grim days are always remembered. (The Association Standard was designed in 1986 and manufactured by the Northamptonshire firm of Zephyr Flags & Banners. The blue silk banner depicts the winged arrow motif of 10 Squadron together with four past aircraft flown by the Squadron, and it was dedicated at St Christopher's Church, RAF Brize Norton on 1 May 1987.)

Due to ill health Doug Dent relinquished his Chairmanship of the Association in November 2003 and his place was taken by David Mole. David was not an ex-RAF man, but a sea-going Chief Engineer for the Shell Oil Company. He may have seemed an unlikely person to fulfil this role, but he is however, the son of a former 10 Squadron WW2 Halifax WOP/Air Gunner, Fg Off Douglas John Mole. His aircraft, Halifax LV 818, had collided in mid-air with a Royal Canadian Air Force (RCAF) 432 Squadron Halifax, on a bombing mission to Duisburg on 18 December 1944. David was just seven days old and, later in life, became determined to research his father's last mission and to trace the relatives of those other crew members who had lost their lives near the village of Tailette, French Ardennes, where Fg Off Mole is buried in the village churchyard, alongside his navigator, Fg Off John Waldron. David's research into his father's last flight had lead him to the 10 Squadron Association via contact with an ex-10 Squadron navigator Ray Thomas.

On becoming Chairman of the Association, David was immediately concerned for the future of the organisation. Its membership had been limited to past wartime colleagues, who were not getting any younger. There was a real likelihood that Association membership would not be sustained, but David initially elected, in his own words, "to steer a steady course." However, at the 2004 Reunion, he somewhat shocked the now-ageing members by telling them that since the Association membership was almost totally WW2 orientated, that membership was getting older and also sadly diminishing. He went on to suggest that the Association must attract younger members from, not only those who had served with the Squadron in WW2, but also from those, like himself, who were the descendants of past Squadron members. In addition it was imperative to encourage those of more recent Squadron service to join. By the same token it was vital to generate some new blood on the Association Committee, if only to relieve the burden on those that had borne it for so long. Thankfully the wise old hands could see the value in David's comments and the message was heeded. The Association turned the corner towards survival and longevity.

Apart from arranging the Reunions, David wanted to add more structure to the Association and so the duties of Secretary, Treasurer and Newsletter Editor became more clear-cut. It is worth bringing back to mind, that during all the years from 1984, Tom Thackray had been producing and publishing the Association Newsletter on a typewriter and that much of the secretarial work had been carried out by Doug Evans.

David arranged an overseas visit in 2004 for members to the Inauguration of the Airforces' Memorial at St Omer, Northern France. (10 Squadron passed through the RFC HQ there in July 1915 en-route to its operational location.) David recalls, "This visit was to mark the 90 years of Airforce presence at St Omer – not of the RAF, which of course was formed near the end of the Great War on 1 April 1918. In fact the oldest attendee was 108 year-old Henry Allingham, who until his death in July 2009 was the longest surviving serviceman who had seen active WW1 service and also the last surviving founding member of the RAF. He told me that he had served there with the Royal Naval Air Service (RNAS). I told him that I felt a bit of a fraud, being there, since I was a sailor. He gripped my hand and said, 'So am I, David.'"

It was just a year later when David himself was taken ill and, after relying heavily on others at the end of 2004, he was forced to hand over the chairmanship of the Association to Michael O'Donovan in the June of 2005. (Doug Dent, the founding Chairman of the Association, sadly died in Leeds on 18 March 2005.) Michael O'Donovan was a former 10 Squadron VC10 captain who had left the RAF in 1987 and sadly, his first task in his new role on becoming Chairman in 2005 was to organise the Association's participation in the forthcoming 10 Squadron Disbandment arrangements which occurred in October of that year. However, the outgoing OC 10 Squadron, Wg Cdr Mike Smart then moved across the airfield to become the CO of the other VC10 squadron at Brize Norton, No 101 Squadron. The Association could not have wished for a better ally. A precedent was set and the Association would have a sponsor in 101 Squadron whilst its parent squadron was in disbandment. Having a current, serving officer in the person of Flt Lt Steve Sansford, a navigator on 101 Squadron, certainly helped as well as he carried out the twin duties of Treasurer and Squadron Liaison Officer. It was not expected at the time that this period of disbandment would last for nearly six years, yet Steve, flying on the remaining RAF VC10s in the transport/tanking roles, did sterling work in enabling the Association to retain its links by encouraging his squadron to host the Association's annual Reunions at Brize Norton. The Association owes a substantial debt of gratitude to 101 Squadron for its assistance over those interim years, and was pleased to hear that 101 Squadron would immediately join 10 Squadron after retiring its VC10s in 2013 in the joint operation of the new Voyager aircraft.

Retired ex-10 Squadron Air Loadmaster, Peter Wentworth took over as Secretary in 2007 and diligently executed this role until he passed over the reigns to another ex-ALM, Sandy Barnes/Butler in 2013. With Mike O'Donovan, he was active in keeping the Association relevant as the Association continued to deal with its ageing membership and with the need to come to grips with 21st century. In 2009, a website www.10sqnass.co.uk was set up and from embryo beginnings it gradually grew to be a medium for contacting members all over the world, historical record storage for the Squadron Operations Records Books, a variety of photographs and anecdotal stories and history. Managed by former VC10 captain Dick King, on the Squadron in the 1980s, the website now commands the No 1 spot in the Google search engine for 10 Squadron. In addition, and after considerable discussion, it was finally decided to discontinue the posted, printed version of the Newsletter after the 50th Anniversary issue in 2013. Printing and postage costs had become prohibitive, the Newsletter went online, and all past Newsletters were made available on the website. A Squadron history project was started – and soon overtaken by the preparation of this volume after a presentation at the 2012 Reunion!

Wing Commander Mike Westwood OBE took over as Chairman in 2014. A Squadron Flight Commander from the

late 1970s, he is still an active pilot with the British Parachute Association, recently received a Royal Aero Club award for his work in that field, and has established good contacts on the Association's behalf across RAF Brize Norton.

AIMS OF THE 10 SQUADRON ASSOCIATION

To preserve the spirit of comradeship founded during RAF service with 10 Squadron

To encourage and sustain a special relationship between 10 Squadron and its Association

To provide a focus for maintaining contact with former Squadron colleagues, by encouraging attendance at the Annual Reunion and Remembrance Day Service at the 10 Squadron Memorial

To encourage the collection and preservation of documents, photographs, anecdotes and memorabilia with which the Squadron's history may be enhanced and preserved

Wartime CO AVM & Mrs Don Bennett, with Bill & Marian Day at a 4 Gp Reunion in September 1981. The demise of the 4 Gp Assoc, formed by Bill Day, later inspired the formation of the 10 Sqn Assoc.

The Blacksmith's ('Bombers') Arms at Seaton Ross where Melbourne personnel were always made welcome during WW2.

10 Sqn participants in a 4 Gp Reunion visit their old wartime local in Seaton Ross in the early 1980s.

As then OC10, Wg Cdr Gerry Bunn opens 'The 10 Sqn Lounge' at the Blacksmith's Arms in September 1983, with the pub's tenants Bob and Wendi Johnson. This function gave rise to the Association's formation.

Officially renamed as "The Bombers Inn" in 1993, the new sign was designed by Norman Appleton, architect of the 10 Sqn War Memorial. When the pub closed in 1998 the sign was taken to the renovated Melbourne Control Tower at Melrose Farm.

Founder Chairman Doug Dent & Norman Appleton, artist and architect.

A rare graduation medal issued in 1936 by the Royal Schools of Music to Chairman Doug Dent. Foregoing a musical career, when WW2 broke out, he became an engine fitter at Melbourne. Doug's medal was one minted with the RSM patron being King Edward VIII.

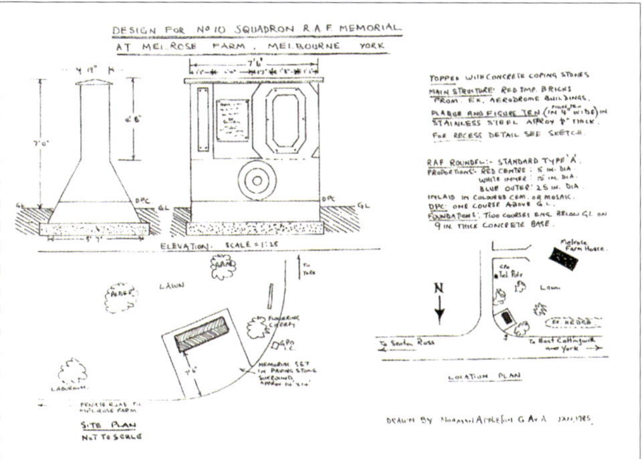

Norman Appleton's original 10 Sqn Memorial Design, to be made from bricks, re-used from RAF Melbourne's derelict WW2 buildings.

Dedication of the Memorial on 15 September 1985. L– Rt : Doug Dent Chairman, 10 Sqn Assoc - Wg Cdr Gerry Bunn, President 10 Sqn Assoc – Rev Jack Armstrong, Rural Dean of York/Chaplain to the ATC - ACM Sir Gus Walker - ADC to the ACM - Wg Cdr Len Marshall, OC10 Sqn

Air Chief Marshal Sir 'Gus' Walker shows his respects with his well-known left-handed salute at the 10 Sqn War Memorial unveiling on 15 September 1985.

Flypast of the VC10 at the Memorial Unveiling: XV 104 flown by Flt Lt Dick King & crew: Flt Lt John Foster - co, Flt Lt Stan Mathews -nav, Flt Lt Paul Woodman – eng.

Marchpast by the 10 Squadron WW2 veterans after the Memorial unveiling.

The 10 Squadron War Memorial on Remembrance Sunday.

The stainless-steel inscription

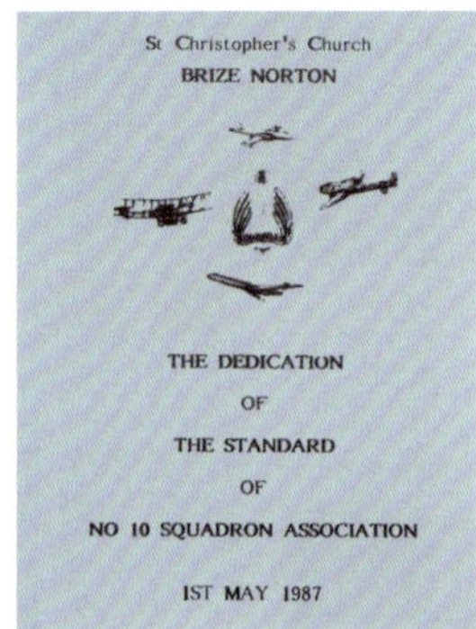

St Christopher's Church
BRIZE NORTON

THE DEDICATION
OF
THE STANDARD
OF
NO 10 SQUADRON ASSOCIATION

1ST MAY 1987

The Association Standard was dedicated on 1 May 1987 in the RAF Brize Norton Station Church.

John Rowbottom of Melrose Farm who generously gave up part of his garden for the War Memorial site and has kindly maintained it ever since.

The Old Melbourne Watch Tower

The Melrose Farm enthusiasts who restored the Tower: l-rt: David Hogyard, Gary Wheeler and Keith Wheeler. (missing from the image is Mike Wood, the driving force behind the Tower refurbishment)

The Tower restored

The original Association Committee WW2 veterans and their wives at a Reunion at Brize Norton in the late 1980s. l-r Tom & Dorothy Thackray, Pam Dent, Dorothy & Doug Evans and Doug Dent end rt. 2nd & 3rd from the right are Vincent and Ghislaine Wuyts-Dennis, two Belgian Resistance members who were involved in helping evaders to escape in WW2: one of whom was Doug Evans brother John, also a Halifax pilot on 158 Sqn who was shot down in 1944.

British Air Services Memorial at St Omer

In September 2004 Association WW2 veterans and ladies attended the unveiling of the Britsh Air Services Memorial at St Omer. 10 Sqn had arrived here in 1915 before moving on to Chocques. St Omer was also the RFC HQ.

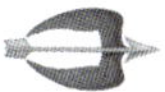

Founder Assoc member Tom Thackray carries the Assoc Standard alongside the 10 Sqn Standard at a Melbourne Remembrance Day Service. His escorts were Peter Jarman(left) and Des Moss (rt). The 10 Sqn Std Bearer was Flt Lt Kim Parkin.

On Remembrance Sunday every year former Sqn members are not forgotten at the 10 Sqn War Memorial.

Wg Cdr Jamie Osborne lays a wreath in November 2014 whilst former 1960's 10 Sqn engine fitter Rev Ian Cooke looks on with Assoc Chairman Mike Westwood.

The 2014 Squadron Standard Party parade for last time at the War Memorial before a new standard was presented in January 2015 by HRH The Princess Royal.

The Remembrance Day service with both the Squadron and Association Standards.

Coffee in the old Melbourne Watch Tower after a Remembrance Sunday service.

The Plough at Allerthorpe where the Remembrance Day lunches take place.

2013 Reunion Members assembled in the Hub at Brize Norton alongside a Voyager.

50 Issues of The Association Newsletter, compiled by former Halifax flight engineer Tom Thackray, were printed and sent to members all over the world. In 2014 the high costs of printing and postage now favoured the use of the Association website and the Newsletters were discontinued.

WW2 aircrew and Founder members of the Association meet a new generation of 10 Sqn at RAF Brize Norton in May 2013. Fred Tiller, Rear Gunner and Doug Evans, Pilot with SACW Aimee Stewart, Voyager Cabin crew.

Remembrance Sunday lunch; Ken Beard, (rt) a Halifax rear gunner who completed a full tour of Ops on 6 June 1944 with Harold Dummer, (left) whose brother, Sgt C.A.R. Dummer was a Halifax mid upper gunner killed in a crash just after takeoff on 16 June 1944 in the same aircraft LV 825 'G – George'.

***Right:** Association members in the 10 Squadron HQ peruse the photo gallery.*

The highlight of the 2015 Centennial Reunion for Association members was the chance to fly on a refuelling flight in a Voyager. Two 11 Squadron Typhoons are seen here after taking on fuel over the southwest of the UK.

THE 10 SQUADRON ASSOCIATION PRESIDENT AND CHAIRMEN PAST AND PRESENT

Gp Cpt Gerry Bunn CBE

Doug Dent

David Mole

Mike O'Donovan

Mike Westwood

'FROM BROOKLANDS TO BRIZE' – THE AUTHORS

Ian Macmillan

Dick King

Art Gallery

Mandy Shepherd:

The eminent wildlife artist Mandy Shepherd kindly painted the unique cover picture specifically for this book, whilst also allowing us to print an image of her painting below. It was commissioned to represent the 39 years of service which the VC10 aircraft had successfully achieved in its various roles before its withdrawl from RAF service in 2013. The original hangs in the 10 Squadron headquarters at RAF Brize Norton.

VC10 – "Shutdown Checks Complete"

David Shepherd:

Mandy's father, David is one of the world's leading wildlife and steam engine artists who is also renowned for his aircraft paintings, many of which hang in the RAF Club, Piccadilly, London. He has kindly allowed us to reproduce his painting of a VC10 at El Adem, Libya. Please support his David Shepherd Wildlife Foundation via www.davidshepherd.org/

El Adem, Libya

Penelope Douglas:

Penelope is a well-known aviation and military artist whose realistic paintings grace many RAF messes. Seen below is the image of a painting commissioned by Wg Cdr Gerry Bunn during the Falklands Conflict of 1982, whilst he was 10 Squadron's Commanding Officer. It depicts a Red Cross painted VC10 during the transit turnround at Ascension Island whilst flying wounded servicemen home to the UK from Montevideo, Uruguay.

'Home from the Conflict'

To form a collection of the three aircraft types flown at RAF Brize Norton in the 1970s Penelope was commissioned by the Brize Norton Officers' Mess to paint the VC10 taking off from Washington on its regular weekly scheduled flight.

'Washington-Dulles Departure'

Penelope hands over her third painting to the Officers' Mess at Brize Norton in January 1977. *The other two paintings depicted Belfast and Britannia aircraft.- from left, Sqn Ldr Sid Adcock 10 Sqn OC A Flt, the then Officers' Mess PMC, and on the right, author Ian Macmillan who now writes, "Whenever I'm back at Brize and in the Mess, I tend to give that painting a mildly proprietorial pat!"*

Bob Murray:

Bob is a self-taught artist who works mainly in watercolours and oils. Having joined the RAF in 1946 as a Halton Apprentice, he passed out as an engine fitter and was subsequently selected for aircrew training as a flight engineer flying on Beverleys and Britannias. Leaving the Air Force in 1975 he continued to fly in civil aviation as a flight engineer on Britannias, DC8s, and Tri-Stars before retiring from flying in 1986. He then devoted his time to painting, specialising in aviation and motoring subjects. Initially many of his paintings were sold by auction houses at their specialist car sales but now Bob's work is bought by collectors in the USA, Europe and Australia. Bob was commissioned by the 10 Squadron Association for the painting below which was presented to the Squadron in May 2015 at the Association's Centenary Reunion Dinner. It depicts all the aircraft types flown by 10 Squadron since its formation on 1 January 1915.

The Association Gift to the Squadron on the occasion of the Centenary

Norman Appleton:

For his work as architect and designer of the 10 Squadron War Memorial, built in 1985 at the entrance to the former RAF Melbourne, East Yorkshire, Norman was made an Honourary Life Member of the 10 Squadron Association. He is also a member of the Guild of Aviation Artists and has continued to support the Association by kindly providing the sketches of the Squadron's aircraft which head each chapter in this book. A selection of his paintings' images and sketches are shown below all of which depict 10 Squadron aircraft although Norman himself was in fact a Lancaster air gunner before serving his career in local government after the war.

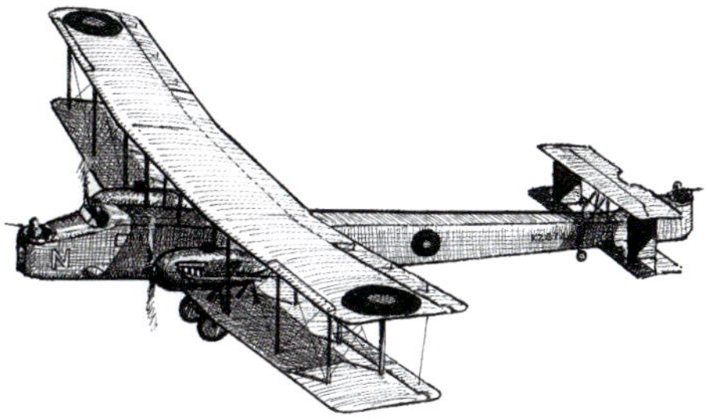

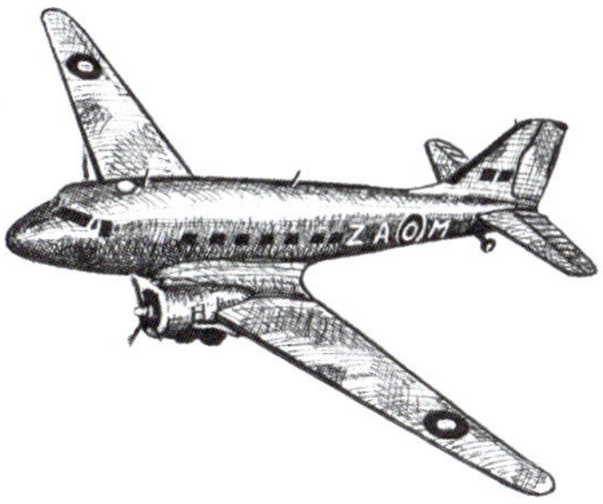

The First Berlin Raid, 1-2 October 1939

10 Squadron owns two distinctive but very different commemorations of the first Berlin raid in October 1939. A memorial plaque for the missing crew was commissioned by Wg Cdr Staton from the Yorkshire workshop of a celebrated wood-carver, 'Mouse Man' Robert Thompson. Although the Squadron was by then at RAF Leeming, on 29 September 1940, the nearest Sunday to the first anniversary, the plaque was dedicated in the station church at Dishforth. It hangs today in the Squadron offices at RAF Brize Norton, with Thompson's distinctive mark, a small mouse, clearly visible above the top right corner. The company carried out a number of wartime commissions from RAF stations and still retains the softwood carving from which the Whitley aircraft on the plaque was developed. These first flights over Berlin were also commemorated by the noted artist and designer, Rex Whistler, who published a drawing in 'Illustrated' magazine in December 1939. In the baroque style and entitled "Flying visit of Truth to Berlin," it showed a winged Britannia sending cherubs to dispense leaflets over a street-map of Berlin, with Hitler and senior Nazis cowering beneath. Wg Cdr Staton approached Whistler to ask if he could buy the original, whereupon Whistler reworked the drawing to change the title to "Flying visit of No 10 Squadron to Berlin" and to add the Squadron badge on Britannia's shield. Wg Cdr Staton was presented with a signed revised version (sold at auction in 2010 for £5200). Smaller unsigned copies were made for each remaining Captain on the raid and one is now held by the Squadron.

Right: In August 2013 with Flt Lt Steve Sansford, 90-year old Hubert Hill and his family see again the plaque that marks the death of his brother, Alfred F. Hill, a gunner in the first 10 Sqn crew lost during WW2. Hubert was to pass away shortly afterwards.

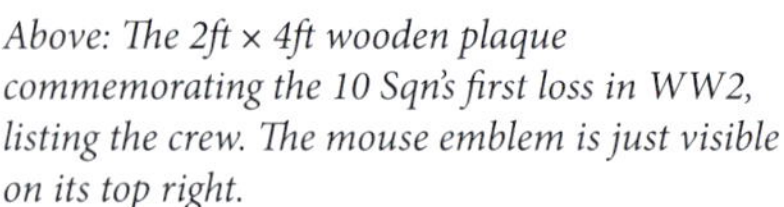

Above: The 2ft × 4ft wooden plaque commemorating the 10 Sqn's first loss in WW2, listing the crew. The mouse emblem is just visible on its top right.

Right: The Rex Whistler print commemorating the first Berlin raid of WW2, modified by him to give the Squadron credit.

WORLD WAR 1 DEATHS IN ACTION AND IN SERVICE

1915

21 September	2Lt S W Caws (P)	
7 November	Lt O V Le Bas (P)	Capt T D Adams (O)

1916

26 April	2Lt J Milner (O)	
22 August	2Lt H C Marnham (P)	Lt C L Tetlow (O)
22 October	2Lt J A Simpson (P – shot down 21 Oct)	
11 December	2Lt G W Dampier (P)	2Lt H C Barr (O)

1917

22 January	2Lt S W Woodley (P)	2Lt W Kellett (O)
10 February	Lt W A Porkess (P)	2Lt E Roberts (O)
18 March	Sgt H La Place (O)	
1 May	2Lt A W Watson (P)	Lt J C Hartney (O)
5 May	Capt H C Lomer (P)	Lt C T Bruce (O)
6/7 May	2Lt R P O Weekes (P)	
24 June	2Lt J W Eyton-Lloyd (P)	Sgt F G Matthews (O)
5 July	Lt F G Pearson (P)	
29 July	Maj O M Conran (P)	Lt H Mitton (O)
14 August	Lt D Gordon (P)	Lt P G Cameron (O)
21 September	Maj G B Ward MC (P)	2Lt W A Campbell (O)
29 September	2Lt E A Barnard (O)	
31 October	2Lt W Davidson (P)	Lt W Crowther (O)
11 November	2Lt K LeGai Mills (P – after an incident on 6 Nov)	
6 December	2Lt A C Ross (P)	Lt C W M Nosworthy (O)
13 December	Lt S W Rowles (O – from wounds on 3 December)	

1918

5 January	2Lt R V Garbett (P)	2Lt B R Raggett (O)
2 April	Lt E D Jones (P)	2Lt W Smith (O)
12 April	Capt A M Maclean (P)	Lt F B Wright (O)
18 April	2Lt E K Harker (O – from wounds on 16 April)	
16 June	Capt E H Comber-Taylor (P)	
20 June	2Lt H L Storrs (P – from wounds on 15 June)	
17 July	Lt N J Forbes (P)	2Lt C F Berry (O)
11 August	2Lt H W Sheard (P)	
4 September	2Lt W J Mills (O – from wounds on 3 September)	
28 October	2Lt D A Mackenzie MM (O)	

Deaths forward with Artillery Batteries

1917

27 October	1AM W M Goldsmith
30 November	1AM R G Streat

1918

14 March	1AM L B Cope
10 July	2AM R J Gibson
13 October	1AM C B Haigh

Other deaths in service

22 March 1915	Capt J A Kane (P)	
14 October 1918	3AM J R Brown	
7 November 1918	2AM I Morgan	
10 November 1918	2AM A W Wells	
17 November 1918	2Lt W K Salton (P)	Lt L H Proctor (O)
12 May 1919	2Lt G H E Kime MM (O – from wounds on 1 October 1918)	

"But all is still – there's not a sound, and I
Remember that somewhere in France you lie
Asleep. In Gonnehem – it is close by.
Oh France, Great Mother – You who watch their sleep.
We thank you, and our gratitude is deep
For the great vigil you so reverently keep."
(The concluding lines of a tribute to Major Ward written by Maurice Baring, a close friend and Staff Officer to Brigadier Trenchard, Commander RFC in France.)

10 SQUADRON AIRCREW KILLED DURING WW2

Listed on the following pages, in date order, are those 10 Squadron aircrew killed whilst on active flying service between September 1939 and December 1945. Deaths of POWs and casualties from flying-training accidents are also included but not those victims of non-flying accidents or who died from natural causes.

The cemeteries listed are the current burial places of the casualties: many may have been initially buried elsewhere, but were exhumed and re-interred after the war. This happened frequently, especially where the maintenance of the original grave could not be guaranteed in perpetuity.

Where a site is listed in brackets, that is the crash site of the aircraft and the casualty may well be buried elsewhere.

The prefix in brackets before a cemetery or crash site indicates the country where it is located:

B – BELGIUM
C – CRETE
D – DENMARK
F – FRANCE
G – GERMANY
H – HOLLAND
I – ITALY
N – NORWAY
U – UK

Those having no known grave are individually commemorated by name on the panels of: *the Royal Air Force Memorial - Runnymede, Surrey.*

Because many of the operations were night raids the actual date of death may be one day either side of the date given in this list.

*****

** - **see in list below** Sgt H. P. Calvert was taken prisoner on 17 August 1941. In 1942 he was involved in a mass escape bid ('The Great Escape' –film) from Stalag 3E in Sagan, now Zagan, Poland and was shot by civil police in Dresden on 20 May 1942.

** - **see in list below** Flt Sgt E. P. Lewis was taken prisoner on 28 June 1941. He escaped from captivity in March 1944, but was recaptured in the July. He was then handed over to the SS and was shot on 1 August 1944.

10 SQN - DATE ORDER LIST OF WW2 CASUALTIES

DATE	NAME	RANK	FORCE	DUTY	BURIED/COMMEMORATED
01-Oct-1939	ALLSOP J. W.	F/L		PILOT	COMM. RUNNYMEDE
01-Oct-1939	SALMON A. G.	P/O		2/PILOT	COMM. RUNNYMEDE
01-Oct-1939	BELL J. R.	AC1		OBS	COMM. RUNNYMEDE
01-Oct-1939	HILL A. F.	AC1		WAG	COMM. RUNNYMEDE
01-Oct-1939	ELLISON F.	LAC		WAG	COMM. RUNNYMEDE
03-Jun-1940	FIELDS A. H.	P/O		WAG	(U) (BATTISFORD, SUFFOLK)
11-Jun-1940	KEAST L. A.	SGT		PILOT	(F) ABBEVILLE, SOMME
11-Jun-1940	BRAHAM D. F.	P/O		2/PILOT	(F) ABBEVILLE, SOMME
11-Jun-1940	MYERS J. J.	SGT		OBS	(F) ABBEVILLE, SOMME
11-Jun-1940	BLACK J. McD.	SGT		WAG	(F) ABBEVILLE, SOMME
11-Jun-1940	NUTTALL R. R. H.	LAC		WAG	(F) ABBEVILLE, SOMME
19-Jun-1940	SMITH H. V.	F/O		PILOT	(U) (AMPTON PARK, SUFFOLK)
13-Aug-1940	PARSONS E. I.	P/O		PILOT	(F) BOULOGNE, PAS-DE-CALAIS
13-Aug-1940	CAMPION A. N.	SGT		2/PILOT	(F) BOULOGNE, PAS-DE-CALAIS
15-Aug-1940	HIGSON K. H.	F/O		2/PILOT	(I) (ITALY)
26-Aug-1940	HOWARD H. H. G.	SGT		PILOT	(I) MILANO
26-Aug-1940	PARVIN J. H. K.	P/O		2/PILOT	(I) MILANO
26-Aug-1940	JOHNSTON N. R.	SGT		OBS	(I) MILANO
26-Aug-1940	STEPHENSON J. W.	SGT		WAG	(I) MILANO
26-Aug-1940	CARTER H. W.	SGT		WAG	(I) MILANO
06-Sep-1940	THOMAS R. H.	F/O		PILOT	COMM. RUNNYMEDE
06-Sep-1940	STEVENS D. J. A.	P/O		2/PILOT	COMM. RUNNYMEDE
06-Sep-1940	HILTON R.	SGT		OBS	COMM. RUNNYMEDE
06-Sep-1940	SEED H. V.	SGT		WAG	COMM. RUNNYMEDE
06-Sep-1940	NEVILLE B. W.	SGT		WAG	COMM. RUNNYMEDE
30-Sep-1940	SNELL V.	SGT		PILOT	(G) REICHSWALD FOREST
30-Sep-1940	ISMAY G. L.	SGT		2/PILOT	(G) REICHSWALD FOREST
14-Oct-1940	WRIGHT D. R.	SGT		PILOT	(U) (WEYBRIDGE, SURREY)
14-Oct-1940	COONEY K.	P/O		2/PILOT	(U) (WEYBRIDGE, SURREY)
14-Oct-1940	CASWELL J. J.	SGT		OBS	(U) (WEYBRIDGE, SURREY)
14-Oct-1940	HENRY B. L.	SGT		WAG	(U) (WEYBRIDGE, SURREY)
14-Oct-1940	DAVIES E.	SGT		WAG	(U) (WEYBRIDGE, SURREY)
14-Oct-1940	DICKINSON R. J.	P/O		OBS	(U) (THIRSK, YORKSHIRE)
14-Oct-1940	NEVILLE L. P.	SGT		WAG	(U) (THIRSK, YORKSHIRE)
21-Oct-1940	PHILLIPS A. S.	F/L		PILOT	(G) DURNBACH

21-Oct-1940	GORDON D. B.	SGT		2/PILOT	(G) DURNBACH
21-Oct-1940	LOFTHOUSE W. A.	SGT		OBS	(G) DURNBACH
21-Oct-1940	MAPPLETHORPE C. P.	SGT		WAG	(G) DURNBACH
21-Oct-1940	WILLS I. G.	SGT		WAG	(G) DURNBACH
05-Nov-1940	JONES A. L.	P/O		PILOT	COMM. RUNNYMEDE
05-Nov-1940	SMITH P. T. H.	SGT		2/PILOT	COMM. RUNNYMEDE
05-Nov-1940	HARDING E. G.	SGT		OBS	COMM. RUNNYMEDE
05-Nov-1940	WATT J. B.	SGT		WAG	COMM. RUNNYMEDE
05-Nov-1940	WALKER W. L. T.	SGT		WAG	COMM. RUNNYMEDE
12-Nov-1940	GOLDSMITH P. D.	SGT		2/PILOT	(U) (RHYMNEY, GLAMORGAN)
13-Nov-1940	FERGUSON K. F.	S/L		PILOT	COMM. RUNNYMEDE
13-Nov-1940	ROGERS C. S.	SGT		2/PILOT	COMM. RUNNYMEDE
13-Nov-1940	FRASER W.	SGT		OBS	COMM. RUNNYMEDE
13-Nov-1940	WATSON A.	SGT		WAG	COMM. RUNNYMEDE
13-Nov-1940	WESTON K. F.	SGT		WAG	COMM. RUNNYMEDE
22-Dec-1940	FLEWELLING R. L.	P/O		2/PILOT	(U) (LEEMING, YORKSHIRE)
16-Jan-1941	SKYRME H. B.	F/O		PILOT	COMM. RUNNYMEDE
16-Jan-1941	ROWLETT J. E.	SGT		2/PILOT	COMM. RUNNYMEDE
16-Jan-1941	SANDLAND G. C.	SGT		OBS	COMM. RUNNYMEDE
16-Jan-1941	POLKINGHORNE E. D.	SGT		WAG	COMM. RUNNYMEDE
16-Jan-1941	BROOKMAN L. P.	SGT		WAG	COMM. RUNNYMEDE
01-Mar-1941	HOARE A. C.	SGT		PILOT	COMM. RUNNYMEDE
01-Mar-1941	WOODBRIDGE P. H.	SGT		2/PILOT	COMM. RUNNYMEDE
01-Mar-1941	FLORIGNY A. A.	P/O		OBS	COMM. RUNNYMEDE
01-Mar-1941	WOODS C. J.	SGT		WAG	COMM. RUNNYMEDE
01-Mar-1941	COATES A.	SGT		WAG	COMM. RUNNYMEDE
18-Mar-1941	WATSON N.	SGT		PILOT	(U) (MASHAM, YORKSHIRE)
16-Apr-1941	SALWAY R. V.	SGT		PILOT	(G) SAGE, OLDENBURG
16-Apr-1941	STRICKLAND V. J.	SGT		2/PILOT	(G) SAGE, OLDENBURG
16-Apr-1941	MORANT N. A.	SGT		OBS	(G) SAGE, OLDENBURG
16-Apr-1941	LYNCH G.	SGT		WAG	(G) SAGE, OLDENBURG
16-Apr-1941	JACKSON N. F.	SGT		WAG	(G) SAGE, OLDENBURG
10-May-1941	GOUGH P. B.	P/O		PILOT	(G) KEIL
10-May-1941	STEWART F. G.	SGT		2/PILOT	(G) KEIL
10-May-1941	MANCHIP L.	SGT		OBS	(G) KEIL
10-May-1941	RICHARDSON R.	SGT		WAG	(G) KEIL
10-May-1941	WATSON D. F. F.	SGT		WAG	(G) KEIL
18-Jun-1941	BRADFORD W. R.	SGT		PILOT	COMM. RUNNYMEDE

18-Jun-1941	CONRY E. J.	SGT	RCAF	2/PILOT	COMM. RUNNYMEDE
18-Jun-1941	GRIFFITHS D. W.	SGT		OBS	COMM. RUNNYMEDE
18-Jun-1941	KELLY W.	F/S		WAG	COMM. RUNNYMEDE
18-Jun-1941	RINTOUL R. O.	SGT		WAG	COMM. RUNNYMEDE
27-Jun-1941	KNAPE A. H.	SGT		PILOT	(G) HAMBURG, OHLSDORF
27-Jun-1941	WALKER D. P.	SGT		2/PILOT	(G) HAMBURG, OHLSDORF
27-Jun-1941	BOLSTER R. V. C.	P/O		OBS	(G) HAMBURG, OHLSDORF
01-Aug-1944	LEWIS E. P. (Note a)	F/S		WAG	(P) MALBORK
27-Jun-1941	CARSON A. S.	SGT		WAG	(G) HAMBURG, OHLSDORF
27-Jun-1941	RICKCORD A. G.	SGT		PILOT	(G) HAMBURG, OHLSDORF
27-Jun-1941	THOMSON C. S.	SGT		2/PILOT	(G) HAMBURG, OHLSDORF
27-Jun-1941	GANE G. L.	SGT		OBS	(G) HAMBURG, OHLSDORF
27-Jun-1941	JONES R. G.	F/S		WAG	(G) HAMBURG, OHLSDORF
27-Jun-1941	HIRD L.	F/S		WAG	(G) HAMBURG, OHLSDORF
27-Jun-1941	SHAW J. S.	SGT		PILOT	(G) BECKLINGEN, SOLTAU
27-Jun-1941	BINGHAM-HALL D. V.	P/O		2/PILOT	(G) BECKLINGEN, SOLTAU
27-Jun-1941	McALONAN J. M.	SGT		OBS	(G) BECKLINGEN, SOLTAU
27-Jun-1941	BANHAM D. W.	SGT		WAG	(G) BECKLINGEN, SOLTAU
27-Jun-1941	LAWLEY E.	SGT		WAG	(G) BECKLINGEN, SOLTAU
27-Jun-1941	WATSON A. K.	P/O		WAG	(G) KIEL
30-Jun-1941	BARRETT J.	P/O		PILOT	(G) REICHSWALD FOREST
30-Jun-1941	RICE W. J. P.	SGT		OBS	(G) REICHSWALD FOREST
30-Jun-1941	BEVERIDGE A. R.	SGT		PILOT	(U) (THETFORD, NORFOLK)
30-Jun-1941	ALCOCK G. A.	SGT		WAG	(U) (THETFORD, NORFOLK)
05-Jul-1941	GOULDING R.	P/O		PILOT	(H) HARDENBURG, LUTTEN
05-Jul-1941	MORRISON D.	SGT		2/PILOT	(H) HARDENBURG, LUTTEN
05-Jul-1941	JORDAN R. H.	SGT		OBS	(H) HARDENBURG, LUTTEN
05-Jul-1941	AIRD R. I. H.	F/S		WAG	(H) HARDENBURG, LUTTEN
07-Jul-1941	BLACK H.	SGT		PILOT	COMM. RUNNYMEDE
07-Jul-1941	HEIGHTON H. R.	SGT		2/PILOT	COMM. RUNNYMEDE
07-Jul-1941	LUCAS O. R.	F/S		OBS	COMM. RUNNYMEDE
07-Jul-1941	THOMPSON R. H.	F/S		WAG	COMM. RUNNYMEDE
08-Jul-1941	LEWIS G. B.	F/S		PILOT	COMM. RUNNYMEDE
08-Jul-1941	WILSON F.	P/O		2/PILOT	COMM. RUNNYMEDE
08-Jul-1941	BUTCHER R. V.	SGT		OBS	COMM. RUNNYMEDE
08-Jul-1941	BEVAN C. J.	F/S		WAG	COMM. RUNNYMEDE
25-Jul-1941	SPIERS W. Mc.N	P/O		PILOT	(B) KOERSAL, LIMBURG
25-Jul-1941	PUTTICK W.	SGT		2/PILOT	(B) KOERSAL, LIMBURG

25-Jul-1941	DANIELS H. J.	P/O		OBS	(B) KOERSAL, LIMBURG
25-Jul-1941	LAWSON C. W. F. D. E.	SGT		WAG	(B) KOERSAL, LIMBURG
25-Jul-1941	BEVERLEY D. B.	SGT		WAG	(B) KOERSAL, LIMBURG
25-Jul-1941	LANDALE P. W. F.	S/L		PILOT	COMM. RUNNYMEDE
25-Jul-1941	PRINGLE G.	P/O		2/PILOT	(G) KEIL
25-Jul-1941	WELLS G.	SGT	RNZAF	OBS	(H) DEN BURG, TEXEL
25-Jul-1941	EARP A. W.	SGT		WAG	COMM. RUNNYMEDE
25-Jul-1941	CHRISTIE G.	F/S		WAG	(G) KEIL
08-Aug-1941	LITTLEWOOD M.	P/O		PILOT	(G) KEIL
08-Aug-1941	BAYLEY E.	SGT		2/PILOT	(G) KEIL
08-Aug-1941	EVANS J. E.	P/O		OBS	COMM. RUNNYMEDE
08-Aug-1941	TIMMS R. T.	SGT		WAG	(G) KEIL
08-Aug-1941	MOORES N.	SGT		WAG	(G) KEIL
08-Aug-1941	MYERS R.	SGT		WAG	COMM. RUNNYMEDE
16-Aug-1941	PEARSON H. H.	P/O		PILOT	(B) HEVERLEE
16-Aug-1941	LOVEDAY S. A.	SGT		2/PILOT	(B) HEVERLEE
16-Aug-1941	WALKER D. W. H.	P/O		OBS	(B) HEVERLEE
16-Aug-1941	WELSH J.	SGT		WAG	(B) HEVERLEE
16-Aug-1941	BOTT S. F. B.	SGT	RCAF	WAG	(B) HEVERLEE
16-Aug-1941	LAGER E. H.	SGT		PILOT	(G) REICHSWALD FOREST
16-Aug-1941	SEWELL V. Y. H.	SGT	RCAF	WAG	(G) REICHSWALD FOREST
20-May-1942	CALVERT H. P. (Note b)	SGT		2/PILOT	(G) 1939-1945, BERLIN
18-Aug-1941	EVILL W. A. S.	P/O		PILOT	(B) REKEM, LIMBURG
18-Aug-1941	TOMKINS K. M.	SGT		2/PILOT	(B) REKEM, LIMBURG
18-Aug-1941	O'DELL C. P.	SGT		OBS	(B) REKEM, LIMBURG
18-Aug-1941	DUFFY D. MacL.	SGT	RCAF	WAG	(B) REKEM, LIMBURG
18-Aug-1941	PARK T. H.	SGT	RCAF	WAG	(B) REKEM, LIMBURG
18-Aug-1941	MANISON E. E.	P/O		2/PILOT	(B) LANAYE, LIMBURG
18-Aug-1941	CRICH H. L.	SGT	RCAF	WAG	(B) HEVERLEE
18-Aug-1941	NORCROSS S.	SGT		WAG	(B) HEVERLEE
22-Aug-1941	LIEBECK K. W.	P/O	RCAF	PILOT	(U) (DENT, WESTMORELAND)
22-Aug-1941	FLETCHER G.	SGT		2/PILOT	(U) (DENT, WESTMORELAND)
06-Sep-1941	STUART W.	SGT	RCAF	PILOT	(U) (ACKLINGTON, NORTHUMBERLAND)
06-Sep-1941	AUSTIN R. S.	P/O	RNZAF	2/PILOT	(U) (ACKLINGTON, NORTHUMBERLAND)
06-Sep-1941	BRYANT P. W.	SGT		OBS	(U) (ACKLINGTON, NORTHUMBERLAND)
06-Sep-1941	POUPARD A.	SGT		PILOT	(H) OLDEBROEK, GELDERLAND
06-Sep-1941	LAURIE-DICKSON J. F.	P/O		2/PILOT	(H) OLDEBROEK, GELDERLAND
06-Sep-1941	WILSON R. H.	F/S	RCAF	OBS	(H) OLDEBROEK, GELDERLAND
06-Sep-1941	GARROD R. J.	SGT		WAG	(H) OLDEBROEK, GELDERLAND

06-Sep-1941	TAKARANGI P. T.	SGT	RNZAF	WAG	(H) OLDEBROEK, GELDERLAND
11-Sep-1941	PURVIS R. P.	P/O		PILOT	COMM. RUNNYMEDE
11-Sep-1941	GREENFIELD J. C.	P/O		2/PILOT	COMM. RUNNYMEDE
11-Sep-1941	FITSELL R. L.	SGT		OBS	COMM. RUNNYMEDE
11-Sep-1941	NIGHTINGALE J.	SGT		WAG	COMM. RUNNYMEDE
11-Sep-1941	IRWIN K. A.	SGT	RCAF	WAG	COMM. RUNNYMEDE
30-Nov-1941	NELSON M. K.	P/O	RNZAF	PILOT	COMM. RUNNYMEDE
30-Nov-1941	ILLINGWORTH G. S.	SGT		2/PILOT	COMM. RUNNYMEDE
30-Nov-1941	KNOWLES J. S.	P/O		OBS	COMM. RUNNYMEDE
30-Nov-1941	WEBSTER W.	SGT		WAG	COMM. RUNNYMEDE
30-Nov-1941	WALES A. F.	SGT		WAG	COMM. RUNNYMEDE
11-Dec-1941	HOSKIN E. C.	SGT		2/PILOT	(U) (EAVESTONE MOOR, YORKSHIRE)
29-Dec-1941	TRIPP W. W.	SGT		PILOT	(U) (LEEMING, YORKSHIRE)
29-Dec-1941	GREEN W. C.	SGT		WOP	(U) (LEEMING, YORKSHIRE)
30-Dec-1941	ROACH R. J.	F/L		AG	COMM. RUNNYMEDE
15-Jan-1942	COWAN T.	SGT		F/E	(U) (NORTHALLERTON, YORKSHIRE)
15-Jan-1942	TAYLOR H. K.	SGT		OBS	(U) (NORTHALLERTON, YORKSHIRE)
15-Jan-1942	VON DADELSZEN M.	P/O	RNZAF	BA	(U) (NORTHALLERTON, YORKSHIRE)
15-Jan-1942	BRICE V. L.	P/O	RCAF	WOP	(U) (NORTHALLERTON, YORKSHIRE)
15-Jan-1942	BRADLEY J. C.	SGT		AG	(U) (NORTHALLERTON, YORKSHIRE)
15-Jan-1942	SAVAGE D.	F/S	RCAF	AG	(U) (NORTHALLERTON, YORKSHIRE)
26-Feb-1942	BISSETT J. A.	SGT	RCAF	PILOT	COMM. RUNNYMEDE
26-Feb-1942	WIELAND E. V. H.	F/S		F/E	COMM. RUNNYMEDE
26-Feb-1942	JEFFRIES J. H.	SGT		OBS	COMM. RUNNYMEDE
26-Feb-1942	WESTLAND J. J.	SGT		BA	COMM. RUNNYMEDE
26-Feb-1942	GLANVILLE W. A. I.	SGT		WOP	COMM. RUNNYMEDE
26-Feb-1942	SIMMONS E. H.	SGT		AG	COMM. RUNNYMEDE
26-Feb-1942	DARWIN C. F.	SGT		AG	COMM. RUNNYMEDE
31-Mar-1942	WEBSTER F. D.	S/L		PILOT	(N) STAVNE, TRONDHEIM
31-Mar-1942	HALL E. A.	SGT		F/E	COMM. RUNNYMEDE
31-Mar-1942	WHEATLEY H. S.	F/S		OBS	COMM. RUNNYMEDE
31-Mar-1942	STEVENS-FOX A. C.	F/L		BA	COMM. RUNNYMEDE
31-Mar-1942	HAGUE A.	SGT		WOP	COMM. RUNNYMEDE
31-Mar-1942	HALL W.	SGT		AG	COMM. RUNNYMEDE
31-Mar-1942	LENEY S. R.	P/O		AG	COMM. RUNNYMEDE
31-Mar-1942	BLUNDEN N. R.	P/O	RNZAF	PILOT	COMM. RUNNYMEDE
31-Mar-1942	DAY G. C.	P/O		F/E	(N) STAVNE, TRONDHEIM
31-Mar-1942	RICHARDS R. G. A.	SGT		OBS	(N) STAVNE, TRONDHEIM
31-Mar-1942	EASTWOOD W. B.	SGT		BA	COMM. RUNNYMEDE

31-Mar-1942	FRANKLIN H. R.	SGT	RCAF	WOP	COMM. RUNNYMEDE
31-Mar-1942	MAY K. C.	SGT		AG	(N) STAVNE, TRONDHEIM
31-Mar-1942	HENMAN A. R.	SGT	RCAF	AG	COMM. RUNNYMEDE
15-Apr-1942	HUGHES R. P.	P/O	RCAF	PILOT	(U) (GREATHAM MOOR, SURREY)
28-Apr-1942	ANNABLE E.	SGT		OBS	COMM. RUNNYMEDE
28-Apr-1942	STOTT H. H.	SGT		AG	COMM. RUNNYMEDE
09-May-1942	GUTHRIE N. D.	S/L		PILOT	(G) 1939-1945, BERLIN
09-May-1942	TOWLER A. E.	SGT		F/E	(G) 1939-1945, BERLIN
09-May-1942	Le VACK F. H.	SGT		OBS	(G) 1939-1945, BERLIN
09-May-1942	HALES D. W.	P/O		BA	(G) 1939-1945, BERLIN
09-May-1942	CLEPHANE D. A.	SGT		WOP	(G) 1939-1945, BERLIN
09-May-1942	DAVIS F. M.	F/O		AG	(G) 1939-1945, BERLIN
09-May-1942	KILLINGBECK M. G.	SGT		AG	(G) 1939-1945, BERLIN
20-May-1942	BAKER G. H.	P/O		PILOT	COMM. RUNNYMEDE
20-May-1942	HIGHAM J. C.	SGT		F/E	COMM. RUNNYMEDE
20-May-1942	EVERITT J.	F/S		OBS	COMM. RUNNYMEDE
20-May-1942	RAYNES D. J.	SGT		BA	COMM. RUNNYMEDE
20-May-1942	FAIRWEATHER R. J. G.	SGT		WOP	COMM. RUNNYMEDE
20-May-1942	HOWICK V. C.	SGT		AG	COMM. RUNNYMEDE
20-May-1942	CHAPMAN H. E.	SGT		AG	COMM. RUNNYMEDE
31-May-1942	MOORE A. R,	SGT		PILOT	(H) WOENSAL GENERAL, EINDHOVEN
31-May-1942	ENGLISH M. F.	SGT	RCAF	AG	(H) WOENSAL GENERAL, EINDHOVEN
31-May-1942	WALKER F.	SGT		AG	(H) WOENSAL GENERAL, EINDHOVEN
02-Jun-1942	WHITFIELD J.	SGT		AG	COMM. RUNNYMEDE
02-Jun-1942	JOYCE D. D. P.	P/O	RCAF	PILOT	(G) REICHSWALD FOREST
02-Jun-1942	GARSON J. G. B.	SGT		F/E	(G) REICHSWALD FOREST
02-Jun-1942	PERRY E. W.	SGT		BA	(G) REICHSWALD FOREST
02-Jun-1942	LAKER W. J.	SGT		WOP	(G) REICHSWALD FOREST
02-Jun-1942	ANDERSON A. M.	SGT		AG	(G) REICHSWALD FOREST
02-Jun-1942	EDGELEY W. J.	SGT		AG	(G) REICHSWALD FOREST
02-Jun-1942	CLOTHIER H. G.	P/O	RNZAF	PILOT	(H) CROOSWIJK GENERAL
02-Jun-1942	GANDERTON J. R. F,	F/S	RNZAF	OBS	COMM. RUNNYMEDE
02-Jun-1942	CLAPHAM J. S.	F/S		BA	COMM. RUNNYMEDE
02-Jun-1942	SIMPSON J.	SGT		WOP	COMM. RUNNYMEDE
02-Jun-1942	MIREAU A. O.	SGT	RCAF	AG	COMM. RUNNYMEDE
02-Jun-1942	MORRIS T. A. G.	F/S		AG	COMM. RUNNYMEDE
06-Jun-1942	ROCHFORD P. G.	F/S		PILOT	(G) RHEINBERG
06-Jun-1942	OPENSHAW E. B. H.	F/O	RNZAF	OBS	(G) RHEINBERG

06-Jun-1942	MATTHIAS L. C.	SGT		WOP	(G) RHEINBERG
21-Jun-1942	SENIOR E. R.	P/O		PILOT	COMM. RUNNYMEDE
21-Jun-1942	JACKSON H.	F/L		F/E	COMM. RUNNYMEDE
21-Jun-1942	SARGAN G. A.	SGT		OBS	COMM. RUNNYMEDE
21-Jun-1942	WILLIAMS A. S,	F/O		BA	COMM. RUNNYMEDE
21-Jun-1942	WALLER G. J.	P/O		WOP	COMM. RUNNYMEDE
21-Jun-1942	JONES G. P.	F/S		AG	COMM. RUNNYMEDE
21-Jun-1942	SALISBURY J. L.	F/S		AG	COMM. RUNNYMEDE
25-Jun-1942	WISEMAN W. J. W.	SGT		PILOT	(U) (LEEMING, YORKSHIRE)
25-Jun-1942	CARTER J. A.	SGT		F/E	(U) (LEEMING, YORKSHIRE)
25-Jun-1942	MUTTER J. G.	F/O		OBS	(U) (LEEMING, YORKSHIRE)
25-Jun-1942	DUVALL S. V.	SGT		BA	(U) (LEEMING, YORKSHIRE)
25-Jun-1942	STEWART D. R.	F/S	RCAF	WOP	(U) (LEEMING, YORKSHIRE)
03-Jul-1942	LAWER E. J.	SGT	RAAF	PILOT	COMM. RUNNYMEDE
03-Jul-1942	JONES D. M.	SGT		F/E	COMM. RUNNYMEDE
03-Jul-1942	MARTIN R. T. C.	SGT		OBS	COMM. RUNNYMEDE
03-Jul-1942	GILL H. J.	SGT		BA	COMM. RUNNYMEDE
03-Jul-1942	MARTIN J. J. F.	SGT		WOP	(H) JONKERBOS, NIJMEGEN
03-Jul-1942	GEORGE H. A.	SGT		AG	COMM. RUNNYMEDE
03-Jul-1942	WILLIAMS O. L.	SGT		AG	COMM. RUNNYMEDE
20-Jul-1942	FEGAN J. S. V.	P/O		PILOT	COMM. RUNNYMEDE
20-Jul-1942	BEATTIE A. J.	SGT		F/E	COMM. RUNNYMEDE
20-Jul-1942	GIBSON W. G.	F/S	RCAF	OBS	COMM. RUNNYMEDE
20-Jul-1942	ATCHISON C.	SGT		BA	COMM. RUNNYMEDE
20-Jul-1942	STEWART F. W.	F/S	RCAF	WOP	COMM. RUNNYMEDE
20-Jul-1942	JOHNSTONE E. E. E.	P/O	RCAF	AG	COMM. RUNNYMEDE
20-Jul-1942	BICK F. G. A.	SGT		AG	COMM. RUNNYMEDE
05-Sep-1942	HACKING A. E.	F/L		PILOT	(C) SOUDA BAY
05-Sep-1942	CARSON A. E.	SGT	RAAF	AG	(C) SOUDA BAY
05-Sep-1942	PORRITT R. G.	SGT	RCAF	AG	(C) SOUDA BAY
07-Sep-1942	MORGAN D. W. E.	P/O		PILOT	(G) REICHSWALD FOREST
07-Sep-1942	RICE J.	SGT		F/E	(G) REICHSWALD FOREST
07-Sep-1942	MILNE A. Mc.D	F/S		OBS	(G) REICHSWALD FOREST
07-Sep-1942	BILLINGS L. E.	SGT		BA	(G) REICHSWALD FOREST
07-Sep-1942	MORRIS A.	SGT		WOP	(G) REICHSWALD FOREST
07-Sep-1942	HENRY A. J.	SGT	RAAF	AG	(G) REICHSWALD FOREST
07-Sep-1942	LOVEDAY H. J.	SGT		AG	(G) REICHSWALD FOREST
20-Sep-1942	McDOUGALL H. C.	SGT		NAV	(U) (BLUBBERHOUSE, YORKSHIRE)

02-Oct-1942	NORREGAARD A.	SGT		BA	(G) KEIL
02-Oct-1942	PETTICAN R. S.	SGT		WOP	(G) KEIL
02-Oct-1942	MOLLER J. H.	F/S	RNZAF	PILOT	(G) KEIL
02-Oct-1942	BRETT T. E.	SGT		F/E	(G) KEIL
02-Oct-1942	HUNTER T. McL. M.	F/S	RCAF	OBS	(G) KEIL
02-Oct-1942	MACPHERSON C. H.	F/S	RCAF	BA	(G) KEIL
02-Oct-1942	PICKARD C. E.	F/S	RCAF	WOP	(G) KEIL
02-Oct-1942	JONES J. L.	F/O		PILOT	(G) KEIL
02-Oct-1942	STEVENS R. Q.	SGT		F/E	(G) KEIL
02-Oct-1942	PICKERING M. A.	SGT		OBS	(G) KEIL
02-Oct-1942	PRIOR R. C.	SGT		BA	(G) KEIL
02-Oct-1942	MOWBRAY A. McK.	F/S	RCAF	WOP	(G) KEIL
02-Oct-1942	CURRAN K. R.	SGT		AG	(G) KEIL
02-Oct-1942	CARTER S. C.	SGT		AG	(G) KEIL
02-Oct-1942	CAMPBELL D.	SGT		PILOT	(D) ODENSE
02-Oct-1942	SPOWART G. W.	SGT		F/E	(D) SVINO
02-Oct-1942	SMYTHSON L.	SGT		OBS	(D) ODENSE
02-Oct-1942	SULLIVAN H. E.	SGT		BA	(D) ODENSE
02-Oct-1942	GOURLAY R. F.	SGT	RCAF	WOP	(D) SVINO
02-Oct-1942	IVERS D. A.	SGT	RCAF	AG	(D) ODENSE
02-Oct-1942	MOORE H. S.	SGT		AG	(D) ODENSE
14-Oct-1942	LINDSAY J. D.	P/O	RCAF	PILOT	COMM. RUNNYMEDE
14-Oct-1942	GREGG R.	SGT		F/E	COMM. RUNNYMEDE
14-Oct-1942	DUNAJSKI F. X. J.	F/S	RCAF	OBS	COMM. RUNNYMEDE
14-Oct-1942	HOGAN F. J.	F/S	RCAF	BA	COMM. RUNNYMEDE
14-Oct-1942	BRALEY R. S.	F/O	RCAF	WOP	COMM. RUNNYMEDE
14-Oct-1942	GIRDLESTONE A. C.	SGT		AG	COMM. RUNNYMEDE
14-Oct-1942	LANGTON A. G.	SGT		AG	COMM. RUNNYMEDE
16-Oct-1942	WILDEY R. K.	W/C		PILOT	(G) RHEINBERG
16-Oct-1942	DuBROY J. W.	SGT	RCAF	AG	(G) RHEINBERG
16-Oct-1942	BRINDLEY A.	F/L		AG	(G) RHEINBERG
09-Nov-1942	HALE K. G. M.	SGT		PILOT	(G) SAGE, OLDENBURG
09-Nov-1942	HUMPHREYS J. E.	SGT		F/E	COMM. RUNNYMEDE
09-Nov-1942	WHITE F. G.	SGT		OBS	(G) SAGE, OLDENBURG
09-Nov-1942	WINDRAM F. J.	SGT		BA	COMM. RUNNYMEDE
09-Nov-1942	BROOM F. A.	SGT		WOP	COMM. RUNNYMEDE
09-Nov-1942	GRISEDALE T. W.	SGT		AG	COMM. RUNNYMEDE
09-Nov-1942	TIDBALL F. L.	SGT		AG	COMM. RUNNYMEDE
23-Nov-1942	JENSON S.	SGT	RCAF	AG	(F) (MORTEFONTAINE, OISE)

30-Nov-1942	WILLMOTT E. J.	F/S		PILOT	(U) (SEATON ROSS, YORKSHIRE)
30-Nov-1942	BARRETT G. S.	S/L		2/PILOT	(U) (SEATON ROSS, YORKSHIRE)
30-Nov-1942	SOGGEE D. A. D.	SGT		F/E	(U) (SEATON ROSS, YORKSHIRE)
30-Nov-1942	FLOWER A. F.	SGT		NAV	(U) (SEATON ROSS, YORKSHIRE)
30-Nov-1942	CLARKE F.	SGT		BA	(U) (SEATON ROSS, YORKSHIRE)
30-Nov-1942	MARCHANT E. A.	SGT		WOP	(U) (SEATON ROSS, YORKSHIRE)
30-Nov-1942	BELLERBY D. A.	SGT		AG	(U) (SEATON ROSS, YORKSHIRE)
30-Nov-1942	BREWER A.	F/S		AG	(U) (SEATON ROSS, YORKSHIRE)
11-Dec-1942	JUNEAU A. R.	F/S	RCAF	PILOT	(F) VILLIERS-LE-DUC, ST-PHAL
11-Dec-1942	WARREN P. A.	SGT		F/E	(F) VILLIERS-LE-DUC, ST-PHAL
11-Dec-1942	KILPATRICK G.	SGT		OBS	(F) VILLIERS-LE-DUC, ST-PHAL
11-Dec-1942	McROBERTS C. A,	F/O	RCAF	BA	(F) VILLIERS-LE-DUC, ST-PHAL
11-Dec-1942	DUNLOP J. L.	F/S	RCAF	WOP	(F) VILLIERS-LE-DUC, ST-PHAL
11-Dec-1942	BUDGEN C. F.	SGT		AG	(F) VILLIERS-LE-DUC, ST-PHAL
11-Dec-1942	CROOKS R. W. J.	SGT		AG	(F) VILLIERS-LE-DUC, ST-PHAL
09-Jan-1943	FISH E.	SGT		PILOT	(H) VREDENHOF, SCHIERMONNIKOOG
09-Jan-1943	SMITH A.	SGT		F/E	(H) VREDENHOF, SCHIERMONNIKOOG
09-Jan-1943	FETHERSTON-HAUGH C. B.	SGT	RCAF	NAV	(H) VREDENHOF, SCHIERMONNIKOOG
09-Jan-1943	SMITH J. W.	SGT	RCAF	BA	(H) VREDENHOF, SCHIERMONNIKOOG
09-Jan-1943	WILSON A. W.	SGT	RCAF	WOP	(H) VREDENHOF, SCHIERMONNIKOOG
15-Jan-1943	BOYLE M. J.	SGT		AG	(H) LEEUWARDEN
09-Jan-1943	KING L. C.	SGT	RCAF	AG	(H) VREDENHOF, SCHIERMONNIKOOG
14-Feb-1943	KAY H.	SGT		AG	(H) JONKERBOS, NIJMEGEN
10-Mar-1943	PECK G. F.	SGT		PILOT	(U) (SEATON ROSS, YORKSHIRE)
10-Mar-1943	MUIRHEAD A. A.	SGT	RCAF	F/E	(U) (SEATON ROSS, YORKSHIRE)
10-Mar-1943	BROWNING W.	SGT		NAV	(U) (SEATON ROSS, YORKSHIRE)
10-Mar-1943	HART B. G.	SGT		BA	(U) (SEATON ROSS, YORKSHIRE)
10-Mar-1943	BOREHAM A. N.	SGT		WOP	(U) (SEATON ROSS, YORKSHIRE)
10-Mar-1943	RETTER D. C.	SGT		AG	(U) (SEATON ROSS, YORKSHIRE)
13-Mar-1943	DICKINSON J.	SGT		PILOT	(G) REICHSWALD FOREST
13-Mar-1943	HENDEN H. E.	SGT		F/E	(G) REICHSWALD FOREST
13-Mar-1943	HARRIS J. H.	SGT		NAV	(G) REICHSWALD FOREST
13-Mar-1943	STANNERS F. W.	SGT		BA	(G) REICHSWALD FOREST
13-Mar-1943	SMITH J. E.	SGT		WOP	(G) REICHSWALD FOREST
13-Mar-1943	GAIT L. J.	SGT		AG	(G) REICHSWALD FOREST
13-Mar-1943	CRAWFORD F. P.	SGT		AG	(G) REICHSWALD FOREST
13-Mar-1943	BARKER L.	SGT		PILOT	(H) JONKERBOS, NIJMEGEN
13-Mar-1943	THOMPSON W. L. G.	SGT	RCAF	F/E	(H) JONKERBOS, NIJMEGEN

13-Mar-1943	PAUL R. J.	F/O		NAV	(H) JONKERBOS, NIJMEGEN
13-Mar-1943	MILLS K. A.	SGT		BA	(H) JONKERBOS, NIJMEGEN
13-Mar-1943	THOMAS L. E.	SGT		WOP	(H) JONKERBOS, NIJMEGEN
13-Mar-1943	FREEL J.	SGT		AG	(H) JONKERBOS, NIJMEGEN
13-Mar-1943	HYATT G. A.	SGT		AG	(H) JONKERBOS, NIJMEGEN
26-Mar-1943	WARING W.	SGT			COMM. RUNNYMEDE
05-Apr-1943	WANN J. A.	F/O		PILOT	COMM. RUNNYMEDE
05-Apr-1943	MAISENBACHER W. M.	SGT	RCAF	F/E	COMM. RUNNYMEDE
05-Apr-1943	BERTRAM N.	P/O		NAV	COMM. RUNNYMEDE
05-Apr-1943	SCANLON W. E.	SGT		BA	COMM. RUNNYMEDE
05-Apr-1943	WHEEN H.	P/O		WOP	COMM. RUNNYMEDE
05-Apr-1943	JAGGER D. C. T.	P/O		AG	COMM. RUNNYMEDE
05-Apr-1943	FRANKLAND E. V.	P/O		AG	COMM. RUNNYMEDE
15-Apr-1943	HANCOCK J. E. G.	F/S		PILOT	(F) ST.-HILAIRE-LE-PETIT, MARNE
15-Apr-1943	OWEN H. G.	SGT		NAV	(F) ST.-HILAIRE-LE-PETIT, MARNE
15-Apr-1943	EVERITT D. B.	SGT		BA	(F) ST.-HILAIRE-LE-PETIT, MARNE
15-Apr-1943	FUNNELL D. E.	SGT		WOP	(F) ST.-HILAIRE-LE-PETIT, MARNE
15-Apr-1943	GRIGGS F. W.	SGT		AG	(F) ST.-HILAIRE-LE-PETIT, MARNE
15-Apr-1943	CULLERTON J. D,	SGT		AG	(F) ST.-HILAIRE-LE-PETIT, MARNE
05-May-1943	HILL E. B.	SGT		2/PILOT	(U) (SUTTON BANK, YORKSHIRE)
05-May-1943	COX T.	SGT		NAV	(U) (SUTTON BANK, YORKSHIRE)
05-May-1943	TAYLOR H. S.	SGT		BA	(U) (SUTTON BANK, YORKSHIRE)
05-May-1943	WAY H. H.	SGT		WOP	(U) (SUTTON BANK, YORKSHIRE)
05-May-1943	WARD G. F.	SGT		AG	(U) (SUTTON BANK, YORKSHIRE)
14-May-1943	McCOY A. F.	W/O	RCAF	AG	COMM. RUNNYMEDE
14-May-1943	MILLS J. F.	F/S	RAAF	PILOT	COMM. RUNNYMEDE
14-May-1943	HOWIE J. C.	SGT		F/E	(H) AMSTERDAM NEW EASTERN
14-May-1943	MACADAM J. S.	SGT		NAV	(H) LEMSTERLAND, LEMMER
14-May-1943	AVENT J. W.	SGT		BA	(H) AMSTERDAM NEW EASTERN
14-May-1943	JONES A.	SGT		WOP	(H) AMSTERDAM NEW EASTERN
14-May-1943	MALTBY C.	SGT		AG	(H) AMSTERDAM NEW EASTERN
14-May-1943	HOWARD E. J. C.	SGT		AG	(H) AMSTERDAM NEW EASTERN
24-May-1943	HINE C. J. J.	SGT		PILOT	(H) WONSERADEEL, MAKKUM
24-May-1943	HALL A.	SGT		F/E	(H) WONSERADEEL, MAKKUM
24-May-1943	KING J. W. T.	F/O		NAV	(H) WONSERADEEL, MAKKUM
24-May-1943	ASHTON H. G.	SGT		BA	(H) WONSERADEEL, MAKKUM
24-May-1943	CHURCH E. C.	SGT		WOP	(H) BERGEN OP ZOOM
24-May-1943	NISBET G. D.	SGT	RCAF	AG	(H) WONSERADEEL, MAKKUM

24-May-1943	BAGGALEY R. F. F.	SGT		AG	(H) BERGEN OP ZOOM
24-May-1943	REES J.	SGT		PILOT	COMM. RUNNYMEDE
24-May-1943	OLIVER W. E.	SGT		F/E	COMM. RUNNYMEDE
24-May-1943	ROSE F. C.	SGT		NAV	(G) KIEL
24-May-1943	GAYWOOD S. J.	SGT		BA	COMM. RUNNYMEDE
24-May-1943	BIRKHEAD D.	SGT		WOP	COMM. RUNNYMEDE
24-May-1943	DAVID E. G.	SGT		AG	COMM. RUNNYMEDE
24-May-1943	FARNELL F. W.	SGT		AG	COMM. RUNNYMEDE
24-May-1943	DENTON J. B.	F/S		PILOT	(G) REICHSWALD FOREST
24-May-1943	INGLIS I. B.	SGT		2/PILOT	(G) REICHSWALD FOREST
24-May-1943	HARRISON M.	SGT		F/E	(G) REICHSWALD FOREST
24-May-1943	PLENDERLIETH N. P.	SGT		NAV	(G) REICHSWALD FOREST
24-May-1943	ADAMS D. H. G.	SGT		BA	(G) REICHSWALD FOREST
24-May-1943	GRIMWOOD P.	SGT		WOP	(G) REICHSWALD FOREST
24-May-1943	WALLIS A. E.	SGT		AG	(G) REICHSWALD FOREST
24-May-1943	LAWSON G. H.	SGT		AG	(G) REICHSWALD FOREST
28-May-1943	RAWLINSON G.	F/O		PILOT	(H) SLEEN GENERAL, DRENTHE
28-May-1943	HOWARTH J.	SGT		F/E	(H) SLEEN GENERAL, DRENTHE
28-May-1943	BEATTIE S. G.	SGT		BA	(H) SLEEN GENERAL, DRENTHE
28-May-1943	BUCK E. S.	SGT		AG	(H) SLEEN GENERAL, DRENTHE
28-May-1943	BLACKBARROW E. B.	SGT		AG	(H) SLEEN GENERAL, DRENTHE
28-May-1943	PRICE H. W.	W/O		PILOT	(G) REICHSWALD FOREST
28-May-1943	WILLIAMS F. G.	SGT		F/E	(G) REICHSWALD FOREST
28-May-1943	LEYLAND R.	SGT		NAV	(G) REICHSWALD FOREST
28-May-1943	CURTIS E. R.	F/O		BA	(G) REICHSWALD FOREST
28-May-1943	PARRY E.	P/O		WOP	(G) REICHSWALD FOREST
28-May-1943	HALSTON E. G.	SGT		AG	(G) REICHSWALD FOREST
28-May-1943	WAGGETT W. E.	SGT		AG	(G) REICHSWALD FOREST
30-May-1943	CLARKE J. E.	F/S	RNZAF	PILOT	(G) REICHSWALD FOREST
30-May-1943	CRANHAM J. R.	SGT		F/E	(G) REICHSWALD FOREST
30-May-1943	SCOTT W. R.	F/S		NAV	(G) REICHSWALD FOREST
30-May-1943	PICKLES J. H.	SGT		BA	(G) REICHSWALD FOREST
30-May-1943	HARRIS R. H.	SGT		WOP	(G) REICHSWALD FOREST
30-May-1943	BIRRELL J. W.	SGT		AG	(G) REICHSWALD FOREST
30-May-1943	SAUNDERS J. A.	F/S		AG	(G) REICHSWALD FOREST
13-Jun-1943	INNES G. M.	SGT		PILOT	(G) REICHSWALD FOREST
13-Jun-1943	SHARP W.	SGT		F/E	(G) REICHSWALD FOREST
13-Jun-1943	JENKINS K. A.	SGT		NAV	(G) REICHSWALD FOREST

13-Jun-1943	SENGER J. E.	SGT		BA	(G) REICHSWALD FOREST
13-Jun-1943	MICHELL F. C. W.	SGT		WOP	(G) REICHSWALD FOREST
13-Jun-1943	SMITH E.	SGT		AG	(G) REICHSWALD FOREST
13-Jun-1943	LEPETIT R. G.	SGT		AG	(G) REICHSWALD FOREST
20-Jun-1943	WATSON J.	SGT		PILOT	(F) BRETTEVILLE-SUR-LAIZE, Canadian
20-Jun-1943	ROCKWOOD G. W.	SGT		F/E	(F) BRETTEVILLE-SUR-LAIZE, Canadian
20-Jun-1943	NASH H. C. W.	SGT		NAV	(F) BRETTEVILLE-SUR-LAIZE, Canadian
20-Jun-1943	MCKAY W. J.	SGT		BA	(F) BRETTEVILLE-SUR-LAIZE, Canadian
20-Jun-1943	BROWN W. G.	SGT		WOP	(F) BRETTEVILLE-SUR-LAIZE, Canadian
20-Jun-1943	LEWIS E. J.	SGT	RCAF	AG	(F) BRETTEVILLE-SUR-LAIZE, Canadian
20-Jun-1943	SUTCLIFFE J.	SGT		AG	(F) BRETTEVILLE-SUR-LAIZE, Canadian
23-Jun-1943	PINKERTON R. M.	SGT		PILOT	COMM. RUNNYMEDE
23-Jun-1943	HOLMES F.	SGT		F/E	(H) AMSTERDAM NEW EASTERN
23-Jun-1943	NUTTALL F. T.	SGT		NAV	(H) AMSTERDAM NEW EASTERN
23-Jun-1943	WARING W.	SGT		BA	COMM. RUNNYMEDE
23-Jun-1943	CONWAY J.	SGT		WOP	(H) AMSTERDAM NEW EASTERN
23-Jun-1943	CROWE J. F. K.	SGT	RCAF	AG	(H) NOORDWIJK GENERAL
23-Jun-1943	MACASKILL T. L.	SGT	RCAF	AG	COMM. RUNNYMEDE
29-Jun-1943	PEATE S.	P/O		PILOT	(H) WOENSAL GENERAL, EINDHOVEN
29-Jun-1943	PAPE H.	SGT		F/E	(H) WOENSAL GENERAL, EINDHOVEN
29-Jun-1943	RAKOCZY P. L.	SGT	RCAF	NAV	(H) WOENSAL GENERAL, EINDHOVEN
29-Jun-1943	PEARSON H. H.	F/O		BA	(H) WOENSAL GENERAL, EINDHOVEN
29-Jun-1943	BAILEY A.	SGT		WOP	(H) WOENSAL GENERAL, EINDHOVEN
29-Jun-1943	ERICKSON H.	SGT	RCAF	AG	(H) WOENSAL GENERAL, EINDHOVEN
29-Jun-1943	SWEENEY J. G.	SGT		AG	(H) WOENSAL GENERAL, EINDHOVEN
29-Jun-1943	GEDDES R. H.	F/S	RAAF	PILOT	(H) JONKERBOS, NIJMEGEN
29-Jun-1943	CROSS H. E.	P/O		F/E	(H) JONKERBOS, NIJMEGEN
29-Jun-1943	BROWN D.	SGT		NAV	(H) JONKERBOS, NIJMEGEN
29-Jun-1943	BRADSHAW R. E.	P/O		BA	(H) JONKERBOS, NIJMEGEN
29-Jun-1943	ENTWISTLE C.	F/S		WOP	(H) JONKERBOS, NIJMEGEN
29-Jun-1943	BOOTH A. W.	SGT		AG	(H) JONKERBOS, NIJMEGEN
29-Jun-1943	WHITE R. S.	SGT		2/PILOT	(H) JONKERBOS, NIJMEGEN
04-Jul-1943	MORLEY A.	F/S	RAAF	PILOT	(B) HOTTON
04-Jul-1943	SADLER F.	SGT		AG	(B) HOTTON
16-Jul-1943	MELLOR H. B.	SGT		PILOT	(F) RECEY-SUR-OURCE, COTE D'OR
16-Jul-1943	MORSE R. A. W.	SGT		F/E	(F) RECEY-SUR-OURCE, COTE D'OR
16-Jul-1943	BUNKER J. D. G.	SGT		NAV	(F) RECEY-SUR-OURCE, COTE D'OR
16-Jul-1943	SMITH H. M.	SGT		BA	(F) RECEY-SUR-OURCE, COTE D'OR
16-Jul-1943	COOPER B. G. A.	SGT		WOP	(F) RECEY-SUR-OURCE, COTE D'OR

16-Jul-1943	ARTHUR S.	SGT		AG	(F) RECEY-SUR-OURCE, COTE D'OR
16-Jul-1943	MCKEOWN R. D.	SGT		AG	(F) RECEY-SUR-OURCE, COTE D'OR
16-Jul-1943	PYLE W. F.	F/S		PILOT	COMM. RUNNYMEDE
16-Jul-1943	COOKE E. T.	SGT		F/E	COMM. RUNNYMEDE
16-Jul-1943	STOCKLEY H. H.	SGT		NAV	COMM. RUNNYMEDE
16-Jul-1943	BURRELL R. A.	SGT		BA	COMM. RUNNYMEDE
16-Jul-1943	RICHARDSON F. J.	SGT		WOP	COMM. RUNNYMEDE
16-Jul-1943	ENOCH D.	SGT		AG	COMM. RUNNYMEDE
16-Jul-1943	CLARK W. H. R.	SGT		AG	COMM. RUNNYMEDE
26-Jul-1943	HIGHTOWER C. E.	P/O	RCAF	F/E	(H) GROESBEEK CANADIAN
26-Jul-1943	JONES W.	P/O		NAV	(H) WOENSEL GENERAL, EINDHOVEN
26-Jul-1943	ACKERLEY D. B.	P/O		BA	(H) WOENSEL GENERAL, EINDHOVEN
26-Jul-1943	COLLINS W.	SGT		AG	(H) WOENSEL GENERAL, EINDHOVEN
26-Jul-1943	DOWNEY G.	F/O		AG	(H) WOENSEL GENERAL, EINDHOVEN
18-Aug-1943	LONG A. J. E.	F/S		PILOT	COMM. RUNNYMEDE
18-Aug-1943	BARBEZAT C. L.	F/O		2/PILOT	COMM. RUNNYMEDE
18-Aug-1943	GALLOWAY D. A.	SGT		F/E	COMM. RUNNYMEDE
18-Aug-1943	HEAL J. J. V.	SGT		NAV	COMM. RUNNYMEDE
18-Aug-1943	COOPER J.	SGT		BA	COMM. RUNNYMEDE
18-Aug-1943	SEFTON L. H.	SGT		WOP	COMM. RUNNYMEDE
18-Aug-1943	GOULDEN D.	SGT		AG	COMM. RUNNYMEDE
18-Aug-1943	WILLETTS F.	SGT		AG	COMM. RUNNYMEDE
20-Aug-1943	SMITH J. C.	F/L		PILOT	(U) (MARKET HARBOROUGH, LEICS.)
20-Aug-1943	STAFFORD A. D.	SGT		F/E	(U) (MARKET HARBOROUGH, LEICS.)
20-Aug-1943	ALLEN R. B.	SGT		WOP	(U) (MARKET HARBOROUGH, LEICS.)
20-Aug-1943	TOWNSEND E.	SGT		AG	(U) (MARKET HARBOROUGH, LEICS.)
20-Aug-1943	NAYLOR R. S.	SGT		AG	(U) (MARKET HARBOROUGH, LEICS.)
28-Aug-1943	WARREN G. R. M.	SGT	RCAF	AG	(B) GOSSELIES, CHARLEROI
06-Sep-1943	D'EATH D. M.	P/O		PILOT	(G) RHEINBERG
06-Sep-1943	ASTIN A.	SGT		F/E	(G) RHEINBERG
06-Sep-1943	DEE E. H.	SGT		NAV	(G) RHEINBERG
06-Sep-1943	MCPHERSON C. C.	F/O	RCAF	BA	(G) RHEINBERG
06-Sep-1943	KEARNES R. T. H.	SGT		WOP	(G) RHEINBERG
06-Sep-1943	HEINIG J. P.	SGT	RCAF	AG	(G) RHEINBERG
06-Sep-1943	COOPER W. A.	SGT		AG	(G) RHEINBERG
07-Sep-1943	DOUGLAS A. G.	F/O		PILOT	(G) DURNBACH
07-Sep-1943	OKILL P. C.	SGT		AG	(G) DURNBACH
07-Sep-1943	DAVIES T. S.	SGT		PILOT	COMM. RUNNYMEDE
07-Sep-1943	TATE R. E.	SGT		F/E	COMM. RUNNYMEDE

07-Sep-1943	MURPHY P. T.	SGT		NAV	COMM. RUNNYMEDE
07-Sep-1943	HILL S. H.	SGT		BA	COMM. RUNNYMEDE
07-Sep-1943	FEAR B. A.	SGT		WOP	COMM. RUNNYMEDE
07-Sep-1943	HOLMES I. M. P.	F/S	RAAF	AG	COMM. RUNNYMEDE
07-Sep-1943	NOLAN W. G.	F/S	RAAF	AG	COMM. RUNNYMEDE
16-Sep-1943	DUNLOP J. M.	SGT		PILOT	(F) ECORCEI, ORNE
16-Sep-1943	COWLER F. G.	SGT		F/E	(F) ECORCEI, ORNE
16-Sep-1943	DUNLOP C. J. L.	P/O		NAV	(F) ECORCEI, ORNE
16-Sep-1943	MACKENNA W. F.	SGT	RCAF	AG	(F) ECORCEI, ORNE
27-Sep-1943	WARDMAN N. P.	F/S		PILOT	(G) HANNOVER
27-Sep-1943	HAND D. T.	SGT		F/E	(G) HANNOVER
27-Sep-1943	PARKER G. E. M.	P/O		BA	(G) HANNOVER
27-Sep-1943	WARRELL A. H. W.	SGT		WOP	(G) HANNOVER
27-Sep-1943	MCINTOSH J. R.	F/S		AG	(G) HANNOVER
27-Sep-1943	WOOD E. M.	SGT	RCAF	AG	(G) HANNOVER
27-Sep-1943	COCKREM H. W. G.	P/O	RAAF	PILOT	(G) HANNOVER
27-Sep-1943	STANWORTH H. A.	SGT		F/E	(G) HANNOVER
27-Sep-1943	MARSH E. G. J.	P/O		NAV	(G) HANNOVER
27-Sep-1943	BARKER R. F. G.	P/O		AG	(G) HANNOVER
27-Sep-1943	PANOS P.	F/S	RAAF	AG	(G) HANNOVER
27-Sep-1943	ROSTRON A.	SGT		PILOT	(G) REICHSWALD FOREST
27-Sep-1943	GREEST T. H.	SGT		F/E	(G) REICHSWALD FOREST
27-Sep-1943	BROWN W. A.	SGT		NAV	(G) REICHSWALD FOREST
27-Sep-1943	JARMAN E. J. E.	SGT		BA	(G) REICHSWALD FOREST
27-Sep-1943	CAVIE J. G.	SGT		WOP	(G) REICHSWALD FOREST
27-Sep-1943	PARSONS A. R. J.	SGT		AG	(G) REICHSWALD FOREST
27-Sep-1943	CHILCOTT F. G.	SGT		AG	(G) REICHSWALD FOREST
22-Oct-1943	PLANT H. E. M.	P/O	RAAF	PILOT	(G) HANNOVER
22-Oct-1943	NOTON L. W.	SGT		F/E	(G) HANNOVER
22-Oct-1943	BLANDFORD S. K.	SGT		NAV	(G) HANNOVER
22-Oct-1943	BURTON A. B.	SGT		WOP	(G) HANNOVER
22-Oct-1943	SMITH L. S.	F/S	RAAF	AG	(G) HANNOVER
22-Oct-1943	PAULL R. J.	F/S	RAAF	AG	(G) HANNOVER
22-Oct-1943	WILKINSON H. H. V.	F/L		PILOT	(G) REICHSWALD FOREST
22-Oct-1943	KEARSLEY H.	SGT		F/E	(G) REICHSWALD FOREST
22-Oct-1943	BALL R. L. H.	F/O		NAV	(G) REICHSWALD FOREST
22-Oct-1943	HEWITT F.	F/O		BA	(G) REICHSWALD FOREST
22-Oct-1943	LOMAS K.	SGT		WOP	(G) REICHSWALD FOREST
22-Oct-1943	BURMAN R. T.	SGT		AG	(G) REICHSWALD FOREST

22-Oct-1943	JENKINS R. L.	F/S		AG	(G) REICHSWALD FOREST
22-Oct-1943	HEPPELL J. W.	P/O		PILOT	(G) HANNOVER
22-Oct-1943	LAMBERT D. N.	SGT		2/PILOT	(G) HANNOVER
22-Oct-1943	SHAW A. E.	SGT		F/E	(G) HANNOVER
22-Oct-1943	GILDING L. C.	F/O		NAV	(G) HANNOVER
22-Oct-1943	ADAMS S. B.	F/S		BA	(G) HANNOVER
22-Oct-1943	KEOGH P. G.	F/O		WOP	(G) HANNOVER
22-Oct-1943	GRAINGER J.	SGT		AG	(G) HANNOVER
22-Oct-1943	FAGAN L. W. D.	F/L		AG	(G) HANNOVER
04-Nov-1943	CAMERON R.	P/O		PILOT	(U) (SHIPDAM, NORFOLK)
04-Nov-1943	EYRE S.	SGT		F/E	(U) (SHIPDAM, NORFOLK)
04-Nov-1943	TANN R. A.	F/S		NAV	(U) (SHIPDAM, NORFOLK)
04-Nov-1943	FIELDER R. J.	F/L		BA	(U) (SHIPDAM, NORFOLK)
04-Nov-1943	HUTTON J.	SGT		WOP	(U) (SHIPDAM, NORFOLK)
04-Nov-1943	WILLIAMSON A. N.	SGT		AG	(U) (SHIPDAM, NORFOLK)
05-Nov-1943	WINSTANLEY J.	SGT		AG	(U) (SHIPDAM, NORFOLK)
04-Nov-1943	HARDEN L. A.	F/L		PILOT	(G) REICHSWALD FOREST
04-Nov-1943	HASKINGS G. E.	SGT		F/E	(G) REICHSWALD FOREST
04-Nov-1943	HOOPER G. T.	SGT	RCAF	AG	(G) REICHSWALD FOREST
04-Nov-1943	PETTY A. H.	P/O		AG	(G) REICHSWALD FOREST
18-Nov-1943	LINDSEY A. M.	P/O	RAAF	PILOT	(F) RAPSCORT, MARNE
18-Nov-1943	HORTON C.	SGT		F/E	(F) RAPSCORT, MARNE
18-Nov-1943	MASON H. V.	F/S		NAV	(F) RAPSCORT, MARNE
18-Nov-1943	BIRD T. A.	SGT		BA	(F) RAPSCORT, MARNE
18-Nov-1943	FRASER G. R.	SGT		WOP	(F) RAPSCORT, MARNE
18-Nov-1943	PALMER G. J.	SGT		AG	(F) RAPSCORT, MARNE
19-Nov-1943	HOLDSWORTH B.	F/S		PILOT	(U) (TANGMERE, SUSSEX)
19-Nov-1943	STEEL R. J. H.	SGT		F/E	(U) (TANGMERE, SUSSEX)
19-Nov-1943	TELFER C.	SGT		NAV	(U) (TANGMERE, SUSSEX)
19-Nov-1943	OUDINOT A. J.	SGT		BA	(U) (TANGMERE, SUSSEX)
19-Nov-1943	DOWNS R. V.	SGT		WOP	(U) (TANGMERE, SUSSEX)
19-Nov-1943	HARPER J.	F/S		AG	(U) (TANGMERE, SUSSEX)
19-Nov-1943	SMITH C. E.	SGT		AG	(U) (TANGMERE, SUSSEX)
23-Nov-1943	PONT D. S.	F/O		PILOT	COMM. RUNNYMEDE
23-Nov-1943	MCKEAG T. H. R.	SGT		F/E	COMM. RUNNYMEDE
23-Nov-1943	BUCHAN A.	SGT		NAV	COMM. RUNNYMEDE
23-Nov-1943	MCMILLAN J.	SGT		BA	COMM. RUNNYMEDE
23-Nov-1943	LANCE K.	SGT		WOP	COMM. RUNNYMEDE
23-Nov-1943	MACKENZIE D.	SGT		AG	COMM. RUNNYMEDE

23-Nov-1943	BAXTER M. F.	F/O		AG	COMM. RUNNYMEDE
23-Nov-1943	HALL T.	F/O		PILOT	(G) REICHSWALD FOREST
23-Nov-1943	ZASTROW W. R.	SGT		F/E	(G) REICHSWALD FOREST
23-Nov-1943	SKELTON G. L.	SGT		NAV	(G) REICHSWALD FOREST
23-Nov-1943	ROGERS S. P.	F/S		BA	(G) REICHSWALD FOREST
23-Nov-1943	ASHTON C. J.	SGT		WOP	(G) REICHSWALD FOREST
23-Nov-1943	ANDERSON H. L.	SGT	RCAF	AG	(G) REICHSWALD FOREST
23-Nov-1943	COLEBROOKE J. R.	SGT		AG	(G) REICHSWALD FOREST
04-Dec-1943	WALKER F. J. T.	P/O		PILOT	(G) BECKLINGEN
04-Dec-1943	REDBOURN P. R.	SGT		2/PILOT	(G) BECKLINGEN
04-Dec-1943	HUDSON D.	SGT		F/E	(G) BECKLINGEN
04-Dec-1943	SLAUGHTER L. G.	SGT		BA	(G) BECKLINGEN
04-Dec-1943	MILES L. R.	SGT		AG	(G) BECKLINGEN
04-Dec-1943	SELBY F. G.	SGT		AG	(G) BECKLINGEN
21-Dec-1943	WHITMARSH A. W.	F/L		PILOT	(G) RHEINBERG
21-Dec-1943	HAYES P. M.	SGT		2/PILOT	(G) RHEINBERG
21-Dec-1943	PRIEST C.	P/O		F/E	(G) RHEINBERG
21-Dec-1943	LEAR K. S.	F/O		BA	(G) RHEINBERG
21-Dec-1943	BUCKNER R. G. E.	F/S		WOP	(G) RHEINBERG
21-Dec-1943	BRITTON M. H.	SGT		AG	(G) RHEINBERG
21-Dec-1943	NORTON K. R.	SGT		AG	(G) RHEINBERG
21-Dec-1943	BORTHWICK W.	SGT		PILOT	COMM. RUNNYMEDE
21-Dec-1943	PALING T.	SGT		WOP	COMM. RUNNYMEDE
21-Dec-1943	ROUCHCOUSTE J. M. H.	SGT		AG	(G) RHEINBERG
21-Dec-1943	GRANT D. C.	F/S	RAAF	AG	COMM. RUNNYMEDE
21-Dec-1943	MORRIS D. G.	SGT		PILOT	(G) RHEINBERG
21-Dec-1943	BERRY S. A.	SGT		BA	(G) RHEINBERG
21-Dec-1943	SMITH E.	SGT	RCAF	AG	(G) RHEINBERG
29-Dec-1943	GREEN P. B.	F/S		PILOT	(H) RUINERWOLD, DRENTHE
29-Dec-1943	HALL W. D.	SGT		F/E	COMM. RUNNYMEDE
29-Dec-1943	COLBOURNE A.	SGT		NAV	(H) RUINERWOLD, DRENTHE
29-Dec-1943	ROOS R. E.	F/S	RCAF	BA	(H) RUINERWOLD, DRENTHE
29-Dec-1943	WEBB S.	F/S		WOP	COMM. RUNNYMEDE
29-Dec-1943	APPLEYARD D. R. C.	SGT		AG	(H) RUINERWOLD, DRENTHE
29-Dec-1943	GREENMON P. J.	SGT		AG	COMM. RUNNYMEDE
21-Jan-1944	CROTHERS F.	P/O		PILOT	(G) 1939-1945, BERLIN
21-Jan-1944	STONES T. R.	SGT		F/E	(G) 1939-1945, BERLIN
21-Jan-1944	MASON C.	SGT		NAV	(G) 1939-1945, BERLIN
21-Jan-1944	EATON H. C.	F/L		BA	(G) 1939-1945, BERLIN

21-Jan-1944	SCOTLAND R. R.	SGT		WOP	(G) 1939-1945, BERLIN
21-Jan-1944	LOGUE D.	SGT		AG	(G) 1939-1945, BERLIN
21-Jan-1944	GOODFELLOW W. A.	SGT		AG	(G) 1939-1945, BERLIN
21-Jan-1944	ARTHUR D. A.	F/S		PILOT	(G) 1939-1945, BERLIN
21-Jan-1944	BRANCHFLOWER R. W.	SGT		F/E	(G) 1939-1945, BERLIN
21-Jan-1944	DRYER L.	SGT		WOP	(G) 1939-1945, BERLIN
21-Jan-1944	BOLTON D. C,	SGT		AG	(G) 1939-1945, BERLIN
21-Jan-1944	GILDARE A. J. B.	SGT		AG	(G) 1939-1945, BERLIN
22-Jan-1944	DIXON R. R.	F/L		PILOT	(G) HANNOVER
22-Jan-1944	DUNBAR W. A.	F/S		F/E	(G) HANNOVER
22-Jan-1944	GRACEY H.	P/O		BA	(G) HANNOVER
22-Jan-1944	PERRETT F. C.	SGT		AG	(G) HANNOVER
22-Jan-1944	PAUL H. A.	F/S		AG	(G) HANNOVER
29-Jan-1944	KILSBY N. W.	F/L	RAAF	PILOT	COMM. RUNNYMEDE
29-Jan-1944	DAGGETT S.	P/O		WOP	COMM. RUNNYMEDE
29-Jan-1944	LING D. F.	SGT		PILOT	(G) 1939-1945, BERLIN
29-Jan-1944	FOSTER J. B.	SGT		AG	(G) 1939-1945, BERLIN
29-Jan-1944	SMITH J. C.	SGT		AG	(G) 1939-1945, BERLIN
29-Jan-1944	O'CONNOR A. B.	SGT		PILOT	(D) ABENRA
29-Jan-1944	WATTERS J.	SGT		F/E	(D) ABENRA
29-Jan-1944	MILES V. L.	F/O	RCAF	NAV	(D) ABENRA
29-Jan-1944	TWIGGE G. A.	WO2	RCAF	BA	(D) ABENRA
29-Jan-1944	MAYES A.	F/S		WOP	(D) ABENRA
29-Jan-1944	SAXTY A. J. W.	SGT		AG	(D) ABENRA
29-Jan-1944	DUDLEY T. K.	SGT		AG	(D) ABENRA
29-Jan-1944	LARGE C.	F/O		PILOT	COMM. RUNNYMEDE
29-Jan-1944	CORBETT K. R.	SGT		F/E	COMM. RUNNYMEDE
29-Jan-1944	HODGKINSON J. G.	F/S		NAV	COMM. RUNNYMEDE
29-Jan-1944	MARTIN R.	SGT		WOP	COMM. RUNNYMEDE
29-Jan-1944	ARMSTRONG G.	SGT		AG	COMM. RUNNYMEDE
29-Jan-1944	TREBILCOCK C. E.	RAAF		AG	COMM. RUNNYMEDE
16-Feb-1944	CLARK W. G.	F/O		PILOT	(G) 1939-1945, BERLIN
16-Feb-1944	CLARKE H. J.	SGT		F/E	(G) 1939-1945, BERLIN
16-Feb-1944	WILLIAMS A.	F/S		NAV	(G) 1939-1945, BERLIN
16-Feb-1944	FERGUSON P. J. H.	SGT		BA	(G) 1939-1945, BERLIN
16-Feb-1944	HARRIS R. P.	SGT		WOP	(G) 1939-1945, BERLIN
16-Feb-1944	JENKINS S. H.	SGT		AG	(G) 1939-1945, BERLIN
16-Feb-1944	COFFEY F.	SGT		AG	(G) 1939-1945, BERLIN
20-Feb-1944	DAVENPORT F.	F/S	RCAF	PILOT	(G) BECKLINGEN

20-Feb-1944	MARSHALL J.	SGT		F/E	(G) BECKLINGEN
20-Feb-1944	WATERSTON E. G.	F/S	RCAF	BA	(G) BECKLINGEN
20-Feb-1944	WOODHOUSE R. L.	F/S		WOP	(G) BECKLINGEN
20-Feb-1944	HEPWORTH W. M.	F/S		AG	(G) BECKLINGEN
20-Feb-1944	HILL C. M.	SGT		AG	(G) BECKLINGEN
27-Mar-1944	WILSON T.	P/O		PILOT	(G) RHEINBERG
27-Mar-1944	KENNEDY A.	SGT		F/E	(G) RHEINBERG
27-Mar-1944	COLBORN E. A.	F/S		NAV	(G) RHEINBERG
27-Mar-1944	MOODY F. A.	F/S		BA	(G) RHEINBERG
27-Mar-1944	GIRLING C. E.	F/S		WOP	(G) RHEINBERG
27-Mar-1944	WALKER B. R. G.	SGT		AG	(G) RHEINBERG
27-Mar-1944	HUTCHEON R.	SGT		AG	(G) RHEINBERG
27-Mar-1944	SIMMONS R. A.	P/O		PILOT	(B) HOTTON
27-Mar-1944	HENDRY J.	SGT		F/E	(B) HOTTON
27-Mar-1944	ARNOLD T.	SGT		NAV	(B) HOTTON
27-Mar-1944	CHAPMAN L. A.	F/S		BA	(B) HOTTON
27-Mar-1944	LAUGHLIN I. E.	SGT		WOP	(B) HOTTON
31-Mar-1944	REGAN W. T. A.	F/S		PILOT	(G) HANNOVER
31-Mar-1944	BIRCH E.	F/S	RAAF	F/E	(G) HANNOVER
31-Mar-1944	TINDAL R. W.	SGT		WOP	(G) HANNOVER
31-Mar-1944	SMITH D. L.	SGT		AG	(G) HANNOVER
11-Apr-1944	BARNES W. G.	F/L		PILOT	(U) (CREMATED, BIRMINGHAM)
11-Apr-1944	HOWELL G. C.	SGT		AG	(F) LONGUEVAL, SOMME
27-Apr-1944	ALLEN C. A. L.	F/L		PILOT	(B) HEVERLEE
27-Apr-1944	HAGERTY C. W.	F/S		NAV	(B) HEVERLEE
27-Apr-1944	HENDRY A. G.	F/O		BA	(B) HEVERLEE
27-Apr-1944	DUERDEN C.	SGT		AG	(B) HEVERLEE
27-Apr-1944	JONES W. S.	SGT	RCAF	AG	(B) HEVERLEE
02-May-1944	LASSEY G.	P/O		PILOT	(B) BRUSSELS TOWN
02-May-1944	PRISKE K. W.	F/S	RCAF	2/PILOT	COMM. RUNNYMEDE
02-May-1944	MILLER N. W. C.	SGT		F/E	COMM. RUNNYMEDE
02-May-1944	KNOWLES D. J.	SGT		NAV	COMM. RUNNYMEDE
02-May-1944	HARTLEY T. J. C.	SGT		BA	COMM. RUNNYMEDE
02-May-1944	LUCAS T. K. D.	SGT		AG	(B) BRUSSELS TOWN
02-May-1944	SETTER D. J.	WO2	RCAF	AG	(B) BRUSSELS TOWN
03-Jun-1944	MURRAY A. A.	F/O		PILOT	(F) ST-ANDRE-DE-L'EURE, EURE
03-Jun-1944	WILLIAMS J.	W/O		WOP	(F) ST-ANDRE-DE-L'EURE, EURE
03-Jun-1944	KUMAR V.	SGT		PILOT	(F) HERMERAY, YVELINES
03-Jun-1944	ARCHER J. T.	SGT		F/E	(F) ST-DESIR

03-Jun-1944	HEYWORTH E. O.	F/O	RCAF	NAV	(F) ST-DESIR
03-Jun-1944	TAYLOR C. D.	F/O		BA	(F) HERMERAY, YVELINES
03-Jun-1944	BLACKLOCK T.	SGT		AG	(F) ST-DESIR
03-Jun-1944	O'LEARY G.	SGT		AG	(F) ST-DESIR
10-Jun-1944	VAN STOCKUM W. J.	F/O		PILOT	(F) VALFLEURY, LAVAL, MAYENNE
10-Jun-1944	ELLYATT J.	F/O		F/E	(F) VALFLEURY, LAVAL, MAYENNE
10-Jun-1944	DANIEL G.			NAV	(F) VALFLEURY, LAVAL, MAYENNE
10-Jun-1944	MARSHALL R. K.	F/O		BA	(F) VALFLEURY, LAVAL, MAYENNE
10-Jun-1944	PERKINS A. C.	SGT		WOP	(F) VALFLEURY, LAVAL, MAYENNE
10-Jun-1944	BEALES F.	SGT	RCAF	AG	(F) VALFLEURY, LAVAL, MAYENNE
10-Jun-1944	MASON A.	SGT		AG	(F) VALFLEURY, LAVAL, MAYENNE
10-Jun-1944	HENDERSON T. W.	P/O	RAAF	PILOT	(F) VALFLEURY, LAVAL, MAYENNE
10-Jun-1944	PEAKE S. W.	SGT		F/E	(F) VALFLEURY, LAVAL, MAYENNE
10-Jun-1944	HENDERSON N. R.	F/O		NAV	(F) VALFLEURY, LAVAL, MAYENNE
10-Jun-1944	BRADBURY H.	SGT		BA	(F) VALFLEURY, LAVAL, MAYENNE
10-Jun-1944	GAINES P.	SGT		WOP	(F) VALFLEURY, LAVAL, MAYENNE
10-Jun-1944	BROTHERTON L. A.	SGT		AG	(F) VALFLEURY, LAVAL, MAYENNE
10-Jun-1944	WICKS E. C.	SGT		AG	(F) VALFLEURY, LAVAL, MAYENNE
17-Jun-1944	LEITCH N. C. C.	P/O	RAAF	PILOT	(U) (GOOLE, YORKSHIRE)
17-Jun-1944	CRAWFORD R. A.	SGT		F/E	COMM. RUNNYMEDE
17-Jun-1944	PEARCE R. F.	SGT		NAV	(U) (GOOLE, YORKSHIRE)
17-Jun-1944	McCARROL W. J.	F/S		BA	COMM. RUNNYMEDE
17-Jun-1944	LEWINGTON C. R. B.	SGT		WOP	(U) (GOOLE, YORKSHIRE)
17-Jun-1944	DUMMER C. A. R.	SGT		AG	(U) (GOOLE, YORKSHIRE)
29-Jun-1944	LIVESEY M.	P/O		PILOT	(F) FEIGNEUX, OISE
29-Jun-1944	SISSONS J. R.	SGT		F/E	(F) FEIGNEUX, OISE
29-Jun-1944	RHODES A. G.	F/S		NAV	(F) FEIGNEUX, OISE
29-Jun-1944	EVANS T. G.	WO2	RCAF	BA	(F) FEIGNEUX, OISE
29-Jun-1944	HUGHES L. W.	F/S		WOP	(F) FEIGNEUX, OISE
29-Jun-1944	TUDBERRY W.	F/S	RAAF	AG	(F) FEIGNEUX, OISE
29-Jun-1944	CHANDLER K. E.	SGT		AG	(F) FEIGNEUX, OISE
29-Jun-1944	CUFFEY J.	SGT		AG	(F) PERLES
01-Jul-1944	ROSEN R. A.	F/O		PILOT	(F) POIX-DE-LA-SOMME
01-Jul-1944	DALEY D.	SGT		F/E	(F) POIX-DE-LA-SOMME
01-Jul-1944	WILLIAMSON -RATTRAY H. C.	F/S		NAV	(F) POIX-DE-LA-SOMME
01-Jul-1944	LELLIOTT J. C.	F/O		BA	(F) POIX-DE-LA-SOMME
01-Jul-1944	LIND G. S.	SGT		WOP	(F) POIX-DE-LA-SOMME
01-Jul-1944	FORDHAM A. S.	SGT		AG	(F) POIX-DE-LA-SOMME

21-Jul-1944	TOUGH F.	SGT		AG	(H) WIJK-EN-AALBURG, NOORD-BRABANT
21-Jul-1944	HADLEY N.	F/O		PILOT	(G) REICHSWALD FOREST
21-Jul-1944	FREEMAN K. H.	F/O	RAAF	2/PILOT	(G) REICHSWALD FOREST
21-Jul-1944	FLANAGAN M.	SGT		F/E	(G) REICHSWALD FOREST
21-Jul-1944	RICH T.	F/O		NAV	(G) REICHSWALD FOREST
21-Jul-1944	MOSLEY-LEIGH P. J.	F/S		BA	(G) REICHSWALD FOREST
21-Jul-1944	TOMLINSON W. R.	SGT		WOP	(G) REICHSWALD FOREST
21-Jul-1944	BRITTAIN P. H.	F/S	RCAF	AG	(G) REICHSWALD FOREST
25-Aug-1944	WALTON E.	F/O		PILOT	(F) LONGUENNES, St-OMER
25-Aug-1944	WHITTAKER J.	F/S		NAV	(F) LONGUENNES, St-OMER
25-Aug-1944	MacDOUGALL C. J. R.	SGT		WOP	(F) LONGUENNES, St-OMER
25-Aug-1944	WARE P. H. C.	SGT		AG	(F) LONGUENNES, St-OMER
25-Aug-1944	BYNG C. A.	SGT		AG	(F) LONGUENNES, St-OMER
24-Sep-1944	KITE G. R. G.	F/O	RCAF	PILOT	(B) LEOPOLDSBURG, LIMBURG
24-Sep-1944	SAUNDERS A. G. T.	F/S		WOP	(B) LEOPOLDSBURG, LIMBURG
16-Oct-1944	OWEN C. M.	F/S		PILOT	COMM. RUNNYMEDE
16-Oct-1944	MURPHY A. G.	SGT		F/E	COMM. RUNNYMEDE
16-Oct-1944	SMITH E. W.	SGT		NAV	COMM. RUNNYMEDE
16-Oct-1944	WYNN H. G.	F/S		BA	COMM. RUNNYMEDE
16-Oct-1944	PICKEN J.	SGT		WOP	COMM. RUNNYMEDE
16-Oct-1944	GRIFFITHS S. G.	SGT		AG	COMM. RUNNYMEDE
16-Oct-1944	McKAY D.	SGT		AG	COMM. RUNNYMEDE
16-Oct-1944	HART S. W.	S/L		PILOT	(D) VADUM
16-Oct-1944	SHEARD F. B.	SGT		F/E	(D) VADUM
16-Oct-1944	McHARDY S. A. F.	P/O	RCAF	BA	(D) VADUM
16-Oct-1944	SLATTER A. L.	P/O		WOP	(D) VADUM
16-Oct-1944	PORTER R. H.	SGT		AG	(D) VADUM
16-Oct-1944	JARDINE W.	P/O		AG	(D) VADUM
05-Nov-1944	HARRIS D. K.	P/O	RNZAF	PILOT	(G) REICHSWALD FOREST
05-Nov-1944	BROWN J.	SGT		F/E	(G) REICHSWALD FOREST
05-Nov-1944	SILSON J. A.	SGT		NAV	(G) REICHSWALD FOREST
05-Nov-1944	TOASE G. R.	SGT		BA	(G) REICHSWALD FOREST
05-Nov-1944	HARLEY H. K.	F/S	RAAF	WOP	(G) REICHSWALD FOREST
05-Nov-1944	MARTIN O. H.	W/O		AG	(G) REICHSWALD FOREST
07-Dec-1944	WELCH E. A.	F/O		PILOT	COMM. RUNNYMEDE
07-Dec-1944	WILCOX E	SGT		F/E	COMM. RUNNYMEDE
07-Dec-1944	SCOTT J. N. L.	F/O		NAV	COMM. RUNNYMEDE
07-Dec-1944	HINDLE J.	F/O		BA	COMM. RUNNYMEDE

07-Dec-1944	O'DONNELL B. J.	F/S	RAAF	WOP	COMM. RUNNYMEDE
07-Dec-1944	SLAVEN J. W. M.	SGT		AG	COMM. RUNNYMEDE
07-Dec-1944	SWINBANK R. M.	SGT		AG	COMM. RUNNYMEDE
18-Dec-1944	BODY G. D.	F/L		PILOT	COMM. RUNNYMEDE
18-Dec-1944	TATAM N. C.	F/L		2/PILOT	(B) LEOPOLDSBURG, LIMBURG
18-Dec-1944	NICHOLSON E.	SGT		F/E	(F) CLICHY, PARIS
18-Dec-1944	WALDRON J. H.	F/O		NAV	(F) TAILLETTE, ARDENNES
18-Dec-1944	LEESE W. H.	P/O	RCAF	BA	(B) LEOPOLDSBURG, LIMBURG
18-Dec-1944	MOLE D. J.	F/O		WOP	(F) TAILLETTE, ARDENNES
18-Dec-1944	MATTHEWS K. F.	SGT		AG	(B) LEOPOLDSBURG, LIMBURG
18-Dec-1944	MAWSON W. E.	SGT		AG	(B) LEOPOLDSBURG, LIMBURG
26-Dec-1944	YATES B.	F/O		PILOT	COMM. RUNNYMEDE
26-Dec-1944	MANSELL P. J.	SGT		F/E	(U) (OFF MARGATE, KENT)
26-Dec-1944	MacLEOD W. R. D.	P/O		NAV	COMM. RUNNYMEDE
26-Dec-1944	GOWER J. G.	P/O		BA	COMM. RUNNYMEDE
26-Dec-1944	EDWARDS P. C.	P/O	RCAF	WOP	(U) (OFF MARGATE, KENT)
26-Dec-1944	MURPHY T.	SGT		AG	(U) (OFF MARGATE, KENT)
26-Dec-1944	ADDYMAN D. E.	P/O		AG	(U) (OFF MARGATE, KENT)
01-Jan-1945	WAITE W. A.	F/S		WOP	(U) (MELBOURNE, YORKSHIRE)
01-Jan-1945	NEWLING H. J.	F/S		AG	(U) (MELBOURNE, YORKSHIRE)
06-Jan-1945	SIFTON C. R.	F/O		PILOT	(G) HANNOVER
06-Jan-1945	BAINBRIDGE W.	SGT		2/PILOT	(G) HANNOVER
06-Jan-1945	BRANSGROVE W. E.	P/O		F/E	(G) HANNOVER
06-Jan-1945	DANIEL W. A.	F/S		NAV	(G) HANNOVER
06-Jan-1945	BULLOCK D. A.	F/S		BA	(G) HANNOVER
06-Jan-1945	JACKSON B.	SGT		WOP	(G) HANNOVER
06-Jan-1945	COATES L. D. I.	F/S		AG	(G) HANNOVER
06-Jan-1945	FRESHWATER N. E.	F/S		AG	(G) HANNOVER
07-Jan-1945	HARROW W. D.	F/L		PILOT	COMM. RUNNYMEDE
07-Jan-1945	WALKER F. J.	F/S		F/E	COMM. RUNNYMEDE
07-Jan-1945	HOLLINGS R. C.	P/O		NAV	(G) HANNOVER
07-Jan-1945	CALVERT A. H.	F/S		BA	(G) HANNOVER
07-Jan-1945	DRAPER J.	F/S		WOP	COMM. RUNNYMEDE
07-Jan-1945	SOLOMON M. A.	SGT		AG	COMM. RUNNYMEDE
07-Jan-1945	THOMAS R. W.	F/S		AG	COMM. RUNNYMEDE
16-Jan-1945	MARSHALL A. J.	F/O		PILOT	(G) HANNOVER
16-Jan-1945	GRIFFITHS A. H.	SGT		F/E	(G) HANNOVER
16-Jan-1945	THORNLEY J.	AG		AG	(G) HANNOVER

03-Feb-1945	GIBBS R. A.	F/O		PILOT	(H) JONKERBOS, NIJMEGEN
03-Feb-1945	ASHTON J.	SGT		F/E	(H) JONKERBOS, NIJMEGEN
03-Feb-1945	CHELL L. H.	F/S		NAV	(H) VENRAY
03-Feb-1945	PARHAM W. P.	F/O		WOP	(H) JONKERBOS, NIJMEGEN
03-Feb-1945	SMITH T. C.	SGT		AG	(H) VENRAY
03-Feb-1945	SEABRIDGE W. H.	SGT		AG	(H) VENRAY
15-Feb-1945	GRAYSHAN J.	P/O		PILOT	(D) HOLBAEK EAST, ZEALAND
15-Feb-1945	BERRY A. J.	F/S		NAV	(D) HOLBAEK EAST, ZEALAND
04-Mar-1945	LAFFOLEY J. G. L.	F/L	RCAF	PILOT	(U) (STAVELEY, YORKSHIRE)
04-Mar-1945	FINCH C. H.	SGT		F/E	(U) (STAVELEY, YORKSHIRE)
04-Mar-1945	THORNDYCRAFT L. A.	P/O	RCAF	BA	(U) (STAVELEY, YORKSHIRE)
04-Mar-1945	FIELD P. H.	F/S		WOP	(U) (STAVELEY, YORKSHIRE)
04-Mar-1945	BRADSHAW E. W.	SGT		AG	(U) (STAVELEY, YORKSHIRE)
06-Mar-1945	TASKER H. W.	SGT		2/PILOT	(G) DURNBACH
06-Mar-1945	DAVENPORT R. E.	F/S		F/E	(G) DURNBACH
06-Mar-1945	WEBSTER L. W.	WO2	RCAF	NAV	(G) DURNBACH
06-Mar-1945	HALL L. L.	SGT	RAAF	WOP	(G) DURNBACH
06-Mar-1945	FEARNLEY F.	F/S		AG	(G) DURNBACH
06-Mar-1945	STEPHEN A. D.	F/L		PILOT	(G) 1939-1945, BERLIN
06-Mar-1945	ELLIOT T. T.	SGT		F/E	(G) 1939-1945, BERLIN
06-Mar-1945	REES K. V.	F/S		NAV	(G) 1939-1945, BERLIN
06-Mar-1945	ROBSON B.	F/S		BA	(G) 1939-1945, BERLIN
06-Mar-1945	MARIA H. C,	F/O	RNZAF	WOP	(G) 1939-1945, BERLIN
06-Mar-1945	ROBERTS C. J.	P/O	RCAF	AG	(G) 1939-1945, BERLIN
06-Mar-1945	HEAP R. E.	F/O		AG	(G) 1939-1945, BERLIN
09-Apr-1945	CURRIE J.	P/O	RNZAF	PILOT	(G) REICHSWALD FOREST
09-Apr-1945	SWITZER J. R.	SGT	RCAF	F/E	(G) REICHSWALD FOREST
09-Apr-1945	PARKIN J.	F/S		NAV	(G) REICHSWALD FOREST
09-Apr-1945	SQUIRE L.	SGT		AG	(G) REICHSWALD FOREST
09-Apr-1945	FORTIN M. K.	SGT		AG	(G) REICHSWALD FOREST

10 SQUADRON – WW2 CASUALTY LIST

Shot

In Memory of

Sergeant

Harold Phillip Calvert

903068, 10 Sqdn., Royal Air Force Volunteer Reserve who died on 20 May 1942

Remembered with Honour

Berlin 1939-1945 War Cemetery

Commemorated in perpetuity by

the Commonwealth War Graves Commission

In Memory of

Warrant Officer

Ernest Philip Lewis

970467, 10 Sqdn., Royal Air Force Volunteer Reserve who died on 01 August 1944

Remembered with Honour

Malbork Commonwealth War Cemetery

Commemorated in perpetuity by

the Commonwealth War Graves Commission

But always Remembered

ACKNOWLEDGEMENTS

Other partial histories of 10 Squadron have been written down the years, usually limited by location or an author's tour length, but this is the first known attempt at covering the Squadron's activity across all of its first 100 years. This general overview has been written with the official records, where available, as primary sources. More academic and detailed histories of certain periods would certainly be possible and may yet emerge, but if the approach taken here conveys in some way the unremitting nature of the daily task facing Squadron personnel in both World Wars, in peacetime service and in training, we will be satisfied.

The sources consulted in compiling this history were many. Individual contributions from former Squadron members have been annotated as such in the main text but, should the following list contain omissions, we readily apologise. We start with the Squadron Association, created in the 1980s with Doug and Pam Dent, Tom Thackray and Doug Evans as prime movers, encouraged by the then CO, Wing Commander Gerry Bunn, now its President. We heartily acknowledge the stories and reminiscences that so many Association members have contributed to the Association Newsletters over the years and, indeed, the remarks that subsequent Squadron COs have written as Forewords – there has been much to draw from in all of those to add colour to the official records. That has proved equally true of various letters and other material held in the Squadron archives/memorabilia at RAF Brize Norton – Flt Lt Steve Sansford's assistance in arranging visits to The Hub to pillage these has been invaluable. That access would, of course, not have been possible without the ready cooperation shown throughout this project by both recent COs, Wg Cdrs Dan James and Jamie Osborne, and XO Sqn Ldr Phil Astle. Jon Salmon and Kevin Daws of AirTanker Ltd have also been most helpful.

Returning to the Association: David Mole, a former Association Chairman and WW2 Bomber Offensive specialist – we cannot thank him sufficiently. His researches into the circumstances of his father's loss in 1944 led him to compile complete listings of WW2 losses, POWs and Evaders, and these have proved invaluable. And there are three gentlemen who have helped enormously whenever we have sought advice on points of detail – pilot Doug Evans, who completed his Ops tour with the Halifax II and Halifax III, and flew in the Berlin Airlift with a charter company before a distinguished career with BOAC and British Airways; former Navigation Leader Doug Newham in both Melbourne and India, who later ran the BOAC/BA Ops setup; and navigator Ken Stewart who flew on Ops in the last months of the war and also went out to India, and who effectively represented the Association as the Prime Minister awarded the first Bomber Command Clasps at 10 Downing Street. The Association in turn owes a great deal to John Rowbottom, the owner of Melrose Farm situated on the site of the Squadron's WW2 Melbourne airfield, near York. John donated the site for the Association Memorial that is the focus of each year's Remembrance Sunday services.

There are two previous internal histories that we must acknowledge. The first, a brief narrative compiled by Flying Officers C A Turner and T A Hoare as the Squadron's Canberra era came to a close, has proved particularly useful with its invaluable WW1 contribution by Percy Russell Mallinson, grounded by a head injury after a crash in 1916 and later the Squadron's Reporting Officer from mid-1918. The second, an episodic survey of Squadron history, compiled as the VC10 era came to an end in 2005 by Master Engineer Martin Blythe.

Beyond all of the above lie the official records at the National Archives - the AIR 1 series for WW1, AIR 27 thereafter, and the Combat Reports in AIR 50 – the book could not exist as it does without these. Online resources have been important, and we would single out the ready assistance on WW1 matters received from members of the Great War Forum, especially Trevor Henshaw and Mike Meech and, for detail related to the VC10, Jelle Hieminga's VC10derness website has proved friendly and most helpful. (How interesting that by far the best VC10 website should

be run from the Netherlands!) More generally, Malcolm Barrass' "Air of Authority" site, another labour of love, has been useful on umpteen points of RAF historical detail. Then there are so many published resources:

For WW1: Trevor Henshaw's masterly "The Sky Their Battlefield," recently re-published in an updated and expanded form; the three volumes of RFC and RAF Communiqués, edited by Christopher Cole and Chaz Bowyer; "Arras" by Mike O'Connor in the Airfields and Airmen series; John Ivelaw-Chapman "High Endeavour;" Ira Jones "Tiger Squadron;" E R Hooton "War Over The Trenches." We are obliged to historian Robert Jackson for permission to use the anecdotes narrated to him by his friend, Capt J A 'Joe' Pattern who, years beforehand, had been his chemistry master at Richmond School, Yorkshire. These can be found in his "Air War Flanders - 1918" and "Army Wings."

For WW2: The first six volumes of W R Chorley's "Bomber Command Losses of the Second World War" and "The Bomber Command War Diaries" compiled by Martin Middlebrook and Chris Everitt were simply indispensable for treatment of the Whitley and Halifax years – after dealing with the exploits of just a single squadron one can only wonder at the hours of research involved in their Command-wide publications. Specific periods of 10 Squadron in WW2 are covered by: Larry Donnelly's "The Whitley Boys;" Tom William's "Gunnery Leader" concerning one of the first commissioned Air Gunners, Ken Bastin,; Brian Rapier "Melbourne Ten;" Esdaile Carter "Pilot and Pacifist;" Nigel Smith "Tirpitz, The Halifax Raids;" AVM D C T Bennett "Pathfinder;" Arthur C Smith "Halifax Crew;" and not forgetting that the Squadron's travails in the Whitley era fill a chapter in Max Hastings' "Bomber Command." Volumes dealing with No 4 Group were: Chris Ward "4 Group Bomber Command, An Operational Record;" Ken Marshall "The Pendulum and The Scythe;" Chris Blanchett "From Hull, Hell and Halifax."

General reference works include Wg Cdr C G 'Jeff' Jefford's two magisterial works, "RAF Squadrons" and "Observers and Navigators," with the latter covering so much more than the title implies. To these can be added: "Forged in War", Humphrey Wynn's official history of Transport Command to 1967; C H Barnes "Handley Page Aircraft since 1907;" Alfred Price "Instruments of Darkness;" Peter Hinchcliffe "The Other Battle;" Roy Irons "The Relentless Offensive;" Victor Bingham "Halifax, Second to None;" Patrick Otter "Yorkshire Airfields in the Second World War;" Robert Jackson "Canberra, The Operational Record;" Andrew Brookes "Victor Units of the Cold War;" Jeremy Brown "A South American War."

Thanks also go to Naomi Neathey, Linzee Duncan, Caroline O'Neill, Lesley Hayward-Mudge, Susanne Pescott, and Claire Jordan who have variously carried out interviews, provided information, obtained permissions, or put in time at the National Archives on the project's behalf – all have helped enormously.

The photographs used have come mainly from Squadron Archives and from contributions by families and ex-Squadron personnel. Particular thanks go to Ernie Lack and Dave Bridger for photographs of the Canberra era, and to John Rattenbury for the later Victor years. Voyager photos came as Crown Copyright via Air Tanker, but we also thank Fg Off Nick Zbieranowski of 10 Sqn and Mark Bentley for our striking front and endplates. We were very fortunate in securing the agreement of noted military and wildlife Mandy Shepherd to provide a cover illustration and we are grateful for her father, David Shepherd's, permission to reproduce one of his paintings. We are also in debt to Norman Appleton for many of the chapter heading sketches of Squadron aircraft types. In addition, we must thank Penelope Douglas for her permission to reproduce a number of her significant VC10 paintings. Such appearance as the volume has is due to the generosity of all of those people.

And when all is said and done, we owe considerable thanks to our wives who have put up with our spending hours at computers on a project that passed its predicted completion date long ago!

Ian Macmillan, Flight Commander (Operations), 1975-1978
Richard King, Copilot and Captain, VIP and Display Pilot, 1980-1986

November 2015